The biology of marine fungi

The biology of marine fungi

Edited by

S. T. MOSS
Department of Biological Sciences
Portsmouth Polytechnic

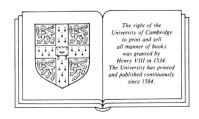

The right of the
University of Cambridge
to print and sell
all manner of books
was granted by
Henry VIII in 1534.
The University has printed
and published continuously
since 1584.

CAMBRIDGE UNIVERSITY PRESS
Cambridge
London   New York   New Rochelle
Melbourne   Sydney

Published by the Press Syndicate of the University of Cambridge
The Pitt Building, Trumpington Street, Cambridge CB2 1RP
32 East 57th Street, New York, NY 10022, USA
10 Stamford Road, Oakleigh, Melbourne 3166, Australia

First published 1986

Printed in Great Britain at the University Press, Cambridge

British Library cataloguing in publication data
The biology of marine fungi
1. Marine fungi
I. Moss, S. T.
589.2'09162   QK618

Library of Congress cataloguing in publication data
The biology of marine fungi
1. Marine fungi.   I. Moss, S. T. (Stephen Thomas), 1943–
QK618.B56   589.2'092   86–12944

ISBN 0 521 30899 2

CONTENTS

CONTRIBUTORS

Dr. L. Adler, Department of Marine Microbiology, Botanical Institute, Carl Skottsbergs Gata 22, S-413 19 Göteborg, Sweden.

Prof. D.G. Ahearn, Laboratory for Microbial and Biochemical Sciences, Georgia State University, Atlanta, Georgia 30303, U.S.A.

Dr. D.J. Alderman, Ministry of Agriculture, Fisheries and Food, Directorate of Fisheries Research, Fish Disease Laboratory, Weymouth DT4 8UB, U.K.

Dr. J.P. Amon, Department of Biological Sciences, Wright State University, Dayton, Ohio 45435, U.S.A.

Dr. G. Bahnweg, Institute für Mikrobiologie, Universität Göttingen, Grisebachstraße 8, 3400 Göttingen, F.R.G.

Dr. T. Booth, Department of Botany, University of Manitoba, Winnipeg, Manitoba, Canada R3T 2N2.

Dr. N.K. Coleman, Department of Biological Sciences, University of Cincinnati, Cincinnati, Ohio 45221-006, U.S.A.

Dr. S.A. Crow, Laboratory for Microbial and Biochemical Sciences, Georgia State University, Atlanta, Georgia 30303, U.S.A.

Dr. R.A. Eaton, School of Biological Sciences, Portsmouth Polytechnic, Portsmouth PO1 2DY, U.K.

Dr. L.V. Evans, Department of Plant Sciences, The University of Leeds, Leeds LS2 9JT, U.K.

Dr. R.D. Fallon, University of Georgia Marine Institute, Sapelo Island, Georgia 31327, U.S.A.

Miss C.A. Farrant, School of Biological Sciences, Portsmouth Polytechnic, Portsmouth PO1 2DY, U.K.

Prof. J.W. Fell, Rosenstiel School of Marine and Atmospheric Science, University of Miami, Miami, Florida 33149, U.S.A.

Dr. R.H. Findlay, Rosenstiel School of Marine and Atmospheric Science, University of Miami, Miami, Florida 33149, U.S.A.

Miss F.M. Gibb, Botany Department, The University, Liverpool L69 3BX, U.K.

Prof. G.C. Hughes, Departments of Botany and Oceanography, University of British Columbia, Vancouver, British Columbia, Canada V6T 2B1.

Dr. K.D. Hyde, School of Biological Sciences, Portsmouth Polytechnic, Portsmouth PO1 2DY, U.K.
(present address: Sains School, Jalan Muara, Bandar S. Begwan, Brunei)

Dr. I. Jäckle, Institute für Mikrobiologie, Universität Göttingen, Grisebachstraße 8, 3400 Göttingen, F.R.G.

Prof. D.H. Jennings, Botany Department, The University, Liverpool L69 3BX, U.K.

Dr. R.G. Johnson, School of Biological Sciences, Portsmouth Polytechnic, Portsmouth PO1 2DY, U.K.
(present address: Research and Development Laboratory, Cyanamid, Gosport PO13 0AS, U.K.)

Prof. E.B.G. Jones, School of Biological Sciences, Portsmouth Polytechnic, Portsmouth PO1 2DY, U.K.

Dr. N. Kenkel, Department of Botany, University of Manitoba, Winnipeg, Manitoba, Canada R3T 2N2.

Dr. D.L. Kingham, Department of Plant Sciences, The University of Leeds, Leeds LS2 9JT, U.K.

Prof. P.W. Kirk, Jr., Department of Biological Sciences, Old Dominion University, Norfolk, Virginia 23508, U.S.A.

Prof. J. Kohlmeyer, Institute of Marine Sciences, University of North Carolina at Chapel Hill, Morehead City, North Carolina 28557, U.S.A.

Dr. J.D. Miller, Chemistry and Biology Research Institute, Agriculture Canada, Ottawa, Ontario, Canada K1A 0C6.

Mr. F.I. Molina, Marine Sciences Centre, University of the Philippines, Diliman, Quezon City 3004, Philippines.
(present address: Departments of Botany and Oceanography, University of British Columbia, Vancouver, British Columbia, Canada V6T 2B1)

Prof. H.P. Molitoris, Institute für Botanik, Universität Regensburg, 8400 Regensburg, F.R.G.

Dr. S.T. Moss, School of Biological Sciences, Portsmouth Polytechnic, Portsmouth PO1 2DY, U.K.

Mr. R. Mouzouras, School of Biological Sciences, Portsmouth Polytechnic, Portsmouth PO1 2DY, U.K.

Dr. W. Mulach, Institute für Botanik, Universität Regensburg, 8400 Regensburg, F.R.G.

Dr. A. Nakagiri, Institute of Biological Sciences, University of Tsukuba, Sakura-mura, Ibaraki 305, Japan.

Dr. S.Y. Newell, University of Georgia Marine Institute, Sapelo Island, Georgia 31327, U.S.A.

Dr. J.L. Polglase, Institute of Aquaculture, University of Stirling, Stirling FK9 4LA, U.K.

Dr. D. Porter, Department of Botany, University of Georgia, Athens, Georgia 30601, U.S.A.

Prof. G.J.F. Pugh, School of Biological Sciences, Portsmouth Polytechnic, Portsmouth PO1 2DY, U.K.

Prof. R.J. Richards, Institute of Aquaculture, University of Stirling, Stirling FK9 4LA, U.K.

Dr. K. Schaumann, Alfred-Wegener-Institut für Polar- und Meeresforschung, Am Handelshafen 12, 2850 Bremerhaven-G, F.R.G.

Dr. C.A. Shearer, Department of Biology, University of Illinois, Urbana, Illinois 61801, U.S.A.

Prof. K. Tubaki, Institute of Biological Sciences, University of Tsukuba, Sakura-mura, Ibaraki 305, Japan.

Dr. J.R. Vestal, Department of Biological Sciences, University of Cincinnati, Cincinnati, Ohio 45221-006, U.S.A.

Dr. J.M. Wethered, Botany Department, The University, Liverpool L69 3BX, U.K.

PREFACE

        This book is based on invited papers presented at the Fourth
International Marine Mycology Symposium convened at Portsmouth
Polytechnic, U.K., in August 1985. A decade has elapsed since a volume
last reviewed the literature on both the lower and higher marine fungi
and seven years since Kohlmeyer and Kohlmeyer's excellent treatise on
the higher marine fungi. Initial studies on marine fungi were
essentially descriptive, taxonomic and biased towards the saprobic
fungi, particularly lignolytic forms. New species continue to be
described as new habitats and geographical regions are investigated but
concomitant with these studies and in an attempt to establish a more
natural classification there has been a reappraisal of characters
considered of taxonomic importance. Criteria considered of phyletic
significance and the techniques for their study in both the lower and
higher fungi have been subjected to critical analysis within the last
six years and this interest is reflected in a number of chapters in this
book.

Modern technology has made the resources of the sea significant
contributors to the world's energy supplies, transport, food and
leisure. Owing to these demands it has become increasingly important to
understand better the marine ecosystem, an environment in which the
fungi are ubiquitous and important members of the biota. The necessity
for a more complete knowledge of the marine environment and its
organisms has stimulated studies into the role of fungi in the sea,
their physiological adaptations, distribution and economic importance
either as pathogens of plants and animals of commercial and/or
environmental importance or degraders of organic materials. This book
considers these aspects in a combination of reviews of the major subject
areas together with original research presentations on topics which
reflect current approaches to the study of marine fungi. Contributors
were selected by the organizing committee, namely, David Jennings,
Gareth Jones, Steve Moss and Geoff Pugh, of the Fourth International
Marine Mycology Congress. Each author is an expert in his field of
research and was requested to review the literature within his
specialized area in order to provide a comprehensive and up-to-date
source of references. References are complete to the end of 1985 and
many chapters cite 1986 literature.

I am indebted to the other members of the organizing committee for
their support and encouragement throughout the preparation of this book.

My sincere thanks and gratitude go to Gillian Perks who diligently and
expertly typed the entire camera-ready copy, including tables, to Lynn
Healey for redrawing Figures 23.1,2 and 27.1 and the authors for their
contributions.

May 1986

S.T. Moss

1 FUNGAL GROWTH IN THE SEA

D.H. Jennings

INTRODUCTION

It is twenty-four years since Johnson & Sparrow (1961) produced their seminal volume on marine fungi, yet there is still debate about those major properties of such fungi which allow them to grow in the sea or for that matter make them less able to grow in fresh water. In a similar vein, there is also uncertainty about the properties of terrestrial fungi which confine them to freshwater. We can be most certain about why the zoosporic fungi, such as *Althornia*, *Dermocystidium*, *Haliphthoros*, *Labyrinthula* and *Thraustochytrium*, grow in the sea. These fungi require sodium for growth and it is this requirement, at concentration which gives the element the status of a macronutrient, that demands they should be marine. It is regrettable for the higher marine fungi that there is yet to be a critical investigation of the presence or lack of a requirement for sodium. Clearly there is no requirement at the macronutrient level but requirement for sodium at the micronutrient level has yet to be ruled out. The demonstration of a requirement for sodium at this latter level by C-4 higher plants (Brownell 1979) provides guidance about how the question might be answered for marine fungi. Nevertheless in spite of the continuing uncertainty about why higher marine fungi and their terrestrial relatives are found in their respective habitats, considerable advances have been made since 1961 in our understanding of the physiology of fungi both terrestrial and marine. This chapter tries to put the physiological information obtained in the context of growth and reproduction of fungi in the sea and suggests those areas of priority demanding further experimental work. Earlier work in this area has been reviewed by Jennings (1983).

VEGETATIVE GROWTH - THE PRESENT POSITION

To date, without exception, studies on the vegetative growth of marine fungi have used radial growth rate on agar and dry matter production in batch culture as measures of growth. It is worth dilating on the inadequacy of these culture techniques, and this is best done from the point of view of considering sea water as the medium in which growth must occur. For a fungus, sea water poses three problems. First, it is a medium of relatively low water potential, second it contains relatively high concentrations of ions, being potentially capable of exerting toxic effects on cell processes and third, it has an alkaline pH. In batch culture, a fungus is able often to nullify relatively easily these properties of sea water which under natural conditions would pose problems for growth. Thus in culture the low water potential presents less of a problem, since the fungus is aided in the generation of the turgor necessary for growth by the relatively high concentration of the carbon source, from which organic solutes can be synthesized within the cytoplasm and these can make a significant contribution to the solute potential and therefore turgor potential. Thus the need to absorb ions to generate turgor is kept to a minimum. This same concentration of carbon source also leads to a very significant production of hydrogen ions, such that the mycelium is soon growing in an acid medium. Finally most studies have involved estimating growth as dry matter production, measurements being made frequently when the mycelium is either entering or in the stationary phase which reduces the value of the data obtained for comparative purposes.

Thus most culture studies have provided little information of ecological significance. There is a clear need for studies using continuous culture although there may be difficulties in growing filamentous species under such conditions. Yeasts therefore become prime candidates for those studies which attempt to probe some of the fundamental physiological features of fungi which allow them to grow in the sea. Indeed even in batch culture, yeasts can have advantages over filamentous forms as demonstrated by Hobot & Jennings (1981) who were able to obtain specific growth rates in a comparative study of the effects of salinity on growth of the marine yeast *Debaryomyces hansenii* (Zopf) Lodder et Kreger-van Rij and *Saccharomyces cerevisiae* Hansen.

Given that data obtained from culture studies have been unsatisfactory, what can we say about the physiology of fungi which will allow us to hypothesize better about why fungi grow in the sea? While the hypotheses presented below are to a degree speculative, they indicate nevertheless the kind of experimental work required to produce that better understanding of marine fungi.

STUDIES ON DEBARYOMYCES HANSENII

The best starting point is the study of Hobot & Jennings (1981) who followed up the observations of Norkrans (1966) and Norkrans & Kylin (1969) which showed that *D. hansenii* can tolerate far higher concentrations of sodium chloride than *S. cerevisiae* and that the former yeast is better able to extrude sodium in exchange for potassium. Hobot & Jennings showed that in acid media *D. hansenii* behaved like *S. cerevisiae* with respect to growth and potassium-sodium exchange. On the other hand, in alkaline media, in which *S. cerevisiae* was unable to grow, irrespective of whether or not sodium chloride was present or not, *D. hansenii* became better able to select for potassium against sodium and there was a much lower rate of proton extrusion than in acid media. The ability of bicarbonate to stimulate growth of *D. hansenii*, but not that of *S. cerevisiae* in acid pH and the low level of proton extrusion in alkaline media suggested that for *D. hansenii*: i) the production of an alkaline cytoplasm removes hydrogen ions from competing with sodium for exchange across the plasma membrane; ii) there may be a bicarbonate pump at the plasma membrane, which is not present in *S. cerevisiae*.

The supposition that cells of *D. hansenii* possess membrane transport systems which function in alkaline media and which are not present in *S. cerevisiae* was confirmed by studies by Comerford *et al.* (1985). They isolated from *D. hansenii* a membrane fraction, relatively free from mitochondrial membrane contamination, which contained a Mg^{2+}-ATPase activity which possessed a pH optimum of 6-6.5 when the cells were grown in non-saline conditions but an additional optimum at pH 8 when 1.5 M sodium chloride was present in the growth medium. One presumes from the fact that growth in such a medium occurs only after a lag (Hobot & Jennings 1981) that synthesis of the alkaline activity is stimulated by sodium chloride. Irrespective of whether the cells were grown in the presence or absence of sodium chloride, the activity functioning at alkaline pH was much less sensitive than the activity functioning at acid pH to 4-acetamido-4'-*iso*-thiocyanatostilbene-2,2'-disulphonate (SITS) (Figure 1.1) and to concentrations of sodium chloride greater

Figure 1.1 Plasma membrane Mg^{2+}-ATPase activity of *Debaryomyces hansenii*, after growth in complete medium together with 1.5 M sodium chloride, as a function of pH (100 mM Tris/maleic acid buffer) in the assay medium in the absence, ● ; and presence of 1.0 mM SITS, ■ . Also present in the assay medium 50 mM potassium chloride (Comerford *et al.* 1985).

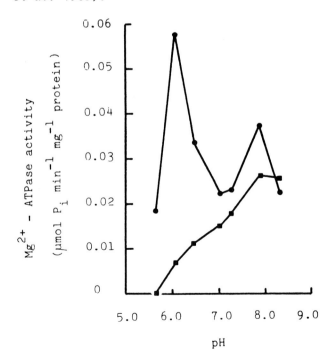

than 100 mM. Indeed the alkaline activity was stimulated by sodium chloride in the range 5-100 mM. The exact location of the ATPase activity extracted from *D. hansenii* was not clear; comparison of the results obtained with those from studies on other yeasts indicated that the activity may have come from both the plasma membrane and the tonoplast. With respect to what has been found for *D. hansenii*, it is interesting to note that Galpin & Jennings (1980) showed for the marine hyphomycete *Dendryphiella salina* (Sutherland) Pugh et Nicot that what was believed to be plasma membrane ATPase has a pH optimum of 8.5-9.5, though the properties of the enzyme were much less extensively studied.

In spite of the uncertainty about the location of the ATPase activity isolated from *D. hansenii*, it is not unreasonable to conclude that when cells are grown in high salinity there is increased activity of a membrane-bound ATPase operating at high pH which helps to maintain a favourable potassium/sodium balance in the cytoplasm. The stimulation by sodium of the alkaline activity is in keeping with this conclusion. Thus the observations described above indicate, in a very preliminary way, an explanation at the molecular level for the observations of Hobot & Jennings (1981) that *D. hansenii* and *S. cerevisiae* respond differently to high concentrations of sodium chloride in the medium.

THE IMPORTANCE OF THE VACUOLE
The indication that there might be a pump at both the plasma membrane and the tonoplast raises the question as to whether or not there is not only extrusion of sodium into the external medium but also into the vacuole. This cannot be assessed properly without a study of ion fluxes and electrical potential measurements, the latter particularly presenting technical difficulties.

With respect to accumulation of ions in the vacuole, it is pertinent to refer to studies by Wethered *et al.* (1985) on ion concentrations within mycelium of *D. salina* grown for 48 h in media containing high concentrations (0.8 osmol [kg H_2O]$^{-1}$) of either sodium chloride or sodium sulphate. Estimates of the protoplasmic sodium concentration were in the order of 280 mM in mycelium grown in the presence of sodium chloride and 180 mM in the presence of sodium sulphate. Studies on the effects of salts on glucose 6-phosphate, glycerol and malic dehydrogenases reported elsewhere in this volume (Gibb *et al.*, Chapter 4) suggest that the above concentrations of sodium in the cytosol are unlikely to have a significant effect on primary metabolism, though a certain minimum level of polyol may be necessary to exert a protective effect on certain enzymes. However, no experiments have yet been carried out to determine the effect of the above concentrations of sodium on the protein synthesis machinery of a marine fungus. It is this part of the metabolic apparatus within the cytoplasm which is likely to be most sensitive to an unfavourable potassium/sodium balance (Wyn Jones *et al.* 1979). Thus, although studies on enzymes of primary metabolism do not indicate a need to sequester sodium in the vacuole away from the metabolic machinery of the cell, further biochemical studies may in fact indicate the contrary.

Indeed, irrespective of whether or not there is a metabolic need, there is indeed evidence that sodium can be located in vacuoles. This evidence comes from a study of the efflux under quasi-steady state conditions of ^{24}Na from mycelium preloaded with the isotope (Wethered et al. 1985). The mycelium behaved as if it contained at least three compartments with respect to ^{24}Na. The fastest exchanging compartment was believed, on the basis of a similarity between the amount of sodium within it and the excess cation content of the mycelium, to be outside the plasma membrane. This being so, it would not be unreasonable to suppose that the other two compartments were cytoplasm and vacuole respectively. However, without a value for the vacuolar volume it is impossible to estimate the concentration of sodium bounded by the tonoplast. It needs to be noted that, while vacuoles appear to be readily discernible in cells of D. hansenii (Gezelius & Norkrans 1970; Lindman & Norkrans 1982), vacuoles do not appear very frequently in the younger hyphae of D. salina. Clearly more work is needed to clarify the situation.

SPORE GERMINATION – A QUESTION MARK

If sodium did need to be sequestered in vacuoles in the mature mycelium of a growing marine fungus, what must be the situation in a germinating spore, where vacuoles are absent? This is a significant question not only because there will be no compartment within the cytoplasm to sequester salt but we know from the x-ray microanalytical studies of the hyphal apex of D. salina that the plasma membrane there is less able to select for potassium against sodium than in older parts of the hypha (Galpin et al. 1978). Further, from histo-chemical staining it would appear that the plasma membrane ATPase is only present some distance from the apex; there is certainly no apparent ATPase activity to be seen in the hyphae as it emerges from the spore (Galpin & Jennings 1975). It would seem that the properties of the apex conspire against the maintenance of a low sodium concentration in the cytoplasm. How then does the germinating fungal spore cope with unfavourable external ionic environment? As far as the generation of turgor for growth is concerned, one can presume that this could be accomplished by the breakdown of insoluble material laid down in the spore when it develops. But the substantial question which remains unanswered is how the internal ionic environment is controlled as the spore germinates. This is a matter which requires urgent attention.

THE ROLE OF POLYOLS IN THE GROWING MYCELIUM OF MARINE FUNGI

Ions appear to contribute something like 60% of the solute potential in 48 h-old mycelium of D. salina grown in the presence of high concentrations of sodium chloride (Wethered et al. 1985). Polyols contribute 30%, the bulk of the remainder being amino nitrogen (which is perhaps not surprising in view of the use of tryptone as the nitrogen-source in the growth medium). Presumably if salt is sequestered in vacuoles, polyols will make a greater contribution to the cytoplasmic solute potential.

There is good evidence that the polyol content of fungi increases with salinity (Jennings 1983; Wethered *et al.* 1985). There is considerable emphasis on the increased synthesis of glycerol by fungi growing under saline conditions and in media of low water potential (Jennings 1983). However, it is clear from studies on *D. salina* that glycerol must not be singled out as playing a special role under the above conditions. These studies (Wethered *et al.* 1985) showed that, while there was certainly increased synthesis of total polyol in growing mycelium of *D. salina* in response to increased medium salinity of different kinds and decreased medium water potential, the proportions of the individual polyols differed depending upon the medium (Table 1.1). Glycerol was certainly the polyol at highest concentration in mycelium grown in high concentrations of sodium chloride but the place of glycerol was taken by mannitol when magnesium chloride or sodium sulphate were at high concentration in the medium.

It has been assumed that polyols, while contributing to the solute potential of a fungus growing in a medium of high salt concentration, or low water potential, are also synthesized because they protect enzymes from the deleterious effects of any high concentration of salt which is generated within the cytoplasm of the fungus. There is certainly good evidence that polyols can exert such a protective effect and it is this evidence that has led to polyols being categorised amongst the so-called compatible solutes (Brown 1978). However, the data of Gibb *et al.* presented in chapter 4 of this volume indicate that we must be careful about taking a too simplistic view of the protective role of glycerol (and indeed other polyols). Likewise Jennings (1983) has urged caution over accepting the assumption that glycerol is the major osmoregulatory solute in fungi.

It has been striking to find that when *D. salina* is grown in the presence of relatively high concentrations of either sodium chloride or magnesium chloride or sodium sulphate or the relatively slowly metabolizable carbohydrate inositol at a given osmolality, the total soluble carbohydrate concentration in the mycelium is constant (Wethered *et al.* 1985). The same constancy of total concentration can result after feeding the non-metabolizable carbohydrate 3-*O*-methyl glucose to non-growing mycelium (Jennings & Austin 1973; McDermott & Jennings 1976). As indicated above, the concentrations of the individual polyols and other carbohydrates present, e.g. inositol and 3-*O*-methyl glucose, can differ very considerably between the treatments. These observations need to be considered alongside the view of Jennings (1985) that polyols have an important role in fungi in the control of cytoplasmic pH. This view is based: a) on the observation of Sanders & Slayman (1982) that, while the cytoplasmic pH of *Neurospora crassa* Shear et Dodge can be regulated by the extrusion of protons via the proton pump in the plasma membrane, it seems more likely under many conditions that cytoplasmic pH is regulated by primary metabolism; b) that the processes involved in the production and breakdown of polyols are not only oxidation-reduction reactions but reactions which lead both to the removal of protons (reduction of sugar to polyol) and their production (oxidation of polyol to sugar). Direct synthesis of polyol from sugar, as is the case for the synthesis of mannitol from glucose via fructose

Table 1.1 Concentration (mol l⁻¹) of solutes in mycelium of *Dendryphiella salina* after 48 h growth in media containing various osmotica to give a final water potential of 0.4 or 0.8 osmol [kg/H₂O] −1. The media were initially buffered to pH 7.2 with 50 mM Tris/HCl. Results are the means of four replicates (Wethered *et al.* 1985)

External osmoticum	Sodium chloride		Magnesium chloride		Sodium sulphate		Inositol	
osmol [kg H₂O]⁻¹	0.4	0.8	0.4	0.8	0.4	0.8	0.4	0.8
glycerol	0.098	0.197	0.038	0.121	0.039	0.062	–	–
erythritol	–	0.006	0.006	–	0.002	0.002	–	–
arabitol	0.007	0.048	0.034	0.065	0.033	0.051	–	0.013
mannitol	0.086	0.102	0.142	0.128	0.126	0.129	0.082	0.082
inositol	–	–	–	–	–	–	0.091	0.214
total	0.191	0.353	0.220	0.314	0.200	0.244	0.173	0.309

– = not present.

could help to increase cytoplasmic pH. On the other hand, the production of a pentitol such as arabitol from glucose will decrease the pH, since the overall process is oxidative owing to the involvement of the pentose phosphate pathway. Thus the constancy of soluble carbohydrate concentration referred to above may be a consequence of the fact that pathways for the synthesis and breakdown of the four major polyols (arabitol, erythritol, glycerol and mannitol) are acting in concert so producing a pH-stat.

The presence of such a pH-stat may be particularly significant for higher marine fungi, insofar that sodium extrusion may far outweigh proton extrusion such that the latter process cannot play a dominant role in the control of cytoplasmic pH. The relative constancy of the medium pH when *D. hansenii* is grown in the presence of 1.5 M sodium chloride at pH 8.3 (Hobot & Jennings 1981) would support this view.

THRAUSTOCHYTRIUM

This genus is the only one of those zoosporic genera referred to above which require sodium for growth for which there is any information about the physiology of its species. With respect to what has been said above for higher marine fungi, there are two relevant pieces of information. First, there is good evidence that phosphate uptake in *T. roseum* Goldstein requires sodium (Siegenthaler *et al.* 1967 a,b). This suggests that the anion is co-transported with sodium and one must presume for such a co-transport process to be effective there must be a sodium extrusion pump also present in the outer membrane. Second, while inorganic ions make the major contribution (91-94%) to the solute potential of the cells, the amino acid proline is the major organic solute present and further its concentration in cells increases with increased salinity of the medium (Wethered & Jennings 1985). While we do not have any information as to how proline is synthesized or broken down in *Thraustochytrium*, it is almost certain from information obtained from bacteria and higher plants and particularly from other fungi (Pateman & Kinghorn 1976) that the amino acid is synthesized reductively from glutamate and the process requires NAD(P)H and ATP. Break-down occurs by conversion back to glutamate and is an oxidative process. Glutamate is present at significant concentrations in *T. aureum* Goldstein and *T. roseum* (Wethered & Jennings 1985). I wish to suggest therefore that for *Thraustochytrium*, as for *D. hansenii*, sodium extrusion across the plasma membrane takes precedence over proton extrusion. Since this is so, there is a need for the metabolic control of cytoplasmic pH and this is brought about by the reductive synthesis and oxidative breakdown of proline. With respect to these ideas, it is interesting to note that Pesci & Beffagna (1985) indicate from studies on the effect of isobutyric acid, that ABA-induced changes in proline level in barley leaves are mediated by changes in intra-cellular pH.

CONCLUSION

There is much speculation in the foregoing but hopefully

that speculation should lead to further experiments. My own feeling, as far as the higher marine fungi are concerned, is that we are achieving a better understanding of their physiology to the extent that we can see in a general way how they are able to function vegetatively in the marine environment. But we have a long way to go before we can relate physiological information to the much greater body of ecological information. Thus at a very minimum we are not yet certain of why marine fungi are restricted to the sea. Nor are we as yet certain why terrestrial fungi are not isolated from marine environments, although it does seem likely that the combination of high salinity and alkaline environment, may restrict invasion of the sea by this group of fungi.

REFERENCES

Brown, A.D. (1978). Compatible solutes and extreme water stress in eukaryotic micro-organisms. Adv. Microbial Physiol., $\underline{17}$, 181-242.

Brownell, P.F. (1979). Sodium as an essential micronutrient element for plants and its possible role in metabolism. Adv. Bot. Res., $\underline{7}$, 118-224.

Commerford, J.G., Spencer-Phillips, P.T.N. & Jennings, D.H. (1985). Membrane-bound ATPase activity, the properties of which are altered by growth in saline conditions, isolated from the marine yeast *Debaryomyces hansenii*. Trans. Br. mycol. Soc., $\underline{85}$, 431-438.

Galpin, M.F.J. & Jennings, D.H. (1975). Histochemical study of the hyphae and the distribution of adenosine triphosphatase in *Dendryphiella salina*. Trans. Br. mycol. Soc., $\underline{65}$, 477-483.

Galpin, M.F.J. & Jennings, D.H. (1980). A plasma-membrane ATPase from *Dendryphiella salina*: cation specificity and interaction with fusicoccin and cyclic AMP. Trans. Br. mycol. Soc., $\underline{75}$, 35-46.

Galpin, M.F.J., Jennings, D.H., Oates, K. & Hobot, J.A. (1978). Localisation by X-ray microanalysis of soluble ions, particularly potassium and sodium in fungal hyphae. Exp. Mycol., $\underline{2}$, 258-269.

Gezelius, K. & Norkrans, B.L. (1970). Ultrastructure of *Debaryomyces hansenii*. Arch. Mikrobiol., $\underline{70}$, 14-25.

Hobot, J.A. & Jennings, D.H. (1981). Growth of *Debaryomyces hansenii* and *Saccharomyces cerevisiae* in relation to pH and salinity. Exp. Mycol., $\underline{5}$, 217-228.

Jennings, D.H. (1983). Some aspects of the physiology and biochemistry of marine fungi. Biol. Rev., $\underline{58}$, 423-459.

Jennings, D.H. (1985). Polyol metabolism in fungi. Adv. Microbial Physiol., $\underline{25}$, 149-193.

Jennings, D.H. & Austin, S. (1973). The stimulatory effect of the non-metabolisable sugar 3-*O*-methyl glucose on the conversion of mannitol and arabitol to polysaccharide and other insoluble compounds in the fungus *Dendryphiella salina*. J. Gen. Microbiol., $\underline{75}$, 287-294.

Johnson, T.W. & Sparrow, F.K. (1961). Fungi in Oceans and Estuaries. Weinheim: Cramer.

Lindman, B. & Norkrans, B. (1982). Membranous interrelationships - an ultrastructural study of the halotolerant yeast *Debaryomyces hansenii*. Protoplasma, 110, 66-70.

McDermott, J.C.B. & Jennings, D.H. (1976). The relationship between the uptake of glucose and 3-*O*-methylglucose and soluble carbohydrate and polysaccharide in the fungus *Dendryphiella salina*. J. Gen. Microbiol., 97, 193-209.

Norkrans, B. (1966). Studies on marine occurring yeasts. Growth related to pH, NaCl concentrations and temperature. Arch. Mikrobiol., 54, 374-392.

Norkrans, B. & Kylin, A. (1969). Regulation of potassium to sodium and of the osmotic potential in relation to salt tolerance in yeasts. J. Bacteriol., 100, 836-845.

Pateman, J.A. & Kinghorn, J.R. (1976). Nitrogen metabolism. In The Filamentous Fungi Vol. 2. Biosynthesis and Metabolism, ed. J.E. Smith & D.R. Berry, pp. 159-237. London: Edward Arnold.

Pesci, P. & Beffagna, N. (1985). Effects of weak acids on proline accumulation in barley leaves: a comparison between abscisic acid and isobutyric acid. Plant, Cell & Environ., 8, 129-134.

Sanders, D. & Slayman, C.L. (1982). Control of intracellular pH. Predominant role of oxidative metabolism, not proton transport, in the eukaryotic micro-organism *Neurospora*. J. Gen. Physiol., 80, 377-402.

Siegenthaler, P.A., Belsky, M.M. & Goldstein, S. (1967 a). Phosphate uptake in an obligately marine fungus: a specific requirement for sodium. Science, 155, 93-94.

Siegenthaler, P.A., Belsky, M.M., Goldstein, S. & Menna, M. (1967 b). Phosphate uptake in an obligately marine fungus. II. Role of culture conditions, energy sources and inhibitors. J. Bacteriol., 93, 1281-1288.

Wethered, J.M. & Jennings, D.H. (1985). The major solutes contributing to the solute potential of *Thraustochytrium aureum* and *T. roseum* after growth in media of different salinities. Trans. Br. mycol. Soc., 85, 439-446.

Wethered, J.M., Metcalf, E.C. & Jennings, D.H. (1985). Carbohydrate metabolism in the fungus *Dendryphiella salina*. VIII. The contribution of polyols and ions to the mycelial solute potential in relation to the external osmoticum. New Phytol., 101, 631-649.

Wyn Jones, R.G., Brady, C.J. & Speirs, J. (1979). Ionic and osmotic relations in plant cells. In Recent Advances in the Biochemistry of Cereals, ed. D.L. Laidman & R.G. Wyn Jones, pp. 63-103. London, New York & San Francisco: Academic Press.

D.G. Ahearn

S.A. Crow

INTRODUCTION

Hydrocarbons from various sources (e.g. anthropogenic pollution, marine seeps, marine algae, atmospheric fallout and terrestrial runoff) enter the ocean daily. These complex hydrocarbon mixtures are dispersed and degraded by abiotic and biogenic processes (Floodgate 1984). The rate of degradation and the significance of microbial activities in the fate of oceanic hydrocarbons vary with environmental conditions and the type of hydrocarbon. Most commonly, bacteria are considered the primary degraders, with algae and fungi having minor roles. Although implied in a number of cases, the degradation of complex hydrocarbon mixtures by a successional microflora containing temporally isolated populations of bacteria and fungi, has been inadequately studied.

METABOLISM

The majority of detailed metabolic studies on hydrocarbon metabolism by fungi have been carried out with geophilic yeasts (Shennan & Levi 1974); however, a variety of fungi from marine habitats also possess the capacity to metabolize hydrocarbons (Ahearn *et al*. 1971; Cerniglia 1984). These diverse groups of fungi commonly utilize short-chained n-alkanes (C_8-C_{18}). As in most microorganisms, the primary route of metabolism in the fungi is monoterminal oxidation to the corresponding alcohol, aldehyde and fatty acid. The fatty acid may then be oxidized to acetate through *beta* oxidation or may be incorporated into cellular lipids (Singer & Finnerty 1984). Finnerty and coworkers (Modrzakowski *et al*. 1977; Modrzakowski & Finnerty 1980) have suggested an alternate pathway for alkane oxidation from evidence employing the alkane analogue dioctyl ether. We have been unable to demonstrate growth on dioctyl ether by yeasts, although several species of *Candida* are capable of growth on the C_6 and C_{10} analogues. Diterminal oxidation yielding diterminal fatty acids is known (Rehm & Reiff 1981). Here the monocarboxylic fatty acid is oxidized to the ω-hydroxy fatty acid and then to the dicarboxylic fatty acid. Certain fungi appear to oxidize alkanes exclusively by subterminal oxidation to secondary alcohols and then to ketones (Rehm & Reiff 1981). Alkenes are oxidized at sites of unsaturation as well as terminally. Branched chain hydrocarbons such as pristane (2,6,10,14-tetramethylpentadecane) are oxidized or cooxidized by certain yeasts (Hagihara *et al*. 1977). We have found that all of the 37 isolates of *Candida lipolytica* (Harrison)

Diddens et Lodder in our collection, including 12 isolates of marine
origin, assimilated pristane as a sole source of carbon for growth.
C. lipolytica also grew with squalene (although more slowly than with
pristane) as the substrate (unpublished data). *Candida maltosa*
Komagata, Nakase et Katsuya did not utilize these substrates as a sole
source of carbon for growth but apparently cooxidized pristane in the
presence of tetradecane (Crow *et al.* 1980).

Few fungi metabolize cycloalkanes. Komagata *et al.* (1964) examined 498
yeast strains and found that none was capable of utilizing cyclopentane.
In our unpublished studies, cycloalkanes have not appeared toxic to
representative yeasts. The metabolism of cycloalkanes in the fungi
apparently proceeds via cooxidation. Oxidation of aromatic compounds
such as phenol, salicylate, benzoate and mandelate has been found to be
an inducible property of *Candida tropicalis* (Castellani) Berkout,
Trichosporon spp. and *Rhodotorula* spp., yeasts that represent
hydrocarbonoclastic taxa. Various species of these genera oxidize
mandelate and benzoate via the protocatechuate branch and salicylate by
the catechol branch of the β-ketoadipate pathway (Durham *et al.* 1984;
Mortberg & Neujahr 1985). We have found that isolates of *Rhodotorula*
and *Trichosporon* grow in sediment and water samples from coastal
habitats that have been enriched with low levels of glucose (<2 μM) and
phenol (5–15 μM) but species of *Candida* with at least some fermentative
capacity become predominant more often at glucose levels of 5 μM or
greater or with the addition of 0.1% yeast extract (unpublished data).
Polycyclic aromatic hydrocarbons (PAHs) are oxidized by a variety of
fungi and moulds from the marine environment (Cerniglia & Crow 1981;
Cerniglia 1984). Polycyclic aromatic hydrocarbons are apparently
oxidized via a cytochrome P-450 monooxygenase to arene oxides that
either isomerize to phenols or are enzymatically hydrated to
transdihydridiols.

DISTRIBUTION OF MARINE FUNGI

Fungi that are endemic to the marine environment are found
mostly in the euphotic zone, particularly in littoral regions. A few
cellulolytic species are found to depths of about 1 km; but in general,
marine fungi seem to be rare in the ocean depths (Kohlmeyer & Kohlmeyer
1979). Kohlmeyer (1983) stated that the major factors controlling the
distribution of marine fungi were availability of substrates or hosts,
temperature, hydrostatic pressure, and oxygen. However *in situ*
evidence that fungi express hydrocarbonoclastic activity in the natural
environment is sparse. Mostly, evidence of activity is indirect;
implied on the basis that a variety of filamentous fungi, many of which
are known to degrade hydrocarbons, has been isolated from the oceans.
Marine arenicolous and lignicolous fungi have not been investigated for
hydrocarbon degradation and there is no evidence that associates any of
these fungi with hydrocarbon enrichment.

The ubiquitous presence of yeasts in the oceans has been amply
documented (Hagler & Ahearn 1986). Apparently most yeasts enter the
oceans with terrestrial runoff. *Debaryomyces hansenii* (Zopf) Lodder et

Kreger-van Rij, the *Candida parapsilosis* (Ashford) Langeron et Talice complex, and *Rhodotorula* spp., for example, seem physiologically adapted for survival in low nutrient conditions and increase in density upon their fortuitous encounter with regions of increased nutrients (Ahearn & Meyers 1976; Fell 1976). *Metschnikowia* spp. may be commensals and pathogens for planktonic crustaceans (Seki & Fulton 1969). Yeast densities may be particularly high (10^3-10^5 ml $^{-1}$) in surface films (Crow *et al.* 1975; Crow *et al.* 1976) and in estuarine marshes (>9×10^4 cm $^{-3}$) (Ahearn *et al.* 1976). Increased densities of yeasts in offshore water have been associated with algal blooms (Meyers *et al.* 1967) and it is probable that yeasts are active in the decomposition of algal products (Gunkel *et al.* 1983).

ASSOCIATION OF FUNGI WITH OIL

A few geophilic moulds, with hydrocarbonoclastic capacities such as *Aspergillus versicolor* (Vuillernin) Tiraboschi and *Cladosporium* spp. show widespread oceanic distribution (Roth *et al.* 1964; Muntañola-Cvetkovic´ & Ristanovic´ 1980). Geophilic Hyphomycetes that are known to develop on decaying organic matter in littoral zones (Kirk 1983) include hydrocarbonoclastic species (Bossert & Bartha 1984). The role of these filamentous fungi in the biodegradation of hydrocarbons in the oceans is unclear. In reviewing factors that may account for the paucity of information on the environmental role of moulds in the degradation of hydrocarbons in freshwater, Cooney (1984) indicated that most environmental studies have centred on bacteria with media and other conditions unfavourable for optimal isolation of moulds or for their utilization of hydrocarbons. These observations may be applied to some marine studies. Fedorak *et al.* (1984) reported that only 29% of 224 moulds from a marine environment altered the n-alkane profile of crude oil. Grüttner & Jensen (1983) found that mould populations in a marine coastal habitat tended to be higher with proximity to an oil refinery. Over 90% of the isolates were able to grow with oil as the only carbon source. Fungi in hydrocarbon enriched littoral regions usually seem to develop secondary to bacteria. In samples from certain of our hydrocarbon enriched ecocores (Spain & Somerville 1985), moulds developed only after several months (unpublished data). This delay in fungal development has been noted also for samples collected from the Amoco Cadiz spill (Ahearn & Crow 1980).

Yeasts and species of the genera *Acremonium*, *Aureobasidium* and *Cladosporium*, which may produce single budding cells under certain conditions, have been associated occasionally (but more often than moulds) with oil pollution of marine habitats. LePetit *et al.* (1970) found that two hydrocarbonoclastic species, *Candida tropicalis* and *C. lipolytica*, occurred in increased densities in littoral regions influenced by effluents from an oil refinery. *Candida lipolytica*, one of the most active hydrocarbonoclastic yeasts, was isolated from the vicinity of wells in the North Sea (Crow *et al.* 1977). The rare occurrence of this species in sea water suggests the presence of hydrocarbons (LePetit *et al.* 1970).

In marshlands dominated by *Spartina alterniflora* Loisel, the native yeast species *Pichia spartinae* Ahearn, Yarrow et Meyers and *Kluyveromyces drosophilarum* (Shehata, Mrak et Phaff) van der Walt, which are not hydrocarbonoclastic, were replaced gradually after the addition of oil by species of *Rhodotorula*, *Trichosporon* and *Pichia* that utilized alkanes for growth (Ahearn *et al.* 1976). *Candida tropicalis* was found to be associated with inshore waters of Biscayne Bay, Florida, but it was not observed to occur in increased densities in proximity to occasional oil slicks (Ahearn 1973). In offshore waters, *Rhodotorula glutinis* (Fresenius) Harrison and *Candida parapsilosis* showed increased densities in surface waters influenced by a spewing oil well; however, these increased densities were not observed in a repeated sampling several weeks later even though oil was still present (Ahearn *et al.* 1971). A possible influence of offshore drilling activities on yeast flora in the North Sea was noted in the comparison of species isolated in 1964-66 with species isolated in 1976 (Ahearn & Crow 1980). *Debaryomyces hansenii*, which shows slow growth on short-chained alkanes, was the predominant species in both sets of samples; but after oil production, *Candida guilliermondii* (Castellani) Langeren et Guerra, a hydrocarbonoclastic yeast, was encountered three times more frequently. Successional development of yeast species with more hydrocarbonoclastic species (*Candida tropicalis* and *Candida guilliermondii* replacing the early developing *Rhodotorula*) has been observed in samples of beached oil from the Amoco Cadiz (Ahearn & Crow 1980).

Yeasts were not observed with an enriched culture medium in our laboratory studies of ground water tainted with gasoline leaking from a storage tank; however, with basal salts agar and extended incubation, *Candida guilliermondii* at densities of 10^3-10^4 cells 1 $^{-1}$ was recovered (Cook & Raymond unpublished). The more volatile hydrocarbons may alter the plasma membrane and, dependent upon concentration, may either stimulate or inhibit growth (Ahearn *et al.* 1971; Gill & Ratledge 1972). The emergence of hydrocarbonoclastic fungi also may be dependent on shifts in pH. In enclosed bunkers of oil tankers that employed sea water compensation systems, approximately 20% of 80 systems were dominated by *Cladosporium resinae* (Lindau) de Vries and *Candida* spp. (Neihof & May 1983). *Cladosporium resinae* grows slowly at the pH of sea water, and organic acid accumulation may be necessary for stimulation of growth (Neihof & May 1983). Microhabitats dominated by fungi may occur in water in oil emulsions and may result in transient periods of yeast and mould degradation of crude oil in marine habitats (Figure 2.1).

UPTAKE OF HYDROCARBONS

Singer & Finnerty (1984) stated that uptake of alkane by yeasts may involve: (1) adsorption to the cell via hydrophobic surface projections; (2) movement through the rigid cell wall via pores or channels with initial accumulation on the outer surface of the cell membrane; (3) movement of the alkane via pinocytosis to microbodies and other sites of oxidation (intra-cytoplasmic inclusions may also be formed). The immediate steps in hydrocarbon uptake are probably energy independent, but facilitated by various complexes of fatty acids with protein and (or) carbohydrates. Kappeli and coworkers have extensively examined the uptake and transport of hydrocarbons in *C. tropicalis* (Käppeli & Fiechter 1977; Käppeli *et al.* 1978; Käppeli & Fiechter 1981). The organism produced a mannan-fatty acid complex that rapidly emulsified the hydrocarbon and induced significant uptake of pure hydrocarbon. *Candida lipolytica* has also been shown to produce an extracellular product capable of inducing both emulsification and uptake of hydrocarbons (Singer & Finnerty 1984). These bioemulsifiers vary in nature or quantity with the particular yeast.

CONCLUSION

While the natural populations of yeasts and moulds in open ocean water would seem to be insufficient to produce significant alterations of the composition of crude oils, unique regions such as the surface film, decomposing algae, or tar-balls may contain highly active levels of fungi. The action of the fungi in these habitats may be noteworthy. During the process of growth on crude oil or mixed hydrocarbons, simultaneous transport of hydrocarbons may occur, resulting in enhanced oxidation of essentially nonwater soluble aromatics such as biphenyl and naphthalene or branch chained alkanes such as pristane. Crow & Bell (1981) have observed the uptake of naphthalene by cells grown on tetradecane and naphthalene. Furthermore, significant uptake of pristane was observed in a yeast incapable of growth on this substrate. The uptake and release of compartmentalized hydrocarbons from eukaryotic microorganisms remains an unquantitated factor in the fate of hydrocarbons in the marine environment.

Figure 2.1 Yeast and mould growth in a smear of beached oil collected from the coast of Louisiana, USA. Bar = 10 µm.

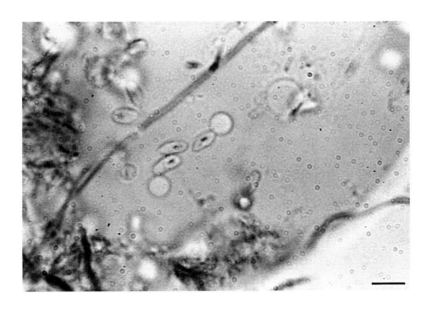

The observed associations do not provide definitive evidence that fungi degrade oils in the marine environment. Nevertheless, the occasional association of certain species with oil enrichment, particularly successional habitation, coupled with knowledge that these developing taxa possess varied hydrocarbonoclastic activities, suggests an active role for fungi. Often fungi do not attain significant densities in an oil-enriched habitat until at least several months after the addition of oil. Since the abiotic parameters of volatilization and aquatic dispersion provide, in the main, for the relatively fast disappearance of oil in open waters, the sites for hydrocarbon degradation by fungi in marine habitats would probably be best found where oil is transported to coastal habitats. Further studies are necessary to elucidate the role of fungi in hydrocarbon degradation in the marine environment.

ACKNOWLEDGEMENT
          The information in this chapter has been funded wholly or in part by the U.S. Environmental Protection Agency under the cooperative agreement number R-809370 awarded to Georgia State University, Atlanta, Georgia.

REFERENCES
Ahearn, D.G. (1973). Effects of environmental stress on aquatic yeast populations. In Estuarine Microbial Ecology, ed. L. Harold Stevenson & R.R. Colwell, pp. 433–439. Columbia, South Carolina: University of South Carolina Press.

Ahearn, D.G. & Crow, S.A. (1980). Yeasts from the North Sea and Amoco Cadiz Oil. Bot. Mar., 23, 123–127.

Ahearn, D.G., Crow, S.A., Berner, N.H. & Meyers, S.P. (1976). Microbiological cycling of oil in estuarine marshlands. In Estuarine Processes, ed. J. Wiley, pp. 483–493. New York: Academic Press.

Ahearn, D.G. & Meyers, S.P. (1976). Fungal degradation of oil in the marine environment. In Recent Advances in Aquatic Mycology, ed. E.B. Gareth Jones, pp. 125–134. London: Elek Science.

Ahearn, D.G., Meyers, S.P. & Standard, P.G. (1971). The role of yeasts in the decomposition of oils in marine environments. Devl. industr. Microbiol., 12, 126–134.

Bossert, I. & Bartha, R. (1984). The Fate of Petroleum in Soil Ecosystems. In Petroleum Microbiology, ed. R. Atlas, pp. 435–474. New York: MacMillan Press.

Cerniglia, C.E. (1984). Microbial metabolism of polycyclic aromatic hydrocarbons. Adv. Appl. Microbiol., 30, 31–71.

Cerniglia, C.E. & Crow, S.A. (1981). Metabolism of aromatic hydrocarbons by yeasts. Arch. Microbiol., 129, 9–13.

Cooney, J.J. (1984). The fate of petroleum pollutants in freshwater ecosystems. In Petroleum Microbiology, ed. R. Atlas, pp. 399–434. New York: MacMillan.

Crow, S.A., Ahearn, D.G., Cook, W.L. & Bouquin, A.W. (1975). Densities of bacteria and fungi in coastal-surface films as determined by a membrane adsorption procedure. Limnol. Oceanogr., 20, 644–646.

Crow, S.A. & Bell, S.L. (1981). Effects of aromatic hydrocarbons on
    growth of *Candida maltosa* and *Candida lipolytica*. Devl.
    Industr. Microbiol., 22, 37-442.

Crow, S.A., Bell, S.L. & Ahearn, D.G. (1980). The uptake of aromatic
    and branched chain hydrocarbons by yeast. Botanica mar., 13,
    117-120.

Crow, S.A., Bourquin, A.W., Cook, W.L. & Ahearn, D.G. (1976).
    Microbiological populations in coastal surface slicks. In
    Proc. Third International Biodegradation Symposium, ed.
    J.M. Sharpley & A.M. Kaplan, pp. 93-98. London: Applied
    Science Publishers.

Crow, S.A., Bowman, P.I. & Ahearn, D.G. (1977). Isolation of atypical
    Candida albicans from the North Sea. Appl. environ.
    Microbiol., 33, 738-739.

Durham, D.R., McNamee, C.G. & Stewart, D.D. (1984). Dissimilation of
    aromatic compounds in *Rhodotorula graminis*: Biochemical
    characterization of pleiotropically negative mutants. J.
    Bact., 160, 771-777.

Fedorak, P.M., Semple, K.M. & Westlake, D.W. (1984). Oil-degrading
    Capabilities of Yeasts and Fungi from Coastal Marine
    environments. Can. J. Microbiol., 30, 565-571.

Fell, J.W. (1976). Yeasts in oceanic regions. In Recent Advances in
    Aquatic Mycology, ed. E.B. Gareth Jones, pp. 93-125.
    London: Elek Science.

Floodgate, G. (1984). The Fate of Petroleum in Marine Ecosystems. In
    Petroleum Microbiology, ed. R. Atlas, pp. 355-398. New
    York: MacMillan.

Gill, C.O. & Ratledge, C. (1972). Effect of *n*-alkanes on the transport
    of glucose in *Candida* sp. strain 107. Biochem. J., 127,
    59-60.

Grüttner, H. & Jensen, K. (1983). Effects of chronic oil pollution from
    refinery effluent on sediment microflora in a Danish coastal
    area. Mar. Poll. Bull., 14, 436-459.

Gunkel, W., Crow, S.A. & Klings, K.W. (1983). Yeast population
    increases during degradation of *Desmarestia viridis*
    (Phaeophyceae) in seawater model microecosystems. Mar.
    Biol., 75, 327-332.

Hagihara, T., Mishina, M., Tanaka, A. & Fukui, S. (1977). Utilization of
    Pristane by a yeast. *Candida lipolytica* fatty acid
    composition of pristane-grown cells. Agric. biol. Chem.,
    41, 1745-1748.

Hagler, A.N. & Ahearn, D.G. (1986). Ecology of aquatic yeasts. In
    The Yeasts, ed. A.H. Rose. London: Academic Press, in
    press.

Käppeli, O. & Fiechter, A. (1977). Component from the cell surface of
    the hydrocarbon-utilizing yeast *Candida tropicalis* with
    possible relation to hydrocarbon transport. J. Bact., 131,
    917-921.

Käppeli, O. & Fiechter, A. (1981). Properties of hexadecane uptake by
    Candida tropicalis. Curr. Microbiol., 6, 21-26.

Käppeli, O., Müller, M. & Fiechter, A. (1978). Chemical and structural
    alterations at the cell surface of *Candida tropicalis*
    induced by hydrocarbon substrate. J. Bact., 133, 952-958.

Kirk, P.W. Jr. (1983). Direct enumeration of marine arenicolous fungi. Mycologia, 75, 670-682.

Kohlmeyer, J. (1983). Geography of marine fungi. Aust. J. Bot. Suppl. Ser., 10, 67-76.

Kohlmeyer, J. & Kohlmeyer, E. (1979). Marine Mycology. The Higher Fungi. New York: Academic Press.

Komagata, K., Nakase, T. & Katsuya, N. (1964). Assimilation of hydrocarbons by yeasts. Preliminary screening. J. gen. Appl. Microbiol., Tokyo, 10, 313-321.

LePetit, J., N'Guyen, M.H. & Deveze, L. (1970). Étude de l'intervention des levures dans la bio-dégradation en mer des hydrocarbures. Annls. Inst. Pasteur, Paris, 118, 709-720.

Meyers, S.P., Ahearn, D.G., Gunke, W. & Roth, F.J. Jr. (1967). Yeasts of the North Sea. Mar. Biol., 1, 118-123.

Modrzakowski, M.C. & Finnerty, W.R. (1980). Metabolism of symmetrical dialkyl ethers by Acinetobacter sp. HO1-N. Arch. Microbiol., 126, 285-290.

Modrzakowski, M.C., Makula, R.A. & Finnerty, W.R. (1977). Metabolism of the alkane analogue n-dioctyl ether by Acinetobacter species. J. Bact., 131, 92-97.

Mortberg, M. & Neujahr, H.Y. (1985). Uptake of phenol by Trichosporon cutaneum. J. Bact., 161, 615-619.

Muntañola-Cvetkovic´, M. & Ristanovic´, B. (1980). A mycological survey of the South Adriatic Sea. J. exp. mar. Biol. Ecol., 43, 193-206.

Neihof, R. & May, M. (1983). Microbial and particulate contamination in fuel tanks on naval ships. Intern. Biodet. Bull., 19, 59-68.

Rehm, H.J. & Reiff, I. (1981). Mechanisms and occurrence of microbial oxidation of long-chain alkanes. Adv. biochem. Engng., 19, 173-215.

Roth, F.J. Jr., Orpurt, P.A. & Ahearn, D.G. (1964). Occurrence and distribution of fungi in a subtropical marine environment. Can. J. Microbiol., 42, 375-383.

Seki, H. & Fulton, J. (1969). Infection of marine copepods by Metschnikowia sp. Mycopath. Mycol. appl., 38, 61-70.

Shennan, J.L. & Levi, J.D. (1974). The growth of yeasts on hydrocarbons. Prog. industr. Microbiol., 13, 1-57.

Singer, M.E. & Finnerty, W.R. (1984). Microbial metabolism of straight-chain and branched alkanes. In Petroleum Microbiology, ed. R.M. Atlas, pp. 1-60. New York: MacMillan.

Spain, J.C. & Somerville, C.C. (1985). Fate and toxicity of high density missile fuels Rd-5 and JP-9 in aquatic test systems. Chemosphere, 14, 239-248.

3 MEASURING FUNGAL-BIOMASS DYNAMICS IN STANDING-DEAD LEAVES OF A SALT-MARSH VASCULAR PLANT

S.Y. Newell

R.D. Fallon

J.D. Miller

INTRODUCTION

Smooth cordgrass (*Spartina alterniflora* Loisel.) experiences little competition from other rooted plants in the intertidal marshes where it grows; thus it is a highly productive plant (e.g. 2840 above-ground g m^{-2} yr^{-1}, coastal Georgia: Schubauer & Hopkinson 1984; compare approx. 2000 g m^{-2} yr^{-1} for tropical rain forests: Tiner 1984). Much of the aerial standing crop of cordgrass consists of dead material, because leaves are not abscised, and stems are generally protected from strong wave action; in Georgia, the annual average dead : live ratio for the standing crop is 1.9 : 1, and the range during a year for standing dead crop in marshes of median height is about 200-1200 kg m^{-2} (Schubauer & Hopkinson 1984).

It has recently been found that this standing-dead material supports a microbial assemblage which is very well adapted to periods of wetness (tides, rain, fog, high humidity) interspersed within periods of extreme dryness (water content <20% of fresh weight) (Newell *et al.* 1986). Rates of carbon dioxide release by standing-dead cordgrass during wet periods are comparable to the maximal rates reported for several types of terrestrial vascular plant litter. Bacterial and algal contributions to the carbon mineralization of the standing-dead cordgrass system were estimated (^{3}H-thymidine incorporation into DNA, and light-stimulated $^{14}CO_2$ fixation, respectively) to amount to a combined 30% or less of the total in some samples (Newell *et al.* 1986). Thus the standing-dead *Spartina* system appears to be one in which saprotrophic activity by fungi and potential fungal production are substantial.

This chapter presents the results of an attempt to measure the changes in fungal biomass content of cordgrass leaves as the leaves aged from the time of senescence. The results of parallel measurements obtained via direct fluorescence microscopy, ergosterol analysis, and immunosorbent assay are compared.

METHODS

Yellow-green, senescent leaves of *S. alterniflora* plants of median height were tagged in place with plastic cable ties (Ty-Rap TY23M-3) on 5 September, 1984 (Hardisky 1980). Samples of these leaves were collected at the time of tagging and at 3-5 week intervals thereafter until the supply of tagged leaves was exhausted (29 November, 1984). Samples of 3 cm length were cut from the basal, midlength, and tip portions of the leaves, washed three times in settled marsh water, and pooled. Each sample leaf area was recorded by photocopying, and dry mass and organic mass per unit area were determined in samples collected for that purpose. Replicate samples consisted of four leaves, and four replicates were analyzed per collection time and type of measurement (except ergosterol, for which 50 leaves per collection were pooled, and 3-4 replicate ground samples analyzed). Only 2 of 4 replicates for immunosorbent assay had been processed at the time of this writing (see note added in proof).

Direct microscopic estimates of fungal biovolume were made by the calcofluor/aniline-blue method of Newell *et al.* (1986). This is a conservative method that results in least sure values for fungal biovolume. Ergosterol analyses were performed by high pressure liquid chromatography (methanol extraction, methanol-potassium hydroxide-refluxing, methylene chloride-methanol solvent, Waters μBondapak RP18 packing) according to Miller *et al.* (1983). Enzyme-linked immunosorbent assays (ELISA) were patterned after the alkaline phosphatase assays of Johnson *et al.* (1982) and Clark & Adams (1977). Immunogen was whole, washed, homogenized mycelium of *Phaeosphaeria typharum* (Desm.) Holm (Newell & Statzell-Tallman 1982) grown in sea water malt extract solution (Casper & Mendgen 1979). Three serial subcutaneous injections of rabbits (Johnson *et al.* 1982) were followed by intravenous injections (Hurn & Chantler 1980).

A conservative factor was chosen to calculate fungal mass from direct microscopically measured biovolumes (200 mg dry cm^{-3}: Newell & Statzell-Tallman 1982; Newell & Fallon 1983). An average factor for calculation of fungal mass from ergosterol values was taken from Miller *et al.* (1984: 246 mg dry mass mg^{-1} ergosterol). The ELISA was experimentally calibrated using mycelium of *P. typharum* which was handled in the same manner as the *S. alterniflora* samples.

RESULTS

The tagged leaves lost organic mass in a smooth, negative exponential fashion (k = -0.014 day^{-1}, r^2 = 0.98) (Figure 3.1). The leaves did not fall from the culms. As the leaves aged, they gradually became shredded lengthwise into long fibres. This was a result of loss of material from the former mesophyll chlorenchyma, which occupied much of the leaf volume between the lengthwise running vascular bundles. After about 30 days, much of the dead mesophyll chlorenchyma appeared, in leaf cross sections, to have been thoroughly pervaded by fungal material, most noticeably the ascocarps of *P. typharum*. These were formed in rows at the abaxial surface, where their orifices reached the exterior of the leaf. By the end of the study period (83 d), leaves were weakened enough that they often bent down and lay partially on the marsh sediment.

Direct microscopic measurements of fungal biovolume and estimates of

fungal mass, designed to be conservative, were extremely so
(Figure 3.1). The mass estimates, as per cent of total mass of the
dead leaf system, rose from 0.03% to 0.25% during the first sampling
interval, then stayed at approximately that level through 83 days. Mean
coefficient of variation (standard deviation as a per cent of the mean)
for hyphal length measurements was 38%, with a range of 30-48%.

Figure 3.1 Organic density (mass per unit original leaf area) and
content of fungal mass for standing leaves of *Spartina alterniflora*.
Fungal mass was measured in parallel by enzyme-linked immunosorbent
assay (ELISA), ergosterol analysis, and conservative direct-count
microscopy. Note scale breaks in the Y axis. (See note added in proof
for revision of ELISA results).

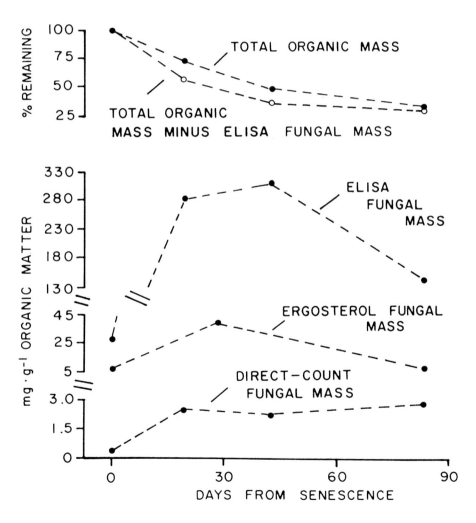

Ergosterol content also showed a rise during the first month after tagging, which indicated a rise in fungal mass content from 0.6 to 3.9% of total leaf mass (Figure 3.1). However, over the subsequent two months of the study period, ergosterol-fungal mass declined to 0.7% of total, near the value for senescent leaves. Mean coefficient of variation for ergosterol content was 7%, with a range of 4-10%.

Fungal content of leaves as determined by ELISA was the highest of the three measures of fungal mass (Figure 3.1). ELISA values were X5 to X19 ergosterol values, and X51 to X138 direct microscopic values. These ratios were rather constant, until the 83-day sampling, when the ELISA : ergosterol factor rose from X7 to X19, and the ELISA : direct microscopy factor fell from X138 to X51. ELISA-fungal content was great enough (up to 31.1% of the total mass of the dead leaf system) that subtracting it from the total mass remaining affected the shape of the curve for leaf weight loss (Figure 3.1, upper). The result of the subtraction was a sharper rate of loss of material over the first 19 days, and a shallower loss rate thereafter, especially after the 42-day sampling. Mean coefficient of variation for ELISA-fungal mass was 49%, with a range of 22-89%.

Specificity testing of the ELISA antibodies showed that response was less than 1% of that to *P. typharum* mycelium for the following materials: marsh-surface sediment; a sediment bacterial assemblage; a bacterial assemblage from the marsh water column; the dominant alga epiphytic on leaves of *S. alterniflora*; young, green leaves of *S. alterniflora*. Eight species of fungi other than *P. typharum*, from *S. alterniflora* leaves or culms, were also tested; six of these gave responses less than 15% of that for *P. typharum*, and one species (*Stagonospora* cfr. species II [see Kohlmeyer & Kohlmeyer 1979]) gave a response 58% of that for *P. typharum*.

DISCUSSION

The very low values (<0.3%) for fungal content found by conservative direct microscopy are probably largely a consequence of two inefficiencies: that of homogenization and that of recognition. Mechanical disruption of vascular plant substrates cannot be expected to result in full release of strongly adherent, pervading mycelia. Thorough breakage of mycelial structures, storage structures, specialized absorbtive structures, and sexual or asexual reproductive structures can be expected to result in partially disguising the fungal nature of these structures. Thus it is not surprising that similarly performed direct microscopic estimations have resulted in similarly low estimates of fungal mass in decaying *S. alterniflora*: about 0.5% (Lee *et al.* 1980); about 4% (Marinucci *et al.* 1983, in Newell & Fallon 1983).

Our ergosterol based estimates of fungal mass showed a pattern of variation with time which was quite similar to that reported by Lee *et al.* (1980): an increase from less than 1% of total mass in senescent material, to about 3-4% early in the decomposition process, followed by a decrease to less than 1% in older decayed material. (Note, however, that Lee *et al*. (1980) used a conversion factor of 500 mg dry mass mg $^{-1}$ ergosterol). Two of the probable principal reasons why the ergosterol estimates are low relative to our ELISA estimates are: a) ergosterol may not be present in dead hyphae which are devoid of cytoplasm; b) the dry mass : ergosterol ratio for *P. typharum* may be considerably higher than 246 : 1 (see discussion in Newell & Fallon 1983). If possibility (a) were true, then the percentage of fungal mass which had not undergone extensive lysis (the 'living' fraction) would range from 5.2% to 20.0%, decreasing with time of decay of *S. alterniflora* leaves. This explanation is attractive, in view of the low percentages for living fungal mass often found using fluorescein diacetate assays (e.g. about 3.5% for a pine forest soil, Bååth 1980; but see Domsch *et al*. 1979, p. 530). However, we used supernatant from centrifugates (700 *g*, 5 min) of leaf homogenate in our ELISA. It seems unlikely that antigens from dead, empty hyphae would be present in these supernatants; if not, then the ELISA would measure only living and recently dead fungal material. Also, the ergosterol mass estimates are all lower than a published estimate for total mass of fungal sexual reproductive structures alone in decaying *S. alterniflora* leaves (about 7% of total mass: Newell & Hicks 1982; Newell & Fallon 1983). We plan to investigate the open question of the likelihood of possibility (b).

Our ELISA values for fungal content in *S. alterniflora* leaves, though five to nineteen times ergosterol values, do not seem unreasonable in comparison to the apparent extensiveness of fungal occupation of the dead mesophyll chlorenchyma of the leaves (see Results). Also, the ELISA values make sense in comparison to the value for mass of sexual structures (previous paragraph), since the sexual structures must be nutritionally supported by a mycelial absorbant mass. When Newell & Hicks (1982) made less conservative direct microscopic estimates of total fungal mass than performed here, they suggested that fungal percentage of total mass for dead *S. alterniflora* leaves lay between 20 and 50%. Finally, the average number of released ascospores of *P. typharum* which we have recorded deposited on surfaces of rinsed senescent leaves of *S. alterniflora* from our tagged samples (about 550 cm $^{-2}$, average for both sides) strongly suggests the presence of a substantial fungal mass in adjacent dead leaves; this concentration of ascospores on leaves translates to about 1.2×10^6 ascospores m $^{-2}$ of marsh, including only those tightly adherent on yellow-green, overstory (shoots >20 cm tall) leaves.

We must note here that questions regarding the accuracy of our ELISA estimates remain, and we are working toward answers. Principal among these questions is that of variability in the antigen : mass ratio.

As a preliminary speculative exercise in interpretation of our ELISA data, let us examine calculated growth efficiency for *P. typharum* based on data from Figure 3.1. We know that microbial CO_2 release from the standing-dead leaves is maximal when leaves are wet owing to rain, high tides, and high humidity (Newell *et al*. 1986). Let us assume that the dead leaves were water-saturated for 30% of the time during the first 19 days after tagging, and that fungal growth occurred only during this time. If this were so, then by the use of CO_2 release rates measured earlier and data used in Figure 3.1, and ascribing all CO_2 release to fungi, one can calculate that fungal specific growth rate would be 0.3 d^{-1}, with a growth efficiency (fungal carbon gain ÷ fungal carbon gain + CO_2 carbon released) of 0.67. Fungal contribution to loss of organic carbon from the leaf, to new fungal mass and fungal CO_2, would be 40% (the remaining loss would be by translocation to rhizomes, as leachate, or as particles). Thus the fungus would rapidly and efficiently convert leaf mass to fungal mass during the first month after leaf senescence. Then for the following two months the fungal material would be gradually lost (k $\cong$ -0.01 to -0.03 day^{-1}), while leaf decay slowed down sharply (see Figure 3.1, upper).

ACKNOWLEDGEMENTS
        We thank M. Johnson, J. Leach, C. Lee and S. Savage for advice, and L. Gassert for the preparation of our illustration. Contribution no. 548 of the University of Georgia Marine Institute.

REFERENCES
Bååth, E. (1980). Soil fungal biomass after clearcutting of a pine forest in central Sweden. Soil Biol. Biochem., 12, 495-500.

Casper, R. & Mendgen, K. (1979). Quantitative serological estimation of a hyperparasite: Detection of *Verticillium lecanii* in yellow rust infected wheat leaves by ELISA. Phytopath. Z., 94, 89-91.

Clark, M.F. & Adams, A.N. (1977). Characteristics of the microplate method of enzyme-linked immunosorbent assay for the detection of plant viruses. J. gen. Virol., 34, 475-483.

Domsch, K.H., Beck, T., Andersen, J.P.E., Söderström, B., Parkinson, D. & Trolldenier, G. (1979). A comparison of methods for soil microbial population and biomass studies. Z. Pflanzenernaehr. Bodenkd., 142, 520-533.

Hardisky, M.A. (1980). A comparison of *Spartina alterniflora* primary production estimated by destructive and nondestructive techniques. In Estuarine Perspectives, ed. V.S. Kennedy, pp. 223-234. New York: Academic Press.

Hurn, B.A.L. & Chantler, S.M. (1980). Production of reagent antibodies. Meth. Enzymol., 70, 104-142.

Johnson, M.C., Pirone, T.P., Siegel, M.R. and Varney, D.R. (1982). Detection of *Epichloë typhina* in tall fescue by means of enzyme-linked immunosorbent assay. Phytopathology, 72, 647-650.

Kohlmeyer, J. & Kohlmeyer, E. (1979). Marine Mycology. The Higher Fungi. New York: Academic Press.

Lee, C., Howarth, R.W. & Howes, B.L. (1980). Sterols in decomposing *Spartina alterniflora* and the use of ergosterol in estimating the contribution of fungi to detrital nitrogen. Limnol. Oceanogr., 25, 290-303.

Marinucci, A.C., Hobbie, J.E. & Helfrich, J.V.K. (1983). Effect of litter nitrogen on decomposition and microbial biomass in *Spartina alterniflora*. Microb. Ecol., 9, 27-40.

Miller, J.D., Young, J.C. & Trenholm, H.L. (1983). *Fusarium* toxins in
        field corn. I. Time course of fungal growth and production
        of deoxynivalenol and other mycotoxins. Can. J. Bot., 61,
        3080-3087.
Miller, J.D., Moharir, Y.E., Findlay, J.A. & Whitney, N.J. (1984).
        Marine fungi of the Bay of Fundy VI: Growth and metabolites
        of *Leptosphaeria oraemaris*, *Sphaerulina oraemaris*,
        Monodictys pelagica, and *Dendryphiella salina*. Proc. N.S.
        Inst. Sci., 34, 1-8.
Newell, S.Y. & Fallon, R.D. (1983). Study of fungal biomass dynamics
        within dead leaves of cordgrass: progress and potential. In
        Proceedings of the International Symposium on Aquatic
        Macrophytes, pp. 150-160. Nijmegen, Netherlands: Catholic
        University.
Newell, S.Y., Fallon, R.D., Cal Rodriguez, R.M. & Groene, L.C. (1986).
        Influence of rain, tidal wetting and relative humidity on
        release of carbon dioxide by standing-dead salt-marsh
        plants. Oecologia (Berl.), in press.
Newell, S.Y. & Hicks, R.E. (1982). Direct-count estimates of fungal and
        bacterial biovolume in dead leaves of smooth cordgrass
        (*Spartina alterniflora* Loisel.). Estuaries, 5, 246-260.
Newell, S.Y. & Statzell-Tallman, A. (1982). Factors for conversion of
        fungal biovolume values to biomass, carbon and nitrogen:
        variation with mycelial ages, growth conditions and strains
        of fungi from a salt marsh. Oikos, 39, 261-268.
Schubauer, J.P. & Hopkinson, C.S. (1984). Above- and belowground
        emergent macrophyte production and turnover in a coastal
        marsh ecosystem, Georgia. Limnol. Oceanogr., 29, 1052-1065.
Tiner, R.W. (1984). Wetlands of the United States: Current Status and
        Recent Trends. Washington, D.C.: U.S. Govt. Printing Office.

Note added in proof: Completion of analysis of all four replicate ELISA
samples has resulted in a change of the mean ELISA-fungal mass values as
follows (see Figure 3.1): 0 days, 22 ± 15 mg g^{-1} organic (mean $\pm$ 1
standard deviation); 19 days, 186 ± 111; 42 days, 188 ± 143; 83 days,
77 ± 121.

4 THE EFFECT OF MONOVALENT IONS ON ENZYME ACTIVITY IN
   DENDRYPHIELLA SALINA

F.M. Gibb

J.M. Wethered

D.H. Jennings

INTRODUCTION

For growth in the sea, a fungus must absorb solutes to generate turgor. The evidence available indicates, as one might anticipate, that ions can make a major contribution to the internal solute potential required for the generation of turgor (Jennings 1983). However, until recently the data for protoplasmic ion concentrations of marine fungi have been ambiguous, owing to uncertainty about the extent of extra-protoplasmic location of ions and the problem of determining the protoplasmic water content (Eamus & Jennings 1986). Nevertheless, recently, Wethered *et al.* (1985) have produced relatively reliable values for the concentrations of ions in mycelium of *Dendryphiella salina* (Sutherland) Pugh et Nicot grown for 48 h in the presence of 0.8 osmol [kg H$_2$O]$^{-1}$ sodium chloride. The protoplasmic sodium concentration was 189 mM, that of potassium 21 mM and chloride 425 mM.

With regard to higher plants, it has been concluded from a large number of *in vitro* studies on their enzymes that if these values were for concentrations within the cytoplasm, then metabolism of the cells would be considerably affected in comparison with those cells containing much lower cytoplasmic concentrations. There is now a consensus amongst higher plant physiologists (Jennings 1976), supported by direct determination either by tracer flux analysis (Jefferies 1973) or x-ray microanalysis (Harvey *et al.* 1981), that the cytoplasmic concentration of monovalent cations in cells of halophytes is maintained at around 150–200 mM. However, to generate turgor, the solute potential of the cell must be much lower than can be generated by this concentration. The shortfall is made up by the synthesis of the so-called compatible solutes, such as glycine-betaine, proline and betaine (Harvey *et al.* 1981). The remainder of the monovalent cations absorbed by the cell of a halophyte are sequestered in the vacuoles as chlorides.

Electron microscope studies of the hyphae of *D. salina* growing in sea water (Metcalf 1980) indicate a paucity of vacuoles in the younger parts of the mycelium. On the other hand, Wethered *et al.* (1985) when determining cytoplasmic ion concentrations also showed that there was a significant total concentration (*ca* 300 mM) of polyols within the cytoplasm. Therefore a question before us is the degree of sensitivity *in vitro* of enzymes extracted from *D. salina* to concentrations of monovalent cation chlorides of the order of 200–400 mM in the presence and absence of polyols.

This chapter is concerned with answering this question. The three
enzymes chosen for study were D-glucose 6-phosphate: NADP oxidoreductase
(EC 1.1.1.49; glucose 6-P dehydrogenase), glycerol: $NADP^+$ oxidoreductase
(EC 1.1.1.72; glycerol dehydrogenase) and L-malate: NAD oxidoreductase
(EC 1.1.1.37; malate dehydrogenase). Glucose 6-P dehydrogenase was
chosen because the enzyme is unquestionably cytoplasmically located and
because the oxidative segment of the pentose pathway has been shown to
be very important in polyol synthesis in *D. salina* (Holligan & Jennings
1972). Glycerol dehydrogenase was chosen because it is almost certain
to be cytoplasmically located and because there is significant synthesis
of glycerol in *D. salina* under saline conditions (Wethered *et al.* 1985).
Malate dehydrogenase was chosen because the enzyme has been extracted
from many higher plant halophytes and its properties studied (Flowers
et al. 1976 a,b).

Only a selection of results are presented here. A much fuller account
including details of procedures used will be given at a later date.

RESULTS

Malate dehydrogenase
        Figure 4.1 shows the effect of varying concentrations of
sodium chloride at three pH values on the activity of the enzyme
isolated from *D. salina.* Essentially the same results were obtained
with potassium chloride. It can be seen that, in the absence of salt,

Table 4.1 K_m and V_{max} values (mean ± s.e.) for malate dehydrogenase
(substrate oxaloacetate) and glucose 6-phosphate dehydrogenase
(substrate glucose 6-phosphate) extracted from mycelium of *Dendryphiella*
grown in media made up either with distilled water or sea water and
assayed at pH 7.4 in the presence or absence of either 200 mM glycerol or
100 mM mannitol

Growth medium	Polyol present in assay medium	K_m (μM)	V_{max} (μM min^{-1} mg^{-1} protein)
malate dehydrogenase			
sea water	–	21.7 ± 3.1	2.1 ± 0.4
sea water	glycerol	11.0 ± 1.1	1.4 ± 0.8
sea water	mannitol	14.3 ± 4.5	1.5 ± 0.7
distilled water	–	8.0 ± 0.0	3.7 ± 3.1
distilled water	glycerol	27	7.4
glucose 6-phosphate dehydrogenase			
sea water	–	147 ± 31.5	668 ± 146
sea water	glycerol	195 ± 25.0	327 ± 79
sea water	mannitol	140 ± 27.5	120

maximum activity is achieved at pH 7.4. At this pH, we have found a variable stimulation of activity at the lower salt concentrations. In Figure 4.1 there is no detectable stimulation. In other experiments, we have seen stimulation by concentrations of sodium chloride as high as 100 mM. However, at 200 mM sodium chloride invariably inhibits enzyme activity, to 36% of the activity when no salts are present as in Figure 4.1, to a lesser extent in other experiments and to 90% in only one instance.

Neither 200 mM glycerol nor 100 mM mannitol, the concentrations at which these two polyols can be found in the mycelium of *D. salina* (Wethered *et al.* 1985), have any significant effect on the activity of the enzyme, irrespective of whether or not the mycelium has been grown in media made up with distilled water or sea water. Table 4.1 shows that these two growth media have no effect on the kinetic properties of the enzyme as exemplified by K_m and V_{max}. No information is available about how the growth media might or might not effect the total amount of enzyme within unit mass of mycelium.

Glucose 6-phosphate dehydrogenase
The activity of this enzyme is relatively little affected by pH. The enzyme is also very much less sensitive to sodium chloride than malate dehydrogenase, particularly at pH 7.4 and 8.5. Figure 4.2 gives a representative set of data. As can be seen, at pH 8.5 the salt stimulates activity at all concentrations up to 400 mM, while at pH 7.4 there is very little effect until 400 mM, at which concentration there is 15% inhibition of activity. At pH 5.5, there is 25% and 50% inhibition with 200 mM and 400 mM sodium chloride respectively. Mannitol (100 mM) has no effect on activity, though glycerol (200 mM) might be slightly inhibitory.

Glycerol dehydrogenase
NAD-linked glycerol dehydrogenase (EC 1.1.1.6) was not detected in either the forward or reverse reactions.

The NADP-linked enzyme was detected using both dihydroxyacetone and D-glyceraldehyde as substrates, the activity with the latter was always less than with the former substrate. Table 4.2 gives the activity of the enzyme extracted from mycelium grown in media made up with distilled water. As can be seen, both sodium chloride and sodium sulphate at 100 mM significantly inhibit enzyme activity. Glycerol at 200 mM was able to reduce the inhibitory effect of the two sodium salts.

DISCUSSION
One must be cautious in extrapolating from data obtained from *in vitro* studies of enzymes to the *in vivo* situation. For example, the proportion of enzyme to substrate *in vivo* may be very different from that in the assay medium. Nevertheless it can be seen that an enzyme which we believe can be unambiguously located in the cytoplasm,

Figure 4.1 The effect of pH and sodium chloride on
activity of malate dehydrogenase from mycelium of
Dendryphiella salina grown in media made up with sea water.
Activity assayed without sodium chloride (▲), and with
10 mM (●), 25 mM (■), 50 mM (△), 100 mM (○) or 200 mM (□)
sodium chloride.

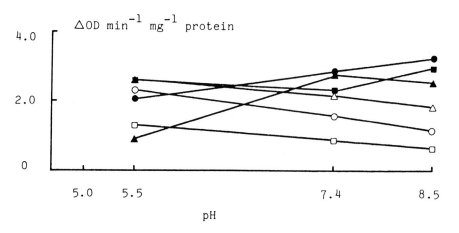

Figure 4.2 The effect of pH and sodium chloride on
activity of glucose 6-phosphate dehydrogenase from mycelium
of *Dendryphiella salina* grown in media made up with
distilled water. Activity assayed without sodium chloride
(▲), and with 10 mM (●), 25 mM (■), 50 mM (△), 100 mM
(○), 200 mM (□), 400 mM (▽) or 500 mM (▼) sodium chloride.

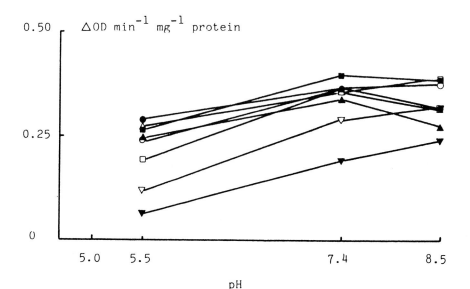

Table 4.2 Activity of glycerol dehydrogenase (μ mol NADP min^{-1} mg protein^{-1} mean ± se) extracted from mycelium of *Dendryphiella salina* grown in media made up with distilled water

Treatment	Activity with dihydroxyacetone as substrate	% inhibition	Activity with glyceraldehyde as substrate	% inhibition
Control, buffer alone*	0.095 ± 0.01	0	0.039 ± 0.005	0
100 mM glycerol	0.106 ± 0.01	0	0.042 ± 0.006	0
100 mM sodium chloride	0.043 ± 0.006	55	0.024 ± 0.001	38
100 mM sodium sulphate	0.043 ± 0.004	55	0.026 ± 0.003	39
100 mM sodium chloride + 200 mM glycerol	0.078 ± 0.009	18	0.036 ± 0.004	8
100 mM sodium sulphate + 200 mM glycerol	0.069 ± 0.003	28	0.032 ± 0.005	18

* 50 mM bicarbonate buffer, pH 10.0.

glucose 6-phosphate dehydrogenase, at pH 7.4 is very little affected by sodium chloride even at concentrations as high as 400 mM. At pH 5.5, activity is much more inhibited. While we have no information about the cytoplasmic pH of *D. salina*, one would expect from the information which is available from other fungi, e.g. *Neurospora crassa* Shear et Dodge (Sanders & Slayman 1982), the value to be around 7.0. If we assume this value, it is clear that the activity *in vivo* of glucose 6-phosphate dehydrogenase is little affected by a relatively high concentration of sodium chloride in the cytoplasm, i.e. of a kind expected from the determinations of protoplasmic concentrations of salt made by Wethered *et al.* (1985).

On the other hand, malate dehydrogenase can be much more sensitive. Here there must be caution in the interpretation of the data because of the uncertainty of the exact location of the enzyme inside the hyphae. In other moulds for instance, there is little doubt that the enzyme is located in the mitochondrion. However, there is also a cytoplasmic enzyme, the synthesis of which is subject to glucose represssion (Benveniste & Munkres 1970). Conceivably, in the experiments described here the mycelium, known to be at the end of the growth phase, may have reduced glucose in the medium to a level where the sugar is less able to repress the synthesis of cytoplasmic malate dehydrogenase. In consequence there might be variable amounts of the cytoplasmic form in the extract.

It was most interesting to find that polyols, at around their putative concentration in mycelium grown in sea water, have little effect on the activity of glucose 6-phosphate dehydrogenase and malate dehydrogenase. On the other hand, glycerol appears to have a protective effect on glycerol dehydrogenase. This enzyme appears to be sensitive to sodium chloride and sulphate at relatively low concentrations (100 mM). However, since the data are only for the high pH of 10.0 and we have made no investigation of the effect of substrate concentration on the activity in the presence of salt, it seems inappropriate to draw many conclusions from the study of this enzyme. Nevertheless it is interesting to see that when the activity of the enzyme was measured *in vitro*, under conditions which approximate somewhat (pH apart) to the ionic milieu of the cytoplasm, activity was not reduced too greatly.

We conclude tentatively from these studies, which must be considered preliminary in nature, that enzymes within the cytoplasm of *D. salina* growing in sea water are able to function at relatively high activity in the presence of sodium chloride at concentrations in the cytoplasm similar to those determined experimentally.

REFERENCES

Benveniste, K. & Munkres, K.D. (1970). Cytoplasmic and mitochondrial malate dehydrogenases of *Neurospora*. Biochim. Biophys. Acta, 220, 161–177.

Eamus, D. & Jennings, D.H. (1986). Water, turgor and solute potentials of fungi. In Water, Fungi and Plants, ed. P.G. Ayres & L. Boddy. Cambridge: Cambridge University Press, in press.

Flowers, T.J., Hall, J.L. & Ward, M.E. (1976 a). Salt tolerance in the
        halophyte *Suaeda maritima*. Further properties of the enzyme
        malate dehydrogenase. Phytochem., 15, 1231-1234.
Flowers, T.J., Ward, M.E. & Hall, J.L. (1976 b). Salt tolerance in the
        halophyte *Suaeda maritima*: some properties of malate
        dehydrogenase. Phil. Trans. Roy. Soc. Lond. B., 273,
        523-540.
Harvey, D.M.R., Hall, J.L., Flowers, T.J. & Kent, B. (1981).
        Quantitative ion localisation within *Suaeda maritima* leaf
        mesophyll cells. Planta, 151, 555-560.
Holligan, P.M. & Jennings, D.H. (1972). Carbohydrate metabolism in the
        fungus *Dendryphiella salina*. III. The effect of the nitrogen
        source on the metabolism of 1-^{14}C- and 6-^{14}C-glucose. New
        Phytol., 71, 1119-1133.
Jefferies, R.L. (1973). The ionic relations of seedlings of the
        halophyte *Triglochin maritima* L. In Ion Transport in Plants,
        ed. W.P. Anderson, pp. 323-335. London & New York: Academic
        Press.
Jennings, D.H. (1976). The effects of sodium chloride on higher plants.
        Biol. Rev., 51, 453-486.
Jennings, D.H. (1983). Some aspects of the physiology and biochemistry
        of marine fungi. Biol. Rev., 58, 423-459.
Metcalf, E.C. (1980). Osmoregulation in the Marine Fungus *Dendryphiella
        salina*. Ph.D. thesis, University of Liverpool.
Sanders, D. & Slayman, C.L. (1982). Control of intracellular pH.
        Predominant role of oxidative metabolism, not proton
        transport, in the eukaryotic microorganism *Neurospora*. J.
        Gen. Physiol., 80, 377-402.
Wethered, J.M., Metcalf, E.C. & Jennings, D.H. (1985). Carbohydrate
        metabolism in the fungus *Dendryphiella salina*. VIII. The
        contribution of polyols and ions to the mycelial solute
        potential in relation to the external osmoticum. New
        Phytol., 101, 631-649.

5 PHYSIOLOGY OF MARINE FUNGI: A SCREENING PROGRAMME FOR GROWTH
AND ENZYME PRODUCTION

H.P. Molitoris

K. Schaumann

INTRODUCTION

Marine mycology was established as a discrete science in the
late 1930's following the work of mycologists such as Höhnk and Sparrow
and papers exemplified by that of Barghoorn & Linder (1944). The
generally accepted definition of marine fungi is that provided by
Kohlmeyer & Kohlmeyer (1979), namely: "obligate marine fungi are those
that grow and sporulate exclusively in a marine or estuarine habitat;
facultative marine fungi are those from fresh water or terrestrial
milieus able to grow (and possibly also to sporulate) in the marine
environment". According to this definition about 500 marine fungi are
known. These comprise about 100 Mastigomycotina and Zygomycotina,
150 filamentous Ascomycotina, 4 filamentous Basidiomycotina, 180 yeasts,
60 Deuteromycotina and 10 lichens. In contrast, about 100 times more
terrestrial fungi, exclusive of the lichens, have been described
(Table 5.1). The relatively low number of marine fungi may be partly
due to the methods of isolation and partly to the fact that large
regions of the oceans have not been thoroughly investigated for their
fungal flora. However, marine fungi do seem to be cosmopolitan in oceans
and estuaries. They are plant and animal parasites, symbionts of algae,
live in lichen-like associations and are also saprophytes on organic
material of plant, animal and anthropogenic origin.

Research on marine fungi has been mainly concerned with their isolation,
cultivation, morphology, taxonomy and systematics. The results of this
research have been presented in few, but excellent, reviews (Johnson &
Sparrow 1961; Kohlmeyer 1974; Hughes 1975; Jones 1976; Kohlmeyer &
Kohlmeyer 1979). Nevertheless, there is an increasing number of papers
on the physiology of marine fungi, in particular enzymatic aspects
(e.g. Nilsson 1974; Schaumann 1974 a; Nilsson & Ginns 1979; Bahnweg &
Bland 1980; Gessner 1980; Leightley 1980; MacDonald & Speedie 1982;
Torzilli 1982; Benner et al. 1984; Breuer & Molitoris 1986). Most of
these papers, however, deal only with single species or isolated
physiological aspects of marine fungi and the methods employed are often
not comparable.

We therefore considered that it was time to begin a broad investigation
of growth and physiology, in particular of enzymology, of marine fungi
based on a larger and more representative number of species from
different taxonomic and ecological groups and, as far as possible, to
compare the results with those of terrestrial fungi.

MATERIALS AND METHODS
Cultures
    With few exceptions the marine fungal species used in this
investigation were obtained from the Kulturensammlung Mariner Pilze,
Bremerhaven, and the terrestrial species from the Botanical Institute,
University of Regensburg.

Culture media
    Two basic media were used (Schaumann 1974 b): a sea water
medium (GPYS: glucose 1.0 g; peptone 0.5 g; yeast extract 0.1 g; agar
16 g; natural sea water or Rila Sea Water Mix 1 l; pH 6.0) and a fresh
water medium (GPYD: glucose 5.0 g; peptone 2.5 g; yeast extract 0.5 g;
agar 16 g; deionized water 1 l; pH 6.0). For growth of species of the
Mastigomycotina the agar concentration was reduced to $8 \, g \, l^{-1}$. Specific
substrates added for enzyme assays are listed in the appropriate
section. All chemicals were of analytical grade unless stated
otherwise. Stock cultures of marine strains and marine isolates were
maintained on a GPYS medium and terrestrial strains on either malt
extract (1.5%) or Moser b (Moser 1958) medium. All stock cultures were
stored at 4°C.

Table 5.1 Number of fungal and lichen species[1] in marine and
terrestrial habitats

Fungal subdivision		Marine	Terrestrial
Myxomycotina			625
Mastigomycotina and Zygomycotina		100	1,935
Ascomycotina		149	28,650
Basidiomycotina		4	16,000
Yeasts[2] ascomycetous	169		
basidiomycetous	3	179	
asporogenous	7		
Deuteromycotina		56	17,000
Lichens		10	13,500
Total		498	77,710

[1] numbers according to Kohlmeyer & Kohlmeyer (1979) and Hawksworth *et*
al. (1983).
[2] terrestrial yeasts included within respective subdivisions.

Growth
Cultures were incubated at 22°C with a diurnal periodicity of twelve hours light and twelve hours dark. Inoculations were made by the transfer of $0.3 mm^3$ pieces of mycelium from the margin of an actively growing culture. In the case of zoosporic fungi an actively growing agar culture was flooded with a few cm^3 of sterile water and inoculum from the resultant zoospore suspension was transferred with a wire loop on to the test plate. Growth of species was measured as increase in colony diameter on agar plates of 9 cm diameter.

Enzyme tests
Production of extracellular enzymes was determined qualitatively and semiquantitatively by incorporation into the media of specific substrates (direct test) or by the addition of substrates and/or reagents after a certain period of growth (indirect test). Cultures were then examined for the intensity and extent of colour, precipitate or clearance zones in order to determine enzyme activity.

Enzyme tests were performed according to the following references: laccase with 1-naphthol, laccase with guaiacol, laccase with benzidine, tyrosinase with p-cresol and glycine and peroxidase with benzidine and H_2O_2 (Lyr 1958; Bresinsky *et al.* 1977); gelatinase (Hankin & Anagnostakis 1975); caseinase (Weyland *et al.* 1970); nitrate reductase (Bresinsky & Schneider 1975); lipase with Tween 80 (Trigiano & Fergus 1979, modified: without saponin); amylase (Trigiano & Fergus 1979); cellulase (Tansey 1971, including preparation of pre-swollen cellulose Avicel); polygalacturonase (medium pH 5.0), pectate transeliminase (medium pH 7.0) and chitinase, including preparation of colloidal chitin (Hankin & Anagnostakis 1975); laminarinase (Chesters & Bull 1963; preparation of insoluble laminarin from phylloids of *Laminaria hyperborea* (Gunn.) Fosl. according to Thiem *et al.* 1977). Lipase activity was determined in both agar plates and screw-capped test tubes, whereas nitrate reductase, cellulase, laminarinase and chitinase activities were determined in screw-capped test tubes only.

All growth and enzyme tests were performed in triplicate on sea water and fresh water media and read at half-weekly or weekly intervals for at least five weeks or until the plates were covered with mycelium.

SCREENING PROGRAMME
Although the oceans cover more than two thirds of the world's surface, our knowledge of marine organisms, in particular marine fungi, is still very superficial. This applies not only to the inventory of marine species (Table 5.1) but also to their horizontal and vertical distribution and even more to their activities *in situ*. In particular, data on substrate degradation and enzyme activities are largely lacking. Information about these activities would allow a better understanding of the essentially unknown role of fungi in the marine ecosystem.

At present taxonomy and systematics of marine fungi are based mainly on morphological characteristics, especially on fruiting structures. Since these structures are often produced only late or not at all in artificial culture, data on the presence and absence of enzymes or electrophoretical enzyme spectra could be helpful in the characterization and identification of marine fungi.

Fungal enzymes are increasingly produced for industrial use; for example up to 300 tons of fungal glucoamylase valued at $30 million are produced annually. Although the fungi employed for enzyme production are usually grown in submerged culture, terrestrial fungi are used almost exclusively for this purpose. By the selection of fungal strains naturally adapted to growth in liquid media, such as marine fungi, it is hoped that species with increased enzyme productivity will be found.

Microorganisms are also used increasingly to clear polluted water and to degrade organic wastes, if possible in combination with the production of valuable biomass. Often, however, too high a concentration of salts in the medium inhibits fungal growth. Here again marine fungi, by definition adapted to higher salt concentrations, seem suitable for these purposes.

The aims of the present study are: 1) to obtain a better understanding of salinity-dependent growth and enzymatic potential of marine fungi in comparison with terrestrial species; 2) to contribute to our understanding of the physiology and ecology of the marine Mycota and their role in nature; 3) to evaluate whether the results obtained are useful for (chemo-)taxonomic purposes; 4) to provide a sound basis for discussion of the economic value of marine fungi; 5) to show ways for industrial exploitation and possible uses of marine fungi and their products.

Selection of species

In order to obtain information on a wide spectrum of fungi, species from different taxonomic and ecological groups of marine fungi and for comparison, of terrestrial fungi were studied. Fungi living in transient zones, such as estuaries and brackish water, were studied by the inclusion of marine, but not necessarily obligate marine, isolates. Table 5.2 shows the taxonomic groups and respective numbers of fungi which have so far been investigated in this programme.

Selection of enzymes

The enzymes investigated in this programme were selected owing to their metabolic and industrial importance, ecological significance (Table 5.3) and the availability of simple, quick, qualitative or semiquantitative tests on agar plates. After the examination of several methods the enzyme assays described in a previous section of this chapter were used.

Phases of the programme
The programme comprised three phases (Table 5.4).   In
phase I a large number of species were tested for growth and the
presence of enzymes, most of which were extracellular. Growth was
determined quantitatively as increase in colony diameter whereas enzyme
activity was determined both qualitatively and semiquantitatively.
Approximately 50 species of marine fungi, which represent about 10% of
the known marine Mycota, have been investigated (Table 5.2) and other
species are presently under investigation. Results on individual
species either have been or will be published elsewhere (Breuer &
Molitoris 1986).

In phase II, as a result of phase I, selected species were grown in
liquid culture and investigated quantitatively for selected enzymes.
For some enzymes electrophoretic isoenzyme spectra were obtained for
chemotaxonomic purposes. This has been done previously for a few
marine and terrestrial fungi with laccase (Molitoris & Prillinger 1986).

In phase III, and based on the results of phases I and II, species and
enzymes which appeared promising for industrial application were
selected. Optimal conditions for growth and enzyme production were
determined and the species adapted to growth in a fermenter. To date
only laccase has been isolated, purified and analyzed (Molitoris 1976).

PRELIMINARY RESULTS
        Representative data from the comparative investigation of
the fungi selected in phase I (Table 5.2) of the programme are
presented below.

Table 5.2 Taxonomic and ecological groups of fungi investigated

Number of species	Taxonomic group	Ecological group						
		Wo	Li	Co	So	Sw	Pa	Rh
	Obligate marine							
12	Mastigomycotina & Zygomycotina	0	1(8)	0	12	10(11)	10	0
11	Ascomycotina	8	6	0	3	0	0	0
2	Basidiomycotina	2	1	0	0	0	0	0
10	Deuteromycotina	10	10	0	3	0	0	0
total 35								
	Marine isolates							
1	Ascomycotina	1	0	0	0	0	0	0
15	Deuteromycotina	2	(2)	0	13(14)	0	0	0
total 16								
	Terrestrial							
3	Mastigomycotina & Zygomycotina	0	0	0	2	0	1	0
10	Ascomycotina	3	4	2	4	0	0	1
34	Basidiomycotina	26	2(3)	1	0	0	1	6
5	Deuteromycotina	(1)	8	0	12	0	5	0
total 52								

Numbers in () include strains with uncertain affiliation to this ecological group.
Wo = wood-degrading, Li = litter-degrading, Co = coprophilic, So = soil or sediment-inhabiting, Sw = free-
floating in sea water, Pa = parasitic, Rh = rhizosphere-inhabiting or mycorrhizal.

Table 5.3 Ecological significance and industrial use of fungal enzymes

Enzyme	Breakdown of natural substrates	Present and potential industrial use
Redox metabolism		
laccase	wood (lignin, cellulose)	pollution control, degradation of waste material
tyrosinase	wood (lignin)	
peroxidase	wood (lignin)	
N metabolism		
proteases		detergents, additives to beer, drugs, cosmetics, pollution control, use of waste material
gelatinase	protein (animals, plants)	
caseinase		
nitrate reductase	NO_3 in water as N source	none
Fat metabolism		
lipase	lipids (animals, plants)	food industry, additives for extractions
C metabolism		
amylase	starch (plants)	sugar, bakery, textile, paper industry
cellulase	cell walls (cellulose, plants, animals)	paper industry, pollution control, use of waste material
pectinases		
polygalacturonase		
pectin transeliminase	cell walls (pectin, plants)	food industry, beverages
chitinase	polysaccharides (chitin, animals, fungi)	biological control of pathogens
laminarinase	polysaccharides (laminarin, marine algae)	fungal biomass from algae

Table 5.4 Screening programme

Species Enzymes	Cultivation	S[1]	D[2]	Determination of growth	Enzyme production and determination	
PHASE I						
many species	agar plates	+	+	qualitative quantitative (diam)	mostly extracellular	qualitative, semi-quantitative
many enzymes	and tubes					
PHASE II						
some species	liquid culture	+	+	quantitative (dry wt)		
some enzymes	mycelium				ME[3] (intracellular enzymes)	quantitative
	culture filtrate				CF[4] (extracellular enzymes)	quantitative (chemotaxonomy)
PHASE III						
single species	fermenter	+	+	optimal conditions	optimal conditions for enzyme production (intracellular/ extracellular)	isolation, purification, (evaluation of industrial value)
single enzyme						

[1] sea water medium.
[2] deionized water medium.
[3] mycelial extract.
[4] culture filtrate.

Growth. When the data on growth of the fungal groups tested are listed according to taxonomic groups (Figure 5.1), it becomes evident that all marine fungi grew on sea water media and all terrestrial fungi grew on fresh water media. Between 60 and 70% of each group were able to grow on both media whereas in the case of marine isolates from transient zones about 90% grew on both media. The two species of marine Basidiomycotina investigated grew on both media, whereas a high percentage of terrestrial Basidiomycotina showed strong inhibition when grown on sea water medium. In contrast, growth of marine species of Mastigomycotina and Zygomycotina was inhibited on fresh water media, whereas terrestrial species of these orders grew on both media. The growth of species of the Deuteromycotina was least influenced by the water quality of the media.

Enzyme activity. The enzymes produced have been found to be correlated with the taxonomic position or ecology of the fungi investigated and whether they are marine or terrestrial. A few representative examples are given.

All obligate marine fungi grown on sea water media produced nitrate reductase and on average more than 90% produced it also on fresh water media (Figure 5.2). Only some of the marine Mastigomycotina and Zygomycotina did not synthesize this enzyme on fresh water media. In terrestrial fungi, on the other hand, this enzyme was less common, with the smallest percentage found in terrestrial Basidiomycotina, an observation in agreement with the findings of Bresinsky & Schneider (1975). With respect to nitrate reductase production the isolates of marine fungi from transient zones resembled the obligate marine fungi, since on sea water media all of them, and on fresh water media 90% of them, produced this enzyme. Production of nitrate reductase enables fungi to use the nitrate in the sea water as a nitrogen source, an element which often limits fungal growth.

The production and quantity of an enzyme is often characteristic of a few ecological groups. For example, a high percentage of fungal species which inhabit and often degrade wood and litter produce phenoloxidases (laccase and tyrosinase) and cellulase (Table 5.5). Further, cellulase production seems to be stimulated by sea water media in marine fungi and by fresh water media in terrestrial fungi.

DISCUSSION
As shown in Table 5.2 approximately 100 marine and terrestrial species have been investigated. Another about 100 species are presently under investigation in order to increase the number of species in certain taxonomic and ecological groups. New groups, such as the Myxomycotina, the Basidiomycotina and the imperfect marine and terrestrial yeasts have also been included. Most of the marine fungi which have been studied in this programme were isolated from temperate zones. However, work currently in progress includes species (provided by J. Kohlmeyer) isolated from subtropical and tropical zones, some species

Figure 5.1 Growth of strains within taxonomic groups of marine fungal isolates, obligate marine and terrestrial fungi incubated for 5 weeks on GPY agar. n = number of strains tested, P = Mastigomycotina and Zygomycotina, A = Ascomycotina, B = Basidiomycotina, D = Deuteromycotina, Ø = average of all taxonomic groups.

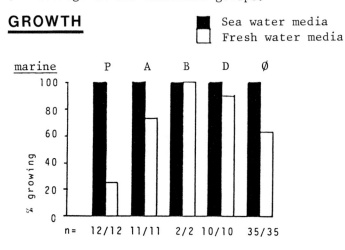

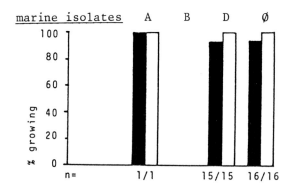

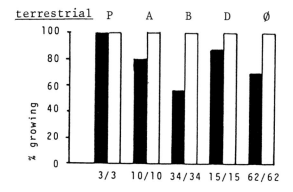

Figure 5.2 Production of nitrate reductase by strains
within taxonomic groups of marine fungal isolates, obligate
marine and terrestrial fungi grown on GPY agar. n = number
of strains tested, P = Mastigomycotina and Zygomycotina,
A = Ascomycotina, B = Basidiomycotina, D = Deuteromycotina,
Ø = average of all taxonomic groups.

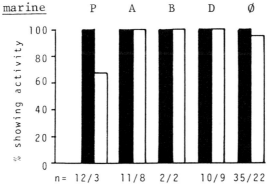

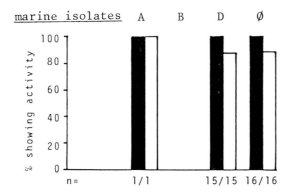

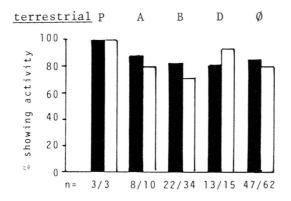

of which have also been examined using isolates from temperate regions. The results of the current work should elucidate any differences in growth and enzyme production which are correlated with geographic and climatic origin of the isolate.

In order to determine whether the presence and the electrophoretic spectra of enzymes are useful for chemotaxonomic purposes, in some genera and species several marine and terrestrial strains were investigated. This data may also reveal differences in growth requirements and enzyme production between marine and terrestrial fungi, as seems to be the case with respect to nitrate reductase. Preliminary experiments have indicated that the production of a number of enzymes, e.g. cellulase, gelatinase, pectate transeliminase and polygalacturonase, is strongly stimulated or inhibited by sea water media. Further experiments are being undertaken to determine whether this is really due to changes in enzyme production, interference of salt ions with enzyme activity or the sensitivity of the test. In addition,

Table 5.5   Tyrosinase and cellulase activity

Enzyme	Ecological group	Medium (S^1/D^2)	Fungal strains					
			Obligate marine		Marine isolates		Terrestrial	
			Number growing	% positive[3]	Number growing	% positive[3]	Number growing	% positive[3]
tyrosinase	Wo	S	20	25	3	0	18	17
		D	16	38	3	33	28	50
	Li	S	18	33	–	–	12	25
		D	16	50	–	–	14	14
	Co	S	–	–	–	–	3	67
		D	–	–	–	–	3	67
	So	S	18	11	13	31	17	18
		D	10	30	13	77	19	26
	Sw	S	10	10	–	–	–	–
		D	2	0	–	–	–	–
	Pa	S	9	0	–	–	6	33
		D	1	0	–	–	8	50
	Rh	S	2	0	–	–	6	33
		D	2	100	–	–	7	57
cellulase	Wo	S	20	50	3	100	25	44
		D	20	35	3	67	28	57
	Li	S	18	61	–	–	13	23
		D	18	44	–	–	14	57
	Co	S	–	–	–	–	2	100
		D	–	–	–	–	3	67
	So	S	18	44	13	54	17	35
		D	18	33	13	54	19	53
	Sw	S	11	18	–	–	–	–
		D	11	18	–	–	–	–
	Pa	S	10	20	–	–	5	20
		D	10	20	–	–	8	25
	Rh	S	2	100	–	–	6	17
		D	2	0	–	–	8	13

[1] sea water medium.
[2] deionized water medium.
[3] per cent of growing strains with enzyme activity.
Wo = wood degrading, Li = litter degrading, Co = coprophilic, So = soil or sediment inhabiting, Sw = free-floating in sea water, Pa = parasitic, Rh = rhizosphere-inhabiting or mycorrhizal.

the number of enzymes to be examined will be increased to include, for example, aminopeptidases, alginase (Schaumann *et al.* Chapter 6), pectin-methylesterase and xylanase. Although the results reported in this chapter concern mainly extracellular enzymes, examination of intracellular enzymes is planned in order to provide a more comprehensive assessment of the complement, role and industrial importance of enzymes produced by marine fungi.

ACKNOWLEDGEMENTS
        The authors thank the "Kulturensammlung mariner Pilze, Bremerhaven (KMPB)" for the majority of the marine fungi, Prof. Bresinsky, Regensburg, for most of the terrestrial strains and Prof. Thiem, Hamburg for a gift of laminarin. Thanks are due to Mr. R. Summers, MA, for critically reading the English manuscript and Mrs. E. Daniel and Mrs. G. Ferstl for excellent technical assistance. The senior author acknowledges a travel grant from Deutsche Forschungsgemeinschaft, Bonn.

REFERENCES
Bahnweg, G. & Bland, E. (1980). Comparative physiology and nutrition of *Lagenidium callinectes* and *Haliphthoros milfordensis*, fungal parasites of marine crustaceans. Botanica mar., **23**, 689-698.
Barghoorn, E.S. & Linder, D.H. (1944). Marine fungi: their taxonomy and biology. Farlowia, **1**, 395-467.
Benner, R., Newell, S.Y., MacCubbin, A.E. & Hodson, R.E. (1984). Relative contributions of bacteria and fungi to rates of degradation of lignocellulosic detritus in salt-marsh sediments. Appl. environm. Microbiol., **48**, 36-40.
Bresinsky, A. & Schneider, G. (1975). Nitratreduktion durch Pilze und Verwertbarkeit des Merkmals für die Systematik. Biochem. Syst. Ecol., **3**, 129-135.
Bresinsky, A., Hilber, O. & Molitoris, H.P. (1977). The genus *Pleurotus* as an aid for understanding the concept of species in Basidiomycetes. In The species concept in hymenomycetes, ed. H. Clémencon. Proc. Herbette Symposium Lausanne, Switzerland, 1976. Biblthca. mycol., **61**, 229-258.
Breuer, S. & Molitoris, H.P. (1986). Anpassungen des aphyllophoralen Pilzes *Halocyphina villosa* an sein marines Habitat. Int. Sympos. Aphyllophorales, Eisenstadt, Austria, 1982, in press.
Chesters, C.G.C. & Bull, A.T. (1963). The enzymic degradation of laminarin. Biochem. J., **86**, 28-31.
Gessner, R.V. (1980). Degradative enzyme production by salt-marsh fungi. Botanica mar., **23**, 130-139.
Hankin, L. & Anagnostakis, S.L. (1975). The use of solid media for detection of enzyme production by fungi. Mycologia, **67**, 597-607.
Hawksworth, D.L., Sutton, B.C. & Ainsworth, G.C. (eds.) (1983). Ainsworth and Bisby's Dictionary of the Fungi. 7. ed. Kew: Commonwealth Mycological Institute.

Hughes, G.C. (1975). Studies of fungi in oceans and estuaries since
        1961. I. Lignicolous, caulicolous and foliicolous species.
        Oceanogr. Mar. Biol. A. Rev., 13, 69-180.
Johnson, T.W. & Sparrow, F.K. (1961). Fungi in Oceans and Estuaries.
        Weinheim: J. Cramer.
Jones, E.B.G. (ed.) (1976). Recent Advances in Aquatic Mycology. London:
        Elek Science.
Kohlmeyer, J. (1974). On the definition and taxonomy of higher marine
        fungi. Veröff. Inst. Meeresforsch. Bremerh., Suppl. 5,
        263-286.
Kohlmeyer, J. & Kohlmeyer, E. (1979). Marine Mycology. The Higher Fungi.
        New York: Academic Press.
Leightley, L.E. (1980). Wood decay activities of marine fungi. Botanica
        mar., 23, 387-395.
Lyr, H. (1958). Über den Nachweis von Oxidasen und Peroxidasen bei
        höheren Pilzen und die Bedeutung dieser Enzyme für die
        Bavendamm-Reaktion. Planta, 50, 359-370.
MacDonald, M.J. & Speedie, M.K. (1982). Cell-associated and
        extracellular cellulolytic activity in the marine fungus
        Dendryphiella arenaria. Can. J. Bot., 60, 838-844.
Molitoris, H.P. (1976). Die Laccasen des Ascomyceten Podospora anserina.
        Beiträge zur Kenntnis von Struktur und Funktion eines
        Systems multipler Enzyme. Biblthca. mycol., 52, 1-81.
Molitoris, H.P. & Prillinger, H.J. (1986). Isoenzyme spectra for
        characterization and identification of fungi. Influence of
        genetical and nongenetical factors. Proc. 2nd Int. Symp.
        Biochemical approaches to the identification of cultivars
        and evaluation of their properties. Braunschweig, 1985, in
        press.
Moser, M. (1958). Die künstliche Mykorrhizaimpfung an Forstpflanzen.
        Forstw. Cbl., 77, 32-40.
Nilsson, T. (1974). Formation of soft rot cavities in various cellulose
        fibres by Humicola alopallonella Meyers & Moore. Stud. For.
        Suec., 112, 1-30.
Nilsson, T. & Ginns, J. (1979). Cellulolytic activity and the taxonomic
        position of selected brown-rot fungi. Mycologia, 71,
        170-177.
Schaumann, K. (1974 a). Experimentelle Untersuchungen zur Produktion und
        Activität cellulolytischer Enzyme bei höheren Pilzen aus dem
        Meer- und Brackwasser. Mar. Biol., 28, 221-235.
Schaumann, K. (1974 b). Zur Verbreitung saprophytischer höherer Pilze in
        der Hochsee. Veröff. Inst. Meeresforsch. Bremerh., Suppl. 5,
        287-300.
Tansey, M.R. (1971). Agar-diffusion assay of cellulolytic ability of
        thermophilic fungi. Arch. Mikrobiol., 77, 1-11.
Thiem, J., Sievers, A. & Karl, H. (1977). Präparative Zugänge zu
        Mannobiose und Laminaribiose. J. Chromat., 130, 305-313.
Torzilli, A.P. (1982). Polysaccharidase production and cell wall
        degradation by several salt marsh fungi. Mycologia, 74,
        297-302.
Trigiano, R.N. & Fergus, C.L. (1979). Extracellular enzymes of fungi
        associated with mushroom culture. Mycologia, 71, 908-917.

Weyland, H., Rüger, H.-J. & Schwarz, H. (1970). Zur Isolierung und Identifizierung mariner Bakterien. Ein Beitrag zur Standardisierung und Entwicklung adäquater Methoden. Veröff. Inst. Meeresforsch. Bremerh., 12, 269-296.

6 COMPARATIVE STUDIES ON GROWTH AND EXOENZYME PRODUCTION OF
DIFFERENT *LULWORTHIA* ISOLATES

K. Schaumann

W. Mulach

H.P. Molitoris

INTRODUCTION

The marine pyrenomycete genus *Lulworthia* was established by
Sutherland in 1916 to accommodate the type species *L. fucicola* Suth.
which was found on the marine brown alga *Fucus vesiculosus* L. collected
near Portsmouth, U.K. Species of *Lulworthia* are ubiquitous and colonize
a wide range of substrates, both dead and living, throughout the world's
oceans and are recognized as among the most common and ecologically most
important marine fungi. Despite this, the present taxonomy of the genus
Lulworthia (Suth.) emend. Kohlmeyer (1980) is unsatisfactory. Koch &
Jones (1984) concurred with Kohlmeyer & Kohlmeyer (1979) on the
inadequate taxonomic status of the genus *Lulworthia* and stated "that
the genus *Lulworthia* is in critical need of revision". These taxonomic
problems are due to the high variability of morphological characters,
particularly those of the ascospores, within the genus *Lulworthia* and
species delimitation based on qualitative and quantitative morphological
data is largely impossible (see Johnson & Sparrow 1961; Meyers *et al.*
1964; Meyers 1966; Cavaliere & Johnson 1966; Hughes 1975; Kohlmeyer &
Kohlmeyer 1979). Speciation within the genus *Lulworthia* may require the
use of techniques less traditional than morphological studies. These
may include ecological, developmental, ultrastructural, cytochemical,
physiological and genetical characters of isolates from different
locations and culture collections.

Owing to the present unsatisfactory taxonomy of the genus *Lulworthia* a
screening programme was initiated in order to study the growth and
exoenzyme production patterns of *Lulworthia* isolates from different
collection sites which had been deposited in the Bremerhaven culture
collection of higher marine fungi. The information obtained from these
studies may not only clarify the taxonomy of the genus *Lulworthia* but
may also provide valuable data on ecophysiological and applied aspects
of the genus *Lulworthia* and the marine fungi in general.

MATERIALS AND METHODS
Samples

The collection sites, dates of collection and the most
important ecological data of eight selected samples which contained
Lulworthia species are listed in Table 6.1. Only wooden substrates,
either driftwood or fixed intertidal wood *sensu* Hughes (1975), were
collected. The methods used for collection, incubation of samples,

microscopic examination and identification of the marine fungi, including the *Lulworthia* species, have been described in detail by Schaumann (1968, 1975). All isolates were deposited in the culture collection at the Institut für Meeresforschung, Bremerhaven at the time of collection.

Isolations
        From the substrates collected at all eight collection sites pure cultures of *Lulworthia* species were established by the transfer of single ascospores on to glucose-peptone-yeast-extract-sea water (GPYS) agar (Schaumann 1974, 1975). When possible more than one strain was isolated from each collection site in order to obtain information on the variability within the *Lulworthia* population. However, the collections from Dakar, Dubay and "Hohe Weg" are represented by only one isolation each. The 37 strains isolated are listed in Figure 6.2. Some of the strains segregated into two different mycelia during cultivation, most probably owing to genetic variation; these are designated by subnumbers. Stock cultures were maintained in screw-capped test tubes on GPYS agar.

Growth and enzyme assays
        The general procedures for growth evaluation and enzyme screening have been outlined by Molitoris and Schaumann in Chapter 5.

Table 6.1 Samples, geographical location, substrate and habitat of the *Lulworthia* collections investigated

Collection no. and site	Geographical location	Substrate	Salinity regime[1]	Date of collection
1 Dakar/W. Africa Ngor	14° 50' N 17° 30' W	driftwood	euhaline zone	22.05.1968
2 Dubay Persian Gulf	25° 51' N 56° 00' E	driftwood	euhaline zone	01.06.1981
3 Helgoland North Sea	54° 11' N 07° 53' E	intertidal wood	euhaline zone	22.10.1980
4 "Hohe Weg" Weser estuary	53° 43' N 08° 15' E	driftwood	mixopoly-haline	30.07.1981
5 Bremerhaven A Weser estuary	53° 32' N 08° 35' E	intertidal wood	mixomeso-haline	06.02.1969
6 Bremerhaven B Weser estuary	53° 32' N 08° 35' E	intertidal wood	mixomeso-haline	17.09.1980
7 Rosfjord S. Norway	58° 04' N 07° 00' E	driftwood (fixed)	mixooligo-haline	20.10.1980
8 Damp Baltic Sea	54° 35' N 10° 01' E	driftwood (fixed)	mesohaline	01.10.1980

[1] Salinity regimes according to Schaumann (1968).

In the present study four additional exoenzymes were incorporated into the screening programme (Table 6.2). The assay of alginase was by precipitation of nondecomposed sodium alginate by $CaCl_2$ (Schaumann & Weide unpublished); pectin methylesterase activity was assayed by the hydroxamic acid method (McComb & McCready 1958), and agarase as well as carrageenase were assayed by liquefaction tests of plain agar and kappa

Figure 6.1 Graphical illustration of ascospore length (a); ascospore width (b); end chamber length (c), and their variability ranges (.....) in *Lulworthia* collections from eight sites together with a differentiation into groups, I, II, III. (For an explanation of the collection site abbreviations see Table 6.1)

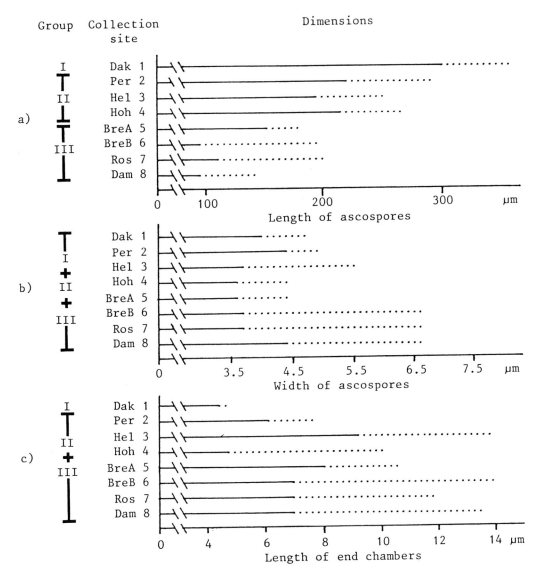

Table 6.2 Enzymes and enzyme substrates *in vitro* and *in situ*

Enzyme	Test substrate	Natural marine substrate	Source
agarase (Aga)	agar-agar	algae (Rhodophyta)	cell wall
alginase (Alg)	Na-alginate Ca-alginate	algae (Phaeophyta)	cell wall
carrageenase (Car)	carrageenan	algae (Rhodophyta)	cell wall
laminarinase (Lam)	soluble laminarin insoluble laminarin	algae (Phaeophyta)	storage product storage product
amylase (Amy)	soluble starch	algae, plant litter	storage product
cellulase (Cel)	NaCMC Avicel	algae, plant litter wood	cell wall
peroxidase (Per)	H_2O_2[1]	wood substrate etc.	cell wall
laccase (Lac)	1-naphthol guaiacol benzidine[1]	wood substrate etc.	cell wall
tyrosinase[2] (Tyr)	p-cresol[1]	wood substrate etc.	cell wall
pectinmethyl- (PME) esterase	pectin A+C	algae, plant litter	cell wall
pectatetrans- (PTE) eliminase	pectin A+C	algae, plant litter	cell wall
polygalac- (P-G) turonase	pectin A+C	algae, plant litter	cell wall
lipase (Lip)	Tween 80[3]	animal & plant-lipids	storage product
protease (Gel) (gelatinase)	gelatine	animal & plant-protein	cellular protein
protease (Cas) (caseinase)	casein	animal & plant-protein	cellular protein
chitinase (Chi)	colloidal chitin	animals & fungi	exoskeleton cell wall
nitrate (Nit) reductase	Na-nitrate	dissolved in sea water	cell-free

[1] growth on GPYS agar (Schaumann 1974).
[2] predominantly intracellular.
[3] Atlas Chemical Division & I.C.I. (= Polyoxyethylensorbitanmonooleate).

carrageenan gels, respectively (Weide 1985). Two sets of strains and of enzymes were screened. In the first set 37 *Lulworthia* strains were tested for growth and production of 13 enzymes; the second set of 7 selected strains which were representative for 6 of the 8 collections was screened for 17 enzymes. All enzyme tests were run with agar plate cultures in Petri dishes, except for nitrate reductase and cellulase where test tubes were used.

Numerical analysis

Cluster analysis was carried out to group the strains and enzymes, respectively, in dendrogram form using the numerical methods of Dice and Jaccard in combination with Ward's method as described by

Table 6.3 Summary of growth and enzyme production data of 37 *Lulworthia* isolates measured after 3 weeks of incubation at 25°C

Substrate	Enzyme	Number of strains showing maximal growth on the substrate in agar plate culture	% of strains exhibiting enzyme activity
agar-agar	agarase	a	29
Na-alginate	alginase	a	0
Ca-alginate	alginase	a	0
carrageenan	carrageenase	a	14
soluble laminarin	laminarinase	a	71
insoluble laminarin	laminarinase	a	71
soluble starch	amylase	14	84
Na-CMC	cellulase	b	100
Avicel	cellulase	b	100
H_2O_2[1]	peroxidase	1	95
1-naphthol	laccase	0	65
guaiacol	laccase	0	65
benzidine[1]	laccase	0	65
p-cresol[1]	tyrosinase[2]	1	3
pectin A+C	pectin methyl-estertase	12	14
pectin A+C	pectate trans-eliminase	12	24
pectin A+C	polygalacturonase	12	32
Tween 80[3]	lipase	2	32
gelatin	protease/gelatinase	12	100
casein	protease/caseinase	0	76
colloidal chitin	chitinase	0	0
Na-nitrate	nitrate reductase	b	22

[1] growth on GPYS agar (Schaumann 1974).
[2] predominantly intracellular.
[3] polyoxyethylensorbitanemonooleate (Atlas Chem.).
a, not yet tested for all the strains.
b, grown in test tube culture.

Steinhausen & Langer (1977). In addition, other methods for cluster analysis were tested, however they all produced similar results. The data were analysed with a Tektronix 4052 computer using a clustering programme developed by J. Dittmer of the Institut für Meeresforschung, Bremerhaven. In general, Ward's method produces a dendrogram on the basis of a dissimilarity coefficient scale, which is dimensionless. For better understanding we added a scale of percentage similarity although the legitimation for doing this is still under discussion.

Figure 6.2 Dendrogram of similarity between 37 strains of *Lulworthia* based on their exoenzyme production patterns. Cluster analysis according to the methods of Jaccard and Ward. (For an explanation of the collection site abbreviations see Table 6.1)

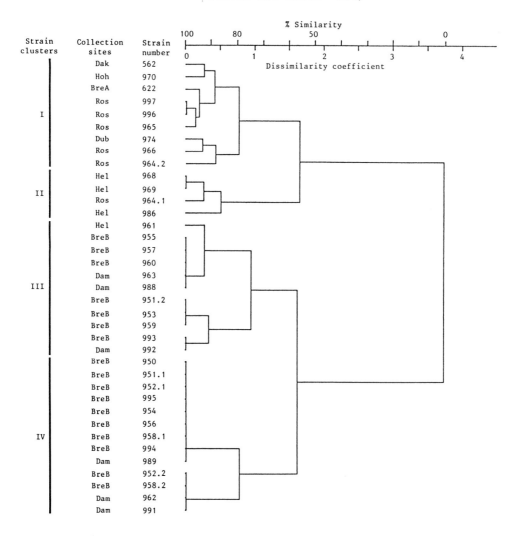

RESULTS
Morphological analysis
        Morphological investigation of the different collections
indicate some degree of identity as well as of diversity among the
different collections. However, the morphological features exhibit a
high degree of variability as is typical for the genus *Lulworthia* and is
the reason for the difficulties which exist in the delimitation of
species. Consequently, neither of the published *Lulworthia* keys by
Kohlmeyer & Kohlmeyer (1979) and Johnson & Sparrow (1961) fits
completely to these data. Of course, some collections can be identified
as being largely identical with one or another described species, e.g.
collection no. 1 from Dakar might be *Lulworthia medusa* var. *biscaynia*
Meyers. However, most of the collections allow no satisfactory
identification. Either the ascospore length fits well and the ascospore
end chambers do not or *vice versa*. With respect to ascospore length
one can detect from Figure 6.1 that three different groups of
collections exist: group I consists exclusively of collection no. 1 from
Dakar, with ascospores longer than 300 µm; Group II comprises the
collection numbers 2 through 4 from Dubay, Helgoland and "Hohe Weg",
respectively, with ascospores shorter than 300 µm but longer than 200 µm;
group III includes collection numbers 5 through 8, which originate from
Bremerhaven, Rosfjord and Damp with ascospores shorter than 200 µm.
This grouping is corroborated only with respect to group I by the
length distribution of the ascospore end chambers and not at all by the
width distribution of the ascospores, even if single collections can be
differentiated against one or a few others with respect to these
features, e.g. collection numbers 4 and 5 against 2 and 8 regarding
ascospore width or collection 2 against 3 and 5 with respect to end
chamber length (Figure 6.1).

Growth
        Analysis of fungal growth on different substrates
incorporated into the basal medium for assaying the exoenzyme production
capacities of *Lulworthia* isolates might provide additional information
for the delimitation of *Lulworthia* species. Moreover, by comparison of
growth data with enzymatic activity some indication of the degree of
substrate utilization by the different strains can be obtained. In
Table 6.3 the growth data after three weeks of incubation are
summarized. These results indicate that starch is a very good
carbohydrate substrate for the growth of *Lulworthia* isolates. It is
much better than glucose. On starch 14 strains exhibited maximum
growth, on glucose only one strain showed maximum growth; this was the
Dakar isolate (562) which occupied a geographically isolated position in
comparison with the other isolates. Pectin and gelatin also proved to
be relatively good substrates for most of the strains, being optimal for
15 and 12 strains, respectively. However, 5 to 6 strains failed totally
in growing on pectin. Guaiacol and 1-naphthol were inhibitory to the
strains as had been expected. Most strains showed growth on colloidal
chitin.

In general, the differentiation of *Lulworthia* strains on the basis of their colony growth patterns on different substrates is very complex and substrate preferences of the strains are obviously heterogeneous. Moreover, some strains generally showed good or poor growth, largely irrespective of the substrate. Consequently, the comparison of colony growth data on different substrates is of no or at best little value for the delimitation of discrete groups or even species in the genus *Lulworthia*. Nonetheless, the data provide valuable information on the design of culture media as well as most probably on the nutritive preferences of the fungi in nature.

Figure 6.3 Dendrogram of co-occurrence of exoenzymes in the genus *Lulworthia*, based on 37 strains from different collection sites. Cluster analysis according to the methods of Dice and Ward. (For an explanation of the enzyme abbreviations see Table 6.2)

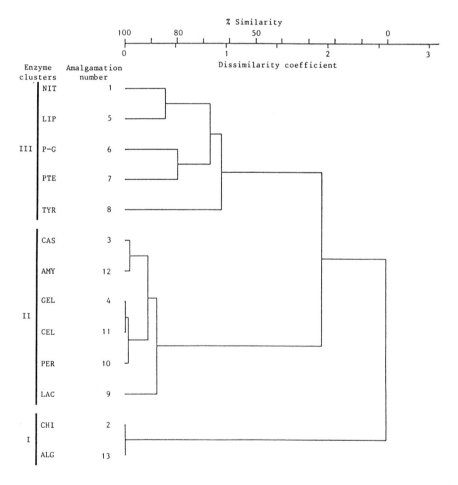

Exoenzyme production
        The semiquantitative results of exoenzyme screening for all
37 *Lulworthia* isolations are summarized in Table 6.3. This work
established that cellulase and gelatinase are produced by all the
strains tested, while alginase and chitinase are absent. Peroxidase is
present in 95% of the strains, amylase in 84%, caseinase in 76%, and
laminarinase in 71%; only one strain, the Dakar isolate (562) exhibited
tyrosinase activity. Carrageenase, agarase and the pectinases were
produced by a limited number of strains; pectinases in only small
amounts. In contrast to the marine fungi tested in our general
screening programme (see Molitoris & Schaumann Chapter 5) only 22% of
the *Lulworthia* strains produced nitrate reductase. However,
characteristically the positive strains, except one strain from the
Rosfjord, originated all from fully marine collection sites, i.e. the
West African coast, the Persian Gulf, the North Sea and the outer Weser
estuary. The remaining isolates from less haline, brackish waters,
except for the strain from the Rosfjord, proved to be nitrate reductase
negative. In a somewhat similar manner the strains reacted to Tween 80
by lipase production. This enzyme was completely absent in all strains
isolated from the mesohaline Bremerhaven B collection. But in
collections from less haline as well as more haline waters most of the
strains were active lipase producers.

Numerical analysis
        The results of exoenzyme production were subjected to a
cluster analysis to determine whether any grouping of the isolates could
be detected and how different these groups would be. Clustering was
done simply on an absence/presence basis without consideration of the
quantitative differences in enzyme production. In the first set all
37 strains were compared with respect to the production of 13 enzymes
and the resulting dendrogram is presented in Figure 6.2. This shows
the existence of different groups of strains. Up to the 75% level of
similarity four separate clusters of strains can be differentiated. In
our opinion these are representative of the material which was
available. If the limits of similarity are set higher, e.g. to the
often used 80% level, seven clusters evolve. However, these are merely
subgroups of the four main clusters mentioned previously.

The original collection sites of the 37 *Lulworthia* strains tested have
also been incorporated into the dendrogram presented in Figure 6.2.
Cluster I comprises the individual isolates from Dakar, "Hohe Weg",
Bremerhaven A and Dubay and also the majority of the Rosfjord isolates.
Cluster II includes three quarters of the Helgoland isolates and one
(segregated) strain from the Rosfjord material. Cluster III comprises
40% of the Bremerhaven B and 50% of the Damp isolates together with a
single supposedly aberrant isolate from Helgoland. Cluster IV includes
the second half of the Damp isolates but is dominated by 60% of the
Bremerhaven B isolates. At present we do not know the reasons for the
differentiation of the Bremerhaven B and Damp material into two
different clusters, although one might speculate that it may be due to a
generally higher genetic variability in these collections. In summary,

however, the clusters predominantly reflect the different collection
sites from which the strains were derived.

This cluster analysis can be inverted to provide valuable information on
enzyme clusters, i.e. enzymes that occur more or less regularly together
among isolates of the genus *Lulworthia*. As is documented by
Figure 6.3, there are three principal clusters: cluster I with the
enzymes alginase and chitinase absent from all isolates; cluster II with
those enzymes which are present in either all or most of the isolates;
cluster III with the enzymes of minor frequency. This latter cluster,
in contrast to cluster II, is relatively heterogeneous and easily
divides into subgroups if the similarity level is raised. In general,
the enzymes cellulase, peroxidase, laccase, amylase and the proteases
caseinase and gelatinase are typical co-occurring exoenzymes in the
genus *Lulworthia*.

DISCUSSION AND SUMMARY

The morphological analysis of the *Lulworthia* isolates
examined illustrates the high degree of plasticity present in this
genus. Although it was possible to define three different groups of
collections on the basis of ascospore length, despite some degree of
variability, we were unable definitely to synonymize these groups with
previously described species, except in one case. These results are in
accordance with the observations of many other authors but in part
contradict the *Lulworthia* concept of Kohlmeyer & Kohlmeyer (1979) which
recognized *Lulworthia fucicola* Suth. as a clearly delimited species
from temperate regions having ascospores with a length of "usually
110 μm or less". However, our collections no. 6 and 8 from Bremerhaven
and Damp have ascospores that are less than 110 μm as well as more than
110 μm in length, they even exceed the variability ranges of this
species as given by Kohlmeyer & Kohlmeyer (1979). Many similarly
problematic *Lulworthia* collections are known throughout the world.
This, and other arguments, forced Cavaliere & Johnson (1966) to reduce
all described *Lulworthia* species to synonymy with *L. medusa* (Ell.et Ev.)
Cribb et Cribb, but this has not been accepted by most marine
mycologists.

In this investigation for new traits to delineate different *Lulworthia*
species or to find additional characters for species delimitation, it
is demonstrated that the comparison of colony growth data is of only
limited taxonomic value. There are, however, indications that
growth-dependent pH changes in certain substrate media might be useful
for differentiation (Schaumann & Molitoris unpublished). Another
physiological approach which looks even more promising is the
differentiation of *Lulworthia* collections by the exoenzyme spectra of
pure cultures derived from natural material. Analysis of the results of
such an enzyme screening by numerical methods is very helpful if not
indispensable as has been demonstrated by this study. It is of interest
that the clustering of *Lulworthia* strains on the basis of their
exoenzyme production capabilities runs parallel primarily with the
different habitats from which the strains were isolated. This result

gives some support to the view of Kirk & Brandt (1980), who stated that water temperature and viscosity may account for the geographical separation of at least some spore groups in *Lulworthia*. Furthermore, we would like to stress the additional influence of salinity in this respect.

Until valid conclusions for the whole genus *Lulworthia* can be drawn, the stability of the strains with regard to exoenzyme production is still an open question and many more strains from different collection sites and substrates must be screened.

ACKNOWLEDGEMENTS
   The authors wish to thank Mrs. M. Sündermann for technical assistance, Mr. J. Dittmer, M.Sc., for aiding in cluster analysis and Mr. M. Gomez, M.Sc., for revision of the English manuscript.

REFERENCES
Cavaliere, A.R. & Johnson, T.W. (1966). Marine Ascomycetes: ascocarp morphology and its application to taxonomy. III. A revision of the genus *Lulworthia* Sutherland. Nova Hedwigia, 10, 425-437.
Hughes, G.C. (1975). Studies on fungi in oceans and estuaries since 1961. I. Lignicolous, caulicolous and foliicolous species. Oceanogr. Mar. Biol., Ann. Rev., 13, 69-180.
Johnson, T.W. & Sparrow, F.K. (1961). Fungi in Oceans and Estuaries. Weinheim: Cramer.
Kirk, P.W. & Brandt, J.M. (1980). Seasonal distribution of lignicolous marine fungi in the lower Chesapeake Bay. Botanica mar., 23, 657-668.
Koch, J. & Jones, E.B.G. (1984). *Lulworthia lignoarenaria*, a new marine pyrenomycete from coastal sands. Mycotaxon, 20, 389-395.
Kohlmeyer, J. & Kohlmeyer, E. (1979). Marine Mycology. The Higher Fungi. New York: Academic Press.
Kohlmeyer, J. (1980). Tropical and subtropical filamentous fungi of the western Atlantic ocean. Botanica mar., 23, 529-544.
McComb, E.A. & McCready, R.M. (1958). Use of the hydroxamic acid reaction for determining pectinesterase activity. Stain Technol., 33, 129-131.
Meyers, S.P. (1966). Variability in growth and reproduction of the marine fungus, *Lulworthia floridana*. Helgoländer wiss. Meeresunters., 13, 436-443.
Meyers, S.P., Kamp. K.M., Johnson, R.F. & Shaffer, D.L. (1964). Thalassiomycetes IV. Analysis of variance of ascospores of the genus *Lulworthia*. Can. J. Bot., 42, 519-526.
Schaumann, K. (1968). Marine höhere Pilze (Ascomycetes und Fungi imperfecti) aus dem Weser-Ästuar. Veröff. Inst. Meeresforsch. Bremerh., 11, 93-117.
Schaumann, K. (1974). Zur Verbreitung saprophytischer höherer Pilze in der Hochsee. Erste quantitative Ergebnisse aus der Nordsee und dem NO-Atlantik. Veröff. Inst. Meeresforsch. Bremerh., Suppl., 5, 287-300.

Schaumann, K. (1975). Ökologische Untersuchungen über höhere Pilze im
 Meer- und Brackwasser der Deutschen Bucht unter besonderer
 Berücksichtigung der holzbesiedelnden Arten. Veröff. Inst.
 Meeresforsch. Bremerh., 15, 79-182.
Steinhausen, D. & Langer, K. (1977). Clusteranalyse: Einführung in
 Methoden und Verfahren der automatischen Klassifikation.
 Berlin, New York: Walter de Gruyter.
Sutherland, G.K. (1916). Additional notes on marine pyrenomycetes.
 Trans. Br. mycol. Soc., 5, 257-263.
Weide, G. (1985). Abbau algenbürtiger Polysaccharide durch marine
 höhere Pilze. Diplomarbeit, Univ. Hamburg.

J.D. Miller

INTRODUCTION

The Office of Technological Assessment of the U.S. congress divides biotechnology into new and old biotechnology. New biotechnology involves genetic engineering and old biotechnology concerns economic activities relating to applied biology, especially microbiology, that man has been doing for thousands of years. Old technology now accounts for the majority of profitable biotechnologies. One area of research that has been of particular interest concerns secondary metabolites of fungi. Everyone is aware of fungal secondary metabolites used in medicine, but there is less awareness of other uses for this sort of biochemical. For a variety of reasons, interest in fungal metabolites useful as fungicides and pesticides has increased (cf. Fischer & Bellus 1983). The Science Council of Canada commissioned a report on biotechnology relating to the forestry industry (Jurasek & Paice 1984) which has subsequently been evaluated by industry experts. One of the most useful areas of research identified in these studies concerns biological control of wood decay. Degradation of wood in chip piles, hydro poles and domestic dwellings alone costs Canada in the order of $CAN 200 million per year. As certain fungicides become unavailable (e.g. pentachlorophenol), coupled with the difficulty of registering such chemicals, biological means (and biochemicals) that prevent decay will become economic. In such cases, the study of lignicolous fungi and their metabolites becomes even more important than is the case now.

Having defined the biotechnological relevance of secondary metabolites of fungi, the next issue is my definition of a fungal secondary metabolite. Other than to note that a secondary metabolite is not a primary metabolite, I would prefer not to contribute to the number of words on that topic (Demain 1972; Bu'Lock 1975; Bennett 1983) because ultimately any definition will break down under close examination. I do believe, however, that secondary metabolites are always useful to the survival of the fungus producing them. Conversely, if fungi are observed in nature, some part of their effects on the substrate and associated biota are likely attributable to secondary metabolites. D.T. Wicklow (pers. com.) showed that the aflatoxin-producing fungus *Aspergillus parasiticus* Speare grows at about one third of the rate of *A. oryzae* (Ahlburg) Cohn. *A. oryzae* is used to make some food products in Japan and is of course not toxigenic. Kurtzman *et al.* (1984) recently showed that *A. flavus* Link ex Fries and *A. oryzae* have 97% nuclear DNA relatedness. The wild organism produces secondary metabolites at least partially at the expense of growth rate, and hence rate of spore production (cf. Rhoades 1979).

Janzen (1977) and Wicklow (1981) described the theoretical basis of interference competition in fungi. Many fungi have been demonstrated to produce antibiotics. Some of these antibiotics contribute to the survival and fitness of the organisms by preventing the growth of other organisms on the colonized substrate. For example, tree stumps colonized by *Trichoderma viride* Persoon ex Fries (a fungus that produces many toxic secondary metabolites) are not invaded by *Fomes annosus* (Fries) Cooke, a phenomenon exploited in the control of *F. annosus* root rots in managed forests. Other fungi produce compounds that play a role in fungal succession on wood. The larch canker fungus *Trichoscyphella willkommii* (Hart.) Nannf. can be displaced by *Cryptosporopsis* species after a period of time (Buczacki 1973). *Cryptosporopsis* species have been shown to produce an antibiotic active against wood decay fungi (Strunz *et al*. 1969). Some secondary metabolites of fungi may play a role in preventing animals from consuming propagules. Various *Aspergillus* toxins have been suggested to exist for this reason (Wicklow & Cole 1982; Wicklow & Shotwell 1983). The presence of competing fungi and other microorganisms in a given econiche affects the kinds and amounts of secondary metabolites produced by a given fungus. Thus the above described processes are dynamic (Wicklow *et al*. 1980; Horn & Wicklow 1983).

The production of secondary metabolites of fungi in quantity in the laboratory is a difficult problem. Only a very general understanding of secondary metabolism exists for filamentous fungi. In general terms, the biosynthesis of a secondary metabolite by a fungus requires a certain sequence of nutrient limitations, as well as particular oxygen, pH, and sometimes temperature conditions (Bu'Lock 1975; Anon. 1983). A knowledge of the type of metabolite desired is helpful, but if this is not known, trial and error experimentation is required. Sea water contains many nutrients and potentially limitation of any one or combination thereof may induce the excretion of a particular secondary metabolite. The substrate that the organism normally grows on should be examined to obtain clues as to the appropriate carbon and nitrogen sources. For example, what works for penicillin production (using mutant strains) will be inappropriate for metabolite production by marine fungi.

INTERFERENCE COMPETITION IN LIGNICOLOUS MARINE FUNGI

In my study of lignicolous marine fungi from the Bay of Fundy, driftwood, intertidal wood and submerged panels (wood from 3 species of trees) were examined for such fungi. The diversity was the same for drift and intertidal wood (26 species) but much reduced on the submerged panels (5 species on birch, 5 species on spruce, 4 species on pine, 9 species in total; Miller & Whitney 1981 a). These data are similar, in general terms, to those of other studies. The panels were submerged for one year and the apparent 'preference' of some fungi for certain types of wood described by Jones (1968) was observed. The low diversity on panels compared to intertidal wood is a well-known phenomenon and that fact prompted a series of investigations concerning the possible role of interference competition in lignicolous marine fungi described in Miller *et al*. (1985) and Strongman *et al*. (1986).

Beech panels of two sizes (10 mm^3 and 50 x 25 x 15 mm) were suspended from a raft in Langstone Harbour, Portsmouth from July to January (5 months). The panels were scraped and incubated in moist chambers at room temperature for 30 days. At that time, the fruiting structures of the marine fungi present were counted on each face (600 faces for both block sizes). *Lulworthia* species was the dominant fungus on 381 of the 600 faces.

On panel faces where *Lulworthia* sp. was the sole ascomycete observed, there was a mean of 137.7 perithecia per 10 mm^2. When perithecia of *Ceriosporopsis halima* Linder were observed on a face (16.8 per 10 mm^2), only 53.4 perithecia of *Lulworthia* were found per 10 mm^2. When cleistothecia of *Amylocarpus encephaloides* Currey (1.3 per 10 mm^2) were found, only 3.3 perithecia of *Lulworthia* were observed per 10 mm^2. Each value of the number of *Lulworthia* per 10 mm^2 in the three cases is statistically different ($p < 0.001$). A similar phenomenon was seen on the larger blocks with some additional species (Miller *et al*. 1985). Wicklow *et al*. (1980) described an *in vitro* method of quantifying antagonism of fungi. In brief, isolates of fungi are paired and allowed to grow together on a Petri dish containing a suitable medium. A number of reaction types are described (i.e. mutual inhibition at a distance; inhibition of one fungus on contact; inhibition of one fungus at a distance; etc.) and points are assigned to each reaction type. The sum of all reactions that a given fungus has to a variety of other fungi yields an "index of antagonism". In a test of 17 higher marine fungi including the three species noted above, the relationship between the three (*Lulworthia* sp., *C. halima*, *A. encephaloides*) was similar *in vitro* and *in vivo*.

Strongman *et al*. (1986) described a similar set of *in vitro* interference competition tests using a modified procedure in which marine fungi were grown on small balsa wood blocks that were placed on ultra pure agarose-artificial sea water medium. An index of antagonism and other factors such as growth rate were determined. These data were used to select isolates for fermentation and bioassays (e.g. antifungal, antibacteria, HeLa 229). Isolates were fermented on artificial sea water medium containing 10 g l^{-1} glucose and organic nitrogen in the form of peptone and yeast extract (0.6 g l^{-1} as N). Two aeration conditions were chosen (50% of saturation and 10% of saturation) and cultures were incubated for 30 d at 28° C. Various extracts of the culture filtrate and hyphae were made and (for the purposes of this discussion) tested using the method of Gueho & Pesando (1982) with a number of marine fungi.

A number of extracts exhibited antifungal activity. By way of example, a fermentation of *Leptosphaeria oraemaris* Linder that had a moderately high index of antagonism produced a solvent extract that showed

antifungal activity. Fractionation of this yielded a number of
compounds with biological activities. One of these was the
sesquiterpene culmorin (Figure 7.1). This compound is soluble enough
to diffuse throughout wood and has antibiotic activity to the marine
fungi tested, which included the "competitors" of *L. oraemaris*
(Strongman *et al*. 1986). This is clear evidence for interference
competition in a lignicolous marine fungus.

A study of the global distribution of lignicolous marine fungi reveals
a group that is cosmopolitan (Jones 1971) and there appear to be species
that are restricted to certain latitudes. Hughes (1975) and Kohlmeyer
(1983) argue that temperature is the most important factor controlling
the biogeography of marine fungi. However, on a local basis,
salinity-temperature-water viscosity have been demonstrated to be
important in controlling the distribution (i.e. a collection) of marine
fungi (Hughes 1969; Kirk & Brandt 1980). Cuomo and co-workers (1979,
1982) reported a number of common North Atlantic marine fungi (and a
lack of tropical forms) from the Mediterranean Sea. Hughes (1969)
reported on the importance of local currents in the distribution of
lignicolous marine fungi but it must be noted that the total number of
reports on marine fungi is still low. It seems possible that speciation
could have occurred within the major current systems of the oceans and
emphasis on current isotheres regarding the global distribution of
lignicolous marine fungi may be displaced (cf. Booth & Kenkel Chapter 25).

Concerning the local distribution of lignicolous marine fungi, Jones
(1976) summarized the observations concerning the colonization of wood
in the sea with respect to salinity, wood preference, "*Phoma*" pattern

Figure   7.1   Structural formula of culmorin.

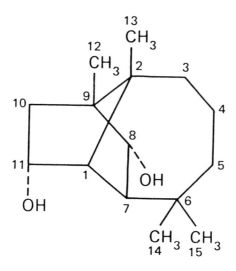

effects and succession. There are on the order of 100 reports of
lignicolous marine fungi in a variety of ecosystems and attempts are
usually made to assign these fungi to a particular econiche. This can
be taken to extremes and complicated hypotheses are sometimes invoked
during this exercise (e.g. mycostatic factors relating to higher marine
fungi, Kohlmeyer 1980).

The local distribution of any organism is governed by the limits of
environmental change that organism can tolerate for growth and
reproduction. The combined effects of temperature and salinity on the
growth of fungi (Richie 1957; Miller & Whitney 1981 b) may be the most
powerful factors that control whether a species can colonize wood in
the sea. Once a fungus lands on that wood, interference competition
likely plays the next most important role in determining whether that
fungus will grow, and in measure, whether a marine mycologist will
record it. If no other fungus is present then the fungus will grow. If
another fungus is present, it may or may not grow, depending on its
ability to resist and/or produce antifungal biochemicals. It will take
many years to establish the significance of these effects as they
relate to the lignicolous marine fungi, but, as is the case for
lignicolous terrestrial fungi, interference competition may be found to
be important in understanding the activities of marine fungi.

Our studies have shown antibiotic extracts from fermentations of marine
fungi. The key to finding more is to listen very carefully to
ecologists and use their observations to understand the biochemical
basis of the field observation. This allows some chance that the
biochemicals found may have some use. Consider the observation of
Miller & Jones (1983) regarding antibacterial activity by algicolous
thraustochytrids or the data of Meyers (1971) in which it appeared that
of 5 marine fungi fed to a nematode, one marine fungus supported only
about one third of the number of nematodes per mg hyphae than the
others. What chemicals may have been present to cause this?

REFERENCES
Anon. (1983). Protection against trichothecene mycotoxins. pp. 27-31.
    Washington, D.C.: National Academy Press.
Bennett, J.W. (1983). Differentiation and secondary metabolism in
    mycelial fungi. In Secondary metabolism and differentiation
    in fungi, eds. J.W. Bennett & A. Ciegler, pp. 1-32. New
    York: Marcel Dekker.
Buczacki, S.T. (1973). A microecological approach to larch Canker
    biology. Trans. Br. mycol. Soc., 61, 315-329.
Bu'Lock, J.D. (1975). Secondary metabolism in fungi and its relationship
    to growth and development. In The filamentous fungi. Vol. 1,
    eds. J.E. Smith & D.R. Berry, pp. 35-58. New York: Wiley.
Cuomo, V., Vanzanella, F. & Grasso, S. (1979). Lignicolous marine fungi
    of the Mediterranean coast of southern Italy. Botanica mar.,
    22, 405.
Cuomo, V., Vanzanella, F., Fresi, E., Cinelli, F. & Mazella, L. (1982).
    Fungal flora Posidonia oceanica and its ecological
    significance. Bull. Br. mycol. Soc., 16 (Suppl. 1), 5.

Demain, A.L. (1972). Cellular and environmental factors affecting
     synthesis and excretion of metabolites. J. appl. Chem.
     Biotechnol., 22, 345-363.
Fischer, H.-P. & Bellus, D. (1983). Phytotoxicants from microorganisms.
     Pestic. Sci., 14, 334-346.
Gueho, E. & Pesando, D. (1982). Antifungal activity of some
     Discomycetes. Biological spectrum of *Ciboria rufo-fusca*.
     Mycopath. Mycol. appl., 72, 123-128.
Horn, B.W. & Wicklow, D.T. (1983). Factors influencing the inhibition
     of aflatoxin production in corn by *Aspergillus niger*. Can.
     J. Microbiol., 29, 1087-1091.
Hughes, G.C. (1969). Marine fungi from British Columbia: occurrence and
     distribution of lignicolous species. Syesis, 2, 121-140.
Hughes, G.C. (1975). Geographical distribution of the higher marine
     fungi. Veröff. Inst. Meeresforsch. Bremerh., Suppl. 5,
     419-441.
Janzen, D. (1977). Why fruits rot, seeds mold and meat spoils. Am.
     Nat., 111, 691-713.
Jones, E.B.G. (1968). The distribution of marine fungi on wood submerged
     in the sea. In Biodeterioration of materials. Vol. 1, eds.
     J.J. Elphick & A.H. Walters, pp. 460-485. New York:
     Elsevier.
Jones, E.B.G. (1971). The ecology and rotting ability of marine fungi.
     In Marine borers, fungi and fouling organisms of wood, eds.
     E.B.G. Jones & S.K. Eltringham, pp. 237-258. Paris: OECD.
Jones, E.B.G. (1976). Lignicolous and algicolous fungi. In Recent
     advances in aquatic mycology, ed. E.B.Gareth Jones,
     pp. 1-49. New York: Wiley.
Jurasek, L. & Paice, M.G. (1984). Biotechnology in the pulp and paper
     industry. Ottawa: Science Council of Canada.
Kirk, P.W. & Brandt, J.M. (1980). Seasonal distribution of lignicolous
     marine fungi in the lower Chesapeake Bay. Botanica mar., 23,
     657-668.
Kohlmeyer, J. (1980). Marine fungi from Martinique. Can. J. Bot., 59,
     1314-1321.
Kohlmeyer, J. (1983). Geography of marine fungi. Aust. J. Bot. Suppl.,
     Ser. 10, 67-76.
Kurtzman, C.D., Smiley, M.J., Robnett, C.J., Axt, A. & Wicklow, D.T.
     (1984). DNA relatedness among the agronomically and
     industrially important fungi *Aspergillus flavus*, *A. oryzae*,
     A. parasiticus and *A. sojae*. Abst. Ann. Meeting Am. Soc.
     Microbiol., 011.
Meyers, S.P. (1971). Developments in the biology of filamentous marine
     fungi. In Marine borers, fungi and fouling organisms of
     wood, eds. E.B.G. Jones & S.K. Eltringham, pp. 217-235.
     Paris: OECD.
Miller, J.D. & Whitney, N.J. (1981 a). Fungi from the Bay of Fundy I:
     Lignicolous marine fungi. Can. J. Bot., 59, 1128-1133.
Miller, J.D. & Whitney, N.J. (1981 b). Fungi from the Bay of Fundy III:
     Geofungi in the marine environment. Mar. Biol., 65, 61-68.
Miller, J.D. & Jones, E.B.G. (1983). Observations on the association of
     thraustochytrid marine fungi with decaying seaweed. Botanica
     mar., 26, 345-351.

Miller, J.D., Jones, E.B.G., Moharir, Y.E. & Findlay, J.A. (1985). Colonization of wood blocks by marine fungi in Langstone Harbour. Botanica mar., 28, 251-257.

Rhoades, D.F. (1979). Evolution of plant chemical defense against herbivores. In Herbivores: their interaction with secondary plant metabolites, eds. G.A. Rosenthal & D.H. Janzen, pp. 3-54. New York: Academic Press.

Richie, D. (1957). Salinity optima for marine fungi affected by temperature. Am. J. Bot., 44, 870-874.

Strongman, D., Miller, J.D., Calhoun, L., Findlay, J.A. & Whitney, N.J. (1986). The biochemical basis of interference competition of some lignicolous marine fungi. Botanica mar., in press.

Strunz, G.M., Court, A.S., Komlossy, J. & Stillwell, M.A. (1969). The structure of cryptosporiopsin, a new antibiotic substance produced by a species of *Cryptosporiopsis*. Can. J. Chem., 47, 2087-2094.

Wicklow, D.T. (1981). Interference competition. In The fungal community, eds. D.T. Wicklow & G.G. Caroll, pp. 351-375. New York: Marcel Dekker.

Wicklow, D.T., Hesseltine, C.W., Shotwell, O.L. & Adams, G.I. (1980). Interference competition and aflatoxin levels in corn. Phytopathology, 70, 761-764.

Wicklow, D.T. & Cole, R.J. (1982). Termorgenic indole metabolites and aflatoxins in sclerotia of *Aspergillus flavus*: an evolutionary perspective. Can. J. Bot., 60, 525-528.

Wicklow, D.T. & Shotwell, O.L. (1983). Intrafungal distribution of aflatoxins among conidia and sclerotia of *Aspergillus flavus* and *Aspergillus parasiticus*. Can. J. Microbiol., 29, 1-5.

J.P. Amon

INTRODUCTION

Rhizophydium littoreum Amon (Amon 1984) was isolated from a green alga (Kazama 1972) in the estuarine environment. An alga can provide copious nutrient to the developing thallus but once the resulting zoospores are released they may be transported to regions of considerably poorer nutrient. The next generation of the organism may be forced to live oligotrophically or starve.

Little is known about the ability of marine fungi to survive in the waters of relatively low nutrient that may often surround them. Laboratory studies of nutrient requirements are designed to test the ability of fungi to use various substrates but rarely addresses whether or not the environment could provide concentrations of the nutrient suitable for continued growth (Jones & Harrison 1976; Amon & Arthur 1980, 1981; Bahnweg & Bland 1980). Since dissolved organic nutrients may exist in marine sediments or in the water column at nannomolar to micromolar concentrations (Henrichs & Farrington 1979; Mopper *et al.* 1980; Seki 1982), it is the intent of this study to demonstrate the ability of the marine chytrid *R. littoreum* to survive and grow in these oligotrophic conditions.

A traditional approach to simulate the natural environment is to use a chemostat. The advantages and shortcomings of this technique have been discussed by Jannasch (1967, 1974) and Harder & Dijkhuizen (1982). In the present study a second approach to continuous culture was used to study *R. littoreum*. A perfusion chamber (PC) with a volume of $0.5\,\text{cm}^3$ was employed to supply continuously nutrient of known concentration to cells attached and growing on its coverglass observation ports. Such a device differs from a chemostat in that a steady state in the PC is provided by a high dilution rate whereas the chemostat depends on a balance between growth rate and dilution rate to achieve steady state. In a chemostat waste accumulates and nutrient is depleted prior to steady state; in the PC waste is continuously diluted and largely removed and nutrients remain at input levels. Cells attach to the glass in the PC technique and very high dilution rates may be employed without loss of the culture to "washout". A system such as the PC should be capable of simulating those environments where, low concentrations of nutrients are available from an infinite sink such as the ocean, or at the sediment water interface where important lower marine fungi are known to exist (Gaertner & Raghu Kumar 1980).

Chemostat studies may be more indicative of the upper layers of
sediment where nutrient and waste are in an equilibrium controlled by
both organismic and diffusion processes. Traditional batch cultures
were used in this study to simulate those transient periods when
abundant nutrient is available and for comparison purposes.

MATERIALS AND METHODS

Rhizophydium littoreum, American Type Culture Collection
number 36100, (known previously as *Phlyctochytrium* sp.) was maintained
by transfers at 48 h intervals on a sea water medium as described
previously (Amon 1976, 1984; Amon & Arthur 1980, 1981; Amon &
Muehlstein 1985).

Perfusion chambers were Sykes—Moore chambers (Belco) or a modification
machined from Lucite and sealed with a flat silicon gasket. Media were
supplied via peristaltic pumps and entered the $0.5\,\text{cm}^3$ growth chamber
through a 23G syringe needle. To assure that nutrient utilization did
not significantly alter the nutrient levels during transit to the
outflow, dilution rates were up to 80 times that estimated to cause
washout of chemostat cultures. Flow through the culture was initiated
only after cells had attached to the bottom coverslip of the chamber
(30-60 min). Cultures were observed 3-8 times daily to record
developmental changes and to measure microscopically the growth of the
cells. All measurements were made at x 400 magnification using a
phase-contrast microscope. Biomass was determined by converting thallus
diameter to volume.

RESULTS

Initial studies using early log phase batch cultures of
R. littoreum indicated that balanced growth could be achieved with a
carbon : nitrogen ratio of approximately 7 : 1. In batch cultures at
glucose concentrations of 0.1 to 10 g 1^{-1} the size of the mature

Table 8.1 Growth of *R. littoreum* in batch culture

Initial glucose concentration (µM)	Size of mature sporangium (µm)	Doubling time (h)	Time to complete life cycle (h)
0	15	4.9	34
556	20	4.1	34
5556	23	4.3	38
13888	33	4.6	48
27778	65+	3.9	53

[1] In a similar test about 4000 µM glucose equivalents of added starch
gave essentially the same result.

sporangium increased with increased glucose level. At the intermediate
concentrations a similar trend was observed in the perfusion chamber.
At lower concentrations, in either batch or perfusion culture, of
glucose (1-100 μM) no further concomitant decrease in the sizes of the
sporangia was noted (Tables 8.1, 8.2). The same trend appeared when
a high level of starch, equivalent to 4000 μM glucose, was added to the
basal medium in batch cultures (Table 8.1). Below 1.0 μM glucose the
maximum size of the sporangium was much reduced (Table 8.2). In
addition, the doubling time was much greater.

Under glucose limitation the doubling time in the PC was relatively
unchanged over the range of 1 to 5000 μM. The time to complete the life

Table 8.2 Carbon limited growth in perfusion continuous cultures of
Rhizophydium littoreum

A. Nitrogen as NH_4Cl at 14000 μM

Glucose concentration in μmoles per litre	Size of mature sporangium (μm)	Doubling time (h)	Time to complete life cycle (h)
0.5	5.7	27.0	72-81
1.0	25	6.1	56
10.0	16	6.6	48
100.0	20	5.8	48
500.0	43	4.3	50
1000.0	57	4.1	52
5000.0	46	5.9	70

B. Effect of Lower Nitrogen Levels (glucose at 10 μM)

Concentration (μM) NH_4Cl	glutamate	Size of mature sporangium (μm)	Doubling time (h)	Time to complete life cycle (h)
1. Nitrogen limited				
0.5	–	6.0	8.3	25
10.0	–	6.2	7.0	22
–	10.0	7.2	5.5	21
2. Carbon limited cultures				
500	–	13.5	3.5	23
–	500	12.5	3.7	23
14000	–	16.0	6.6	48

cycle was consistent over the range of 10 to 1000 μM glucose. The
increase in doubling time, size at maturity, and the time to complete
the life cycle at 0.5 μM glucose was indicative of the starvation
conditions (Figure 8.1). The longer life cycle time at the highest
glucose level resulted from a delay of sporulation which occurred at
high concentrations of glucose (Chin 1981). Reducing the nitrogen

Figure 8.1 Doubling time and time to complete life cycle.
Note extremes. Concentration scale is logarithmic.

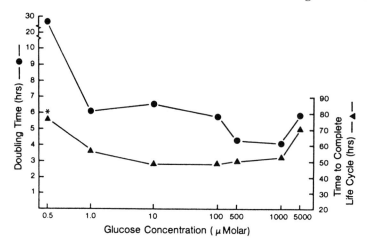

Figure 8.2 Three size responses to glucose concentration.
Below 1 μM cells barely survive.

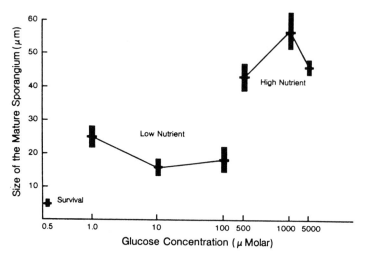

levels, either as ammonium or glutamate-N, gave a decrease in the doubling time and a slight decrease in the size of the sporangium in the PC (Table 8.2).

Effects of nitrogen limitation are shown in Table 8.2 B,1. With glucose at 10 µM the concentration of ammonium had little effect on the time to complete the life cycle, doubling time or the final size of the sporangium. Glutamate-N appeared to be a better source of nitrogen as it somewhat increased sporangium size and decreased doubling time. The overall effect of lower (limiting) levels of nitrogen was a decrease in sporangium size regardless of whether the culture was carbon or nitrogen limited.

It appeared that three types of size response may be caused by glucose concentration (Figure 8.2). At concentrations of 1 to 100 µM glucose the sporangium remained near 20 µm diam. At concentrations of 500 µM glucose and above the sporangium was at least twice as large. Below 1 µM glucose the sporangium was so small that it produced only one or two zoospores at maturity. In addition, the lowest concentration produced a number of individuals (ranging from less than 5% to 50%) which initially grew but never matured. When either glucose or the nitrogen source was provided to PC cultures at less than 1 µM many cells died soon after the zoospores settled. Cell debris from the mortality was penetrated and apparently used by the developing rhizoids of the remaining cells (Figure 8.3a).

In addition to the effect on the sizes of sporangia nutrient concentration affected other aspects of morphology. In low nutrient conditions rhizoids were long and narrow whereas in high nutrient conditions they tended to be shorter and more blunt (Figures 8.3b,c). In low nutrient, distribution of cytoplasmic inclusions tended to be

Figures 8.3a-d *Rhizophydium littoreum*. Light micrographs. Bars = 25 µm.
Figure 8.3a Uniformly granular cell in low nutrient penetrates cell debris (C) with rhizoids (R).
Figure 8.3b Cell with narrow rhizoids (R) in low nutrient.
Figure 8.3c Cell with wider rhizoids (R) in high nutrient.
Figure 8.3d Cell with storage material (S) in high nutrient.

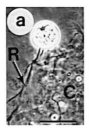 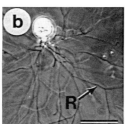 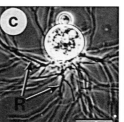 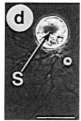

uniform throughout the life cycle (Figure 8.3a) while in high nutrient levels midphase cells had a large granulation-free region (Figure 8.3d) in the upper part of the cell (Amon 1976). Ultrastructural examination (Figure 8.4a) showed that this clear region resembled glycogen (Coulter & Aronson 1977). Under low nutrient conditions sporangia were smaller, thinner walled (Figure 8.4b) and possessed fewer exit papillae than those grown at high nutrient levels (Figure 8.4c).

Chemostat cultures behaved differently from perfusion cultures. Input concentrations of 100, 500, 1000 and 5000 µM glutamic acid (nitrogen limiting) were tested but after numerous attempts it was evident that at

Figures 8.4a-c *Rhizophydium littoreum.*
Figure 8.4a T.E.M. of the thick cell wall (W) of a thallus grown in 500 µm glucose. Note the glycogen-like storage material (S). Bar = 0.2 µm.
Figure 8.4b T.E.M. of the thinner cell wall (W) of a thallus grown in 5 µm glucose. Note the reduced amount of storage material. Bar = 0.2 µm.
Figure 8.4c Thalli grown in 1-10 µm glucose are small and have few exit papillae. L.M. Bar = 20 µm.

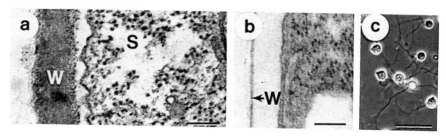

Figures 8.5a-c *Rhizophydium littoreum.*
Figures 8.5a,b Intrasporangial germination. T.E.M. (Fig. 5a) and phase-contrast (Fig. 5b) micrographs of thalli (T) showing rhizoids (R) penetrating mother cell wall (M). Fig. 5a, bar = 1 µm. Fig. 5b, bar = 25 µm.
Figure 8.5c Autoradiograph of thallus (T) and rhizoids (R) indicating uptake of C-14 bicarbonate. Bar = 25 µm.

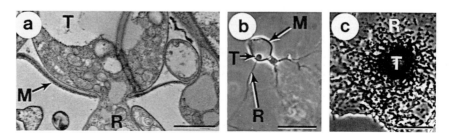

100 µM and 5000 µM consistent results could not be attained. Cultures
with inputs of 500 and 1000 µM glutamic acid produced normal steady
state conditions and could be manipulated to determine maximum growth
rates. Residual substrate concentrations in the growth vessels normally
fell to about 2 to 10% of the initial value. Steady state doubling
times were calculated to be 42 h and 125 h for the 1000 and 500 µM
chemostats respectively. Comparison shows these doubling times to be
orders of magnitude greater than PC cultures (Table 8.2). Sizes of
the sporangia in chemostat cultures varied with the input concentration
of substrate in much the same manner as observed in perfusion chambers.

The morphology of chemostat grown cells was related to the clumping of
individual cells. Some clumping was apparently due to a large degree of
intrasporangial germination. Intrasporangial germination resulted when
zoospores were not released from the sporangia. In some cases this may
have been caused when cells surrounding the mother cell blocked the path
of release. In other cases no apparent blockage could be implicated.
Intrasporangial germination was most frequent at the lower concentration
of limiting substrate in chemostats and was occasionally observed in PC
cultures (Figures 8.5a,b).

Routine monitoring of the level of glucose in nitrogen limited
chemostats revealed an apparent increase in the level of that compound.
The input level of glucose was 10 g l $^{-1}$ but glucose levels occasionally
rose as high as 17 g l $^{-1}$ although the average amount was approximately
11 g l $^{-1}$. Labelling studies using C-14 labelled sodium bicarbonate
showed that cells grown in batch, chemostat, or PC cultures were
capable of fixing significant amounts of bicarbonate carbon (Jones
1983). A large portion of the fixed carbon was quickly transferred to
soluble compounds found in the culture supernatant. There was little
difference between cells grown in lighted and dark conditions but the
presence of glucose did enhance the fixation process. Preliminary
characterization of carbohydrates in the supernatant by thin layer
chromatography demonstrated that labelled glucose was present but in a
mixture with many other carbohydrates and other unknown compounds.
Figure 8.5c shows an autoradiogram which demonstrates the uptake of
labelled bicarbonate.

DISCUSSION
Studies on zoosporic marine fungi have demonstrated their
abundance in the ocean from the tropics to the poles (Johnson & Sparrow
1961; Ulken 1970; Bahnweg & Sparrow 1974). In addition these fungi have
been isolated from both intertidal and benthic environments (Bremer
1976; Gaertner & Raghu Kumar 1980). Given such broad distribution and
the variable nature of nutrient in these habitats, it is important to
understand the ability of zoosporic marine fungi to use the ambient
levels of nutrient. If nutrient levels fall below that necessary to
support growth it is important to know of any survival strategies
present in these fungi.

Marine bacteriologists have noted several characteristics which would

enable a microbe to survive in low nutrient environments. Poindexter
(1981) suggested that the mechanisms for copiotrophs differ from the
true oligotrophs in that they may not have the ability to use the very
low concentrations of nutrient available in some habitats.
R. littoreum does not grow well at concentrations below 1 μM glucose or
glutamic acid which suggests that it is not capable of sustained growth
in the water column of the open ocean where nanomole quantities of
these substrates are common. Concentration of nutrients in marine
sediments is often 10 times higher (sediment is in the 1 to 10 μM range
for amino acid and carbohydrate) than in the overlying waters (Crawford
et al. 1974; Henrichs & Farrington 1979; Mopper et al. 1980; Southward
& Southward 1982) and it appears that R. littoreum may be able to grow
under these slightly elevated nutrient conditions as well as it does in
very rich media.

Attachment of an organism to surfaces may also enhance its survival at
low nutrient levels (Poindexter 1981; Fletcher & Marshall 1982; Paerl
1982). If attachment gives R. littoreum any advantage it may explain
why this and other similar fungi are found either in sediments or
attached to algae. Additionally, if attachment promotes the uptake of
nutrients, it may account for the faster growth rate of R. littoreum in
the PC than cells grown in similar nutrient concentrations suspended in
a chemostat vessel.

Paerl (1982) and Morita (1982) found that cells which survived
oligotrophic conditions often decreased in size and thus effectively
increased their surface to volume ratio. Such an adaptation makes
procuring nutrient more energetically efficient. When R. littoreum is
provided with abundant nutrient it grows rapidly and to great size,
stores carbohydrate, and produces thick rhizoids. The effect of
excessive glucose is to delay maturation and release of zoospores, and
to produce a greater number of progeny per sporangium. Conversely,
scarcity of glucose leads to production of small one or two zoospored
sporangia which mature rather quickly, elimination of storage products
and an increase of surface area (narrow rhizoids). This provides an
advantage because the population has the ability to respond to copious
nutrient with copious growth, or to low nutrient by smaller size and
faster maturation. In either case the zoospores, just like motile
bacteria (Paerl 1982), are the key to establishing the next generation
in favourable nutrient and zoospores can swim for at least 24 hours
(Muehlstein 1985). Zoospores are known to be attracted to various
carbohydrates and mixtures of amino acids (Amon & Muehlstein 1985).

Morita (1982) suggested that organisms with the greatest ability to
survive will be prepared to use any substrate when it becomes
available. Harder & Dijkhuizen (1982) further postulated that the
ability to use mixtures of substrates is a strategy for growth at low
nutrient levels. Yei (1981) in this laboratory, discovered that at
high levels of nitrogen only organic forms gave high growth rates,
while at lower nitrogen levels both inorganic and organic forms produced
similar growth rates. Likewise, Chin (1981) demonstrated that high
levels of glucose repressed the utilization of starch, but low levels

did not.

Storage of reserve materials during growth in excess glucose may aid
microbes to survive periods when nutrient levels fall below those that
normally support growth (Morita 1982; Paerl 1982). With this energy
reserve the starving organisms may be able to complete development and
release zoospores which may swim to another location of more favourable
nutrient. Chytrids are known to store glycogen during their growth
phase (Coulter & Aronson 1977) and the zoospores of this fungus have a
large lipid body (Kazama 1972). Cryptic growth, or growth at the
expense of dead and dying cells in the population has been observed and
this is, in effect, similar to the use of storage materials.
Intrasporangial germination may also tend to conserve loss of reserves
as well as protect the thallus from losses by diffusion.
Intrasporangial germination leads to clumps which provide food through
cryptic growth.

Carbon dioxide fixation may act as an alternative source of carbon to
this chytrid but that can be significant only if a source of energy is
found for the process. Any of these adaptations fit the established
strategies for survival under low nutrient conditions.

To understand the nutritional requirements of *R. littoreum* and similar
fungi in nature it may be necessary to analyze three types of culture.
The batch culture is appropriate to model growth of the fungus on rich
nutrients consumed by growth. Such a situation occurs when fresh plant
or animal material dies and is colonized by these saprobes. The
opposite extreme of environment is found in the open water column. In
the water column nutrient levels are very low, and they do not appear to
change as a result of biological activity. Waters in such an
environment may contain a residual nutrient which is the "steady state"
result of biological activity. This situation can be modelled by either
a chemostat or by the perfusion chamber. Both provide a constant
nutritional environment. In nature metabolites produced by growth of
one organism are degraded by the actions of the community around it. In
a chemostat one or even a few species may reach a steady state
influenced by cumulative, albeit, diluted and only partially processed
wastes. The chemostat situation is not, however, entirely unrealistic.
Populations living as microcolonies or small simple communities may live
in the semiconfinement of marine sediments. As in the chemostat they
may receive a constant input of nutrient, use it, produce waste and
experience dilution, by diffusion from above. The PC provides the
researcher with an opportunity to reconstruct the known chemistry of a
given habitat and determine whether or not it can support the growth or
survival of a particular species. In addition the PC provides surfaces
for attachment which are typically not available in or must be
specifically guarded against in the chemostat.

The perfusion chamber is not a new tool. Pasteur described such a
device in 1879 (Pasteur 1879) and others have used it in culture of
animal cells and fungi (Salkin 1970; Salkin & Robertson 1970). Its use
as a system to approximate the environment has not been broadly

exploited. The advantages of this device should aid in the understanding of microbial nutrition in nature.

ACKNOWLEDGEMENTS
The author recognizes the contributions of P. Jones, S. Yei, S. Chin, D. Jennings, L. Muehlstein, D. Cool, B. Lutz, D. Leffler, and M. Yaggi to the completion of this work, which was supported in part by a grant from the National Science Foundation OCE-8007958.

LITERATURE CITED

Amon, J.P. (1976). An estuarine species of *Phlyctochytrium* (Chytridiales) having a transient requirement for sodium. Mycologia, 68, 470–480.

Amon, J.P. (1984). *Rhizophydium littoreum*: A chytrid from siphonaceous marine algae – an ultrastructural examination. Mycologia, 76, 132–139.

Amon, J.P. & Arthur, R.D. (1980). The requirement for sodium in marine fungi: Uptake and incorporation of amino acids. Botanica mar., 23, 639–644.

Amon, J.P. & Arthur, R.D. (1981). Nutritional studies of a marine *Phlyctochytrium* sp. Mycologia, 72, 1049–1055.

Amon, J.P. & Muehlstein, L.K. (1985). Chemotaxis and phototaxis in *Rhizophydium littoreum*. Poster, Fourth International Marine Mycology Symposium, Portsmouth, 10–17 August.

Bahnweg, G. & Bland, C.E. (1980). Comparative Physiology and nutrition of *Lagenidium callinectes* and *Halipthoros milfordensis*, fungal parasites of marine crustaceans. Botanica mar., 23, 689–698.

Bahnweg, G. & Sparrow, F.K. (1974). Occurrence distribution and kinds of zoosporic fungi in subantarctic and antarctic waters. Veröff. Inst. Meeresforsch. Bremerh., Suppl. 5, 149–157.

Bremer, G.B. (1976). The ecology of marine lower fungi. In Recent Advances in Aquatic Mycology, ed. E.B. Gareth Jones, pp. 313–333. London: Elek.

Chin, S.C. (1981). Effects of glucose limitation on the regulation of cell activity in *Phlyctochytrium* sp. M.S. thesis, Wright State University, U.S.A.

Coulter, D.B. & Aronson, J. (1977). Glycogen and other soluble glucans from Chytridiomycete and Oomycete species. Arch. Microbiol., 115, 317–322.

Crawford, C.C., Hobbie, J.E. & Webb, K.L. (1974). The utilization of dissolved free amino acids by estuarine microorganisms. Ecology, 55, 551–563.

Fletcher, M. & Marshall, K.C. (1982). Are solid surfaces of ecological significance to aquatic bacteria? In Advances in Microbial Ecology, ed. K.C. Marshall, vol. 6, pp. 199–236. New York: Plenum Press.

Gaertner, A. & Raghu Kumar, S. (1980). Ecology of the Thraustochytrids (lower marine fungi) in the Fladen Ground and other parts of the North Sea. Meteor Forch.-Ergebn. A., 22, 165–185.

Harder, W. & Dijkhuizen, L. (1982). Strategies of mixed substrate
     utilization in microorganisms. Phil. Trans. R. Soc. Lond.
     Ser. B, 297, 459-480.
Henrichs, S.M. & Farrington, J.L. (1979). Amino acids in interstitial
     water of marine sediment. Nature, 279, 319-322.
Jannasch, H.W. (1967). Growth of marine bacteria at limiting
     concentrations of organic carbon in seawater. Limnol.
     Oceanogr., 12, 264-271.
Jannasch, H.W. (1974). Steady state and the chemostat in ecology.
     Limnol. Oceanogr., 19, 716-720.
Johnson, T.W. Jr. & Sparrow, F.K. Jr. (1961). Fungi in Oceans and
     Estuaries. Weinheim: J. Cramer.
Jones, P. (1983). Effects of carbon limitation on the growth and
     development of Rhizophydium sp.: a perfusion chamber study.
     Honours thesis, Wright State University, U.S.A.
Jones, E.B.G. & Harrison, J.L. (1976). Physiology of marine
     phycomycetes. In Recent Advances in Aquatic Mycology, ed.
     E.B.Gareth Jones, pp. 261-278. London: Elek.
Kazama, F. (1972). Development and morphology of a chytrid isolated
     from Bryopsis plumosa. Can. J. Bot., 50, 499-505.
Mopper, K., Dawson, R., Leibzeit, G. & Ittekkot, V. (1980). The
     monosaccharide spectra of natural waters. Mar. Chem., 10,
     55-66.
Morita, R.Y. (1982). Starvation - Survival of heterotrophs in the
     marine environment. In Advances in Microbial Ecology, ed.
     K.C. Marshall. Vol. 6, pp. 171-198. New York: Plenum Press.
Muehlstein, L.K. (1985). Phototaxis and chemotaxis in the marine
     fungus Rhizophydium littoreum Amon. M.S. thesis, Wright
     State University, U.S.A.
Paerl, H.W. (1982). Factors limiting productivity of freshwater
     ecosystems. In Advances in Microbial Ecology, ed.
     K.C. Marshall. Vol. 6, pp. 75-110. New York: Plenum Press.
Pasteur, L. (1879). The physiological theory of fermentation. Transl.
     by F. Faulkner & D.C. Robb. In The five foot shelf of
     books. Vol. 38, ed. C.W. Eliot. New York: P.F. Collier &
     Son.
Poindexter, J.S. (1981). Oligotrophy: Feast and famine existence. In
     Advances in Microbial Ecology. Vol. 5. ed. M. Alexander,
     pp. 63-89. New York: Plenum Press.
Salkin, I.F. (1970). Allochytrium expandens, gen. et sp. n.: Growth and
     morphology in continuous culture. Am. J. Bot., 57, 649-658.
Salkin, I.F. & Robertson, J.A. (1970). Use of a tissue culture chamber
     for developmental studies of aquatic phycomycetes. Arch.
     Mikrobiol., 70, 157-160.
Seki, H. (1982). Organic materials in aquatic ecosystems. Boca Raton,
     Florida: CRC Press.
Southward, A.J. & Southward, E.C. (1982). The role of dissolved organic
     matter in the nutrition of deep-sea benthos. Am. Zool., 22,
     647-659.
Ulken, A. (1970). Phycomyceten aus der mangrove bei Cananeia (Sao
     Paulo, Brasilien). Veröff. Inst. Meeresforsch. Bremerh., 12,
     313-319.

Yei, S. (1981). Effects of low L-glutamic acid conditions on growth, development and survival of the marine fungus *Phlyctochytrium* sp. in continuous culture. M.S. thesis, Wright State University, U.S.A.

9   PHYSIOLOGICAL AND BIOCHEMICAL CHARACTERISTICS OF THE YEAST
    DEBARYOMYCES HANSENII IN RELATION TO SALINITY

L. Adler

INTRODUCTION
        Marine-occurring yeasts differing in halotolerance have been
isolated by Norkrans (1966). One of the yeast species found,
Debaryomyces hansenii (Zopf) van Rij, was strongly salt-tolerant, with
growth in the range 0-24% NaCl (w/v). At high salinities the internal
salt level is not sufficient to balance the osmotic potential of the
environment (Norkrans & Kylin 1969; Hobot & Jennings 1981) and other
osmolytes have to be accumulated. Evidence indicates that this type of
regulation is mediated by the accumulation of polyhydroxy alcohols in
D. hansenii, as reported below. Cell osmolytes that are accumulated to
high intracellular levels have to provide an environment that does not
adversely affect macromolecular structure and function, which has led to
the introduction of the concept of compatible solutes (Brown 1976,
1978); a concept extended and presented with an evolutionary perspective
by Yancey *et al.* (1982).

In this chapter glycerol accumulation and its role in osmoregulation of
D. hansenii are discussed. Based on mutant studies a possible scheme
for the glycerol pathways is presented.

MATERIAL AND METHODS
        Debaryomyces hansenii strain 26 (Norkrans 1966), and two
glycerol nonutilizing strains (strain 26-3 and 26-6) isolated as
described by Adler *et al.* (1985) were used. Media, growth conditions,
extraction and analysis of glycerol were as described by Adler &
Gustafsson (1980) and Adler *et al.* (1985). Preparation of crude
extracts and measurements of enzyme activities were as given by Adler
et al. (1985).

For experiments (Table 9.1 & Figure 9.2) which were not described by
Adler & Gustafsson (1980) and Adler *et al.* (1985), cells were grown in
Fernbach flasks containing 500 cm^3 of a 0.5% glucose mineral-salts
medium (Adler & Gustafsson 1980) to a cell density of 0.6-0.7 mg cell
dry wt cm^{-3}. Experiments were started by diluting the exponentially
growing culture with an equal volume of fresh medium containing the
stress-solute. Culture volumes of 500 cm^3 were immediately
re-established, and at time intervals, glycerol was extracted and
analyzed as described by Adler *et al.* (1985). Cell dry weight was
estimated from measurements of absorbance at 610 nanometres using

standard curves.

RESULTS
Polyol accumulation
        The production and accumulation of polyols in *D. hansenii*
during a growth cycle in batch cultures at 0 and 16% NaCl (w/v) is shown
in Figure 9.1. The accumulation of glycerol and arabinitol each
follows a unique time course and the intracellular levels are markedly
increased at 16% NaCl. Intracellular glycerol increases sharply during
the lag phase, peaks during the log phase, and decreases and eventually
disappears in the early stationary phase. Arabinitol remains
essentially constant during the lag and early log phase, but increases
in the late log and early stationary phase concomitant with the decrease
of glycerol (Adler & Gustafsson 1980).

The two polyols were the only low-molecular weight organic compounds
found to increase in salt-stressed cells; amino acid and gas
chromatographic analyses of trimethylsilyl derivatives from cell
extracts revealed no marked changes of other compounds.

Figure 9.2 shows the effect of a sudden increase of salinity on growth
(A) and glycerol content (B) of exponentially growing cells. Following
this "upshock" intracellular glycerol increased after 2-4 h to a plateau
level, proportional to the salt concentration of the medium. The level

Figure 9.1 Changes in polyol levels during a growth cycle
of *D. hansenii* wild-type in media of 0% and 16% NaCl.
Intra- (■) and extracellular (□) glycerol; intra- (●)
and extracellular (○) arabinitol. Log dry wt of yeast
(△). Redrawn from Adler & Gustafsson (1980).

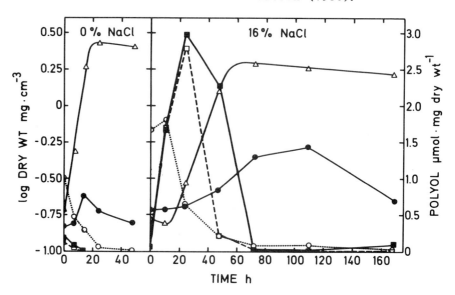

reached is very close to that seen in *D. hansenii* grown in media of comparable salinity (Adler *et al.* 1985). Assuming a cell water content of 2 µl mg dry wt $^{-1}$ (cf. Meredith & Romano 1977) the glycerol concentration reached after transfer to 8% NaCl was about 1.25 M. There was no obvious effect on growth by salinities of 0.25-1% NaCl, whereas 2 and 4% caused a short lag phase, extended at 8% NaCl to last approximately equally long as the time needed to build-up the internal plateau level of glycerol.

To examine whether the glycerol accumulation is specifically triggered by NaCl or a result of a more generalized response to the water activity (a_W) of the medium, the cells were exposed to several different stress-solutes added in concentrations to yield an a_W of 0.9964 or 0.9927. (The concentration of NaCl needed for this adjustment was 0.64 and 1.31% (w/v), respectively). Essentially the same results were observed when the a_W of the medium was adjusted with different salts or with sucrose (Table 9.1).

Figure 9.2 Growth (A) and glycerol accumulation (B) by *D. hansenii* wild-type after exposure to selected salinities. The concentrations of NaCl (%, w/v) were as indicated in the figure.

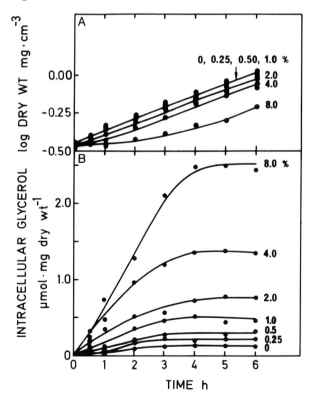

During characterization of a number of glycerol non-utilizing mutants, it was observed that one of the mutants isolated, strain 26-6, showed a reduced ability to produce glycerol when subjected to salt-stress. The rate of glycerol accumulation was slower than for the wild-type and the maximum levels attained in media of 2, 4 and 8% NaCl were *ca* 0.3, 0.6, and 1.1 µmol mg dry wt $^{-1}$, respectively (Adler & Blomberg unpublished), i.e. about half the amount accumulated by the parental strain under similar

Table 9.1 Glycerol accumulation by *D. hansenii* wild-type after exposure to various stress-solutes in isotonic concentrations. The solute concentrations used to adjust a_w were as given in the tables of Harris (1981). The glycerol levels shown represent the plateau levels reached 2-4 h after addition of the stress-solute. The control value obtained with no stress-solute added was 0.09 µmol mg dry wt $^{-1}$.

Stress-solute	Intracellular glycerol µmol mg dry wt $^{-1}$	
	a_w 0.9964	a_w 0.9927
KCl	0.24	0.44
NaCl	0.25	0.43
Na_2SO_4	0.25	0.44
Sucrose	0.27	0.49

Figure 9.3 Growth of *D. hansenii*, mutant strain 26-6, on 1% glucose media containing 14% NaCl (w/v). The media were supplemented as follows: (upper row, left-right) no addition, 1 mM arabinitol, 1 mM mannitol; (lower row, left-right) 1 mM betaine, 1 mM glycerol, 1 mM proline. Each plate was inoculated with approximately 400 colony-forming units.

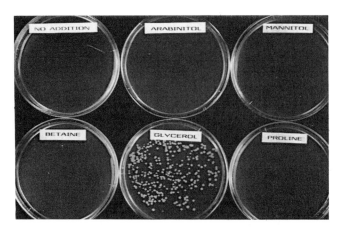

conditions (cf. Figure 9.2). Unlike the wild-type, the mutant did not grow when plated on solid 1% glucose media containing >10% NaCl. However, as illustrated in Figure 9.3, the mutant grows well at 14% NaCl if the medium is supplemented with 1 mM glycerol. Addition of other well-known osmoprotectants (mannitol, arabinitol, proline and betaine; cf. Le Rudulier *et al.* 1984) to the same concentration did not promote growth of the mutant.

Glycerol metabolism

With the aim of studying the glycerol metabolism in *D. hansenii* one of the glycerol non-utilizing mutants, strain 26-3, that grow on gluconeogenic carbon sources other than glycerol, was characterized further (Adler *et al.* 1985). At increased salinity the total glycerol production was greater for the mutant than for the wild-type and the mutant was unable to reutilize the produced glycerol. However, the internal levels of glycerol were similar to those found in the parental strain; the excess glycerol produced was lost to the extracellular medium. Thus, the ability to adjust internal levels of glycerol in response to salt-stress was unaffected by the block in the dissimilatory pathway.

Table 9.2 compares enzyme activities of the parental strain to those of the mutant strain 26-3. The pattern is similar for all enzymes of the glycerol metabolism examined except that the activity of the mitochondrial glycerol 3-phosphate (G3P) dehydrogenase was lacking in the mutant under the assay conditions used. A more detailed examination of the mutant enzyme revealed a 330-fold increase in the apparent value for K_m, for the substrate G3P (Adler *et al.* 1985).

DISCUSSION

In *D. hansenii* polyols are accumulated in response to salt-stress (Gustafsson & Norkrans 1976; Adler & Gustafsson 1980). During a growth cycle at high salinity, the polyol pool shows a highly dynamic behaviour and growing cells have qualitatively and quantitatively a significantly different polyol composition than stationary cells (Figure 9.1). In the exponential phase glycerol is the main polyol accumulated while arabinitol predominates in the stationary phase.

When exponentially growing cells were subjected to a sudden increase of salinity (Figure 9.2), the intracellular content of glycerol increased during a period of 2-4 h to a level proportional to the salinity of the medium (0.25-8% NaCl). Thus, the cells have ability to sense gradual differences in salinity and adapt the intracellular contents of glycerol to a wide range of external salinities. Since isotonic concentrations of several different stress-solutes (three salts and a non-electrolyte) induce glycerol accumulation to similar levels (Table 9.1), it is likely that the glycerol accumulation, at least during mild stress, is a generalized response to osmotic stress. The response of glycerol accumulation, the high intracellular levels that can be achieved, and the fact that changes of internal salts (Norkrans & Kylin 1969; Hobot &

Jennings 1981), amino acids and other low-molecular weight compounds are not sufficient to adjust osmotic parity with the environments, suggest that glycerol has a major role in the osmoregulation of growing cells. For organisms tolerating low a_w, cellular osmolytes can make important contributions to the intracellular composition, requiring that the osmolytes accumulated are widely compatible with macromolecular functions. The nonpertubing characteristics of glycerol are well-known and supported by the observed effects of glycerol on kinetic parameters

Table 9.2 Specific activities of enzymes of glycerol metabolism in cell-free extracts of *D. hansenii* wild-type and mutant strain 26-3. Enzyme activities are expressed as nmol min^{-1} mg protein^{-1} ± standard deviation. From Adler *et al.* (1985)

Enzyme	Specific activity	
	wild-type	mutant strain 26-3
Glycerol kinase	25.5 ± 1.3	27.3 ± 1.7
Mitochondrial G3P DH*	33.6 ± 5.3	0.0
NADP-dependent glycerol DH*	2.0 ± 0.6	1.8 ± 0.7
Dihydroxyacetone kinase	3.2 ± 0.9	6.0 ± 1.0
NAD-dependent G3P DH*	144 ± 7.0	166 ± 15

* DH = dehydrogenase

Figure 9.4 A model for the glycerol metabolism of *D. hansenii*. (1) Glycerol kinase; (2) Mitochondrial G3P dehydrogenase; (3) NADP-dependent glycerol dehydrogenase; (4) Dihydroxyacetone kinase; (5) NAD-dependent G3P dehydrogenase; (6) Phosphatase; (7) Glycerol transport protein(s). GAP: glyceraldehyde phosphate; DHA(P): dihydroxyacetone (phosphate).

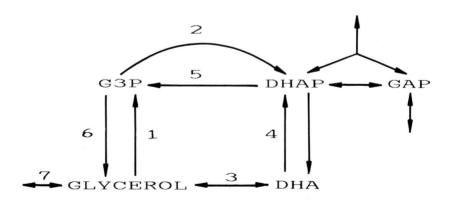

of a number of enzymes (Ruwart & Suelter 1971; Myers & Jakoby 1973; Adler 1978; Brown 1978). By virtue of its large number of hydrogen bonding groups/volume (Crowe *et al.* 1984), glycerol has a high potential for interaction with cellular macromolecules and may even during severe stress substitute for water. Nuclear magnetic resonance studies of packed cells of *D. hansenii* grown at 8% NaCl (Adler *et al.* 1981) indicated a significant restriction in motional freedom of the intracellular glycerol, interpreted as being due to interaction with macromolecular surfaces. The ability of glycerol to confer protection against severe salt-stress is supported by the observation on the growth of the mutant strain 26-6 of *D. hansenii* which has a reduced ability to accumulate glycerol. Small amounts of glycerol added to the medium permitted growth at salt concentrations otherwise inhibitory to the mutant (Figure 9.3). No such response was exhibited by other osmolytes added, such as polyols (mannitol and arabinitol), an amino acid (proline) and a methylamine (betaine). Since a transport system for glycerol is reported for *D. hansenii* (Adler *et al.* 1985) one can assume that the glycerol present in the medium was concentrated by the cells and used as a compatible solute inside the cells. A need for appropriate glycerol regulation for growth initiation is supported by results presented in Figures 9.1 and 9.2, which show that during increased stress glycerol accumulation precedes growth of the wild-type strain.

To approach the mechanism of glycerol regulation, the pathways of the glycerol metabolism have been studied. Our present concept of the metabolism of glycerol in *D. hansenii* is presented in Figure 9.4. The enzyme activities in crude extract (Table 9.2) indicate the existence of two pathways for glycerol dissimilation; one way involving glycerol kinase and mitochondrial glycerol 3-phosphate (G3P) dehydrogenase (G3P pathway), and another way involving glycerol dehydrogenase and dihydroxyacetone kinase (DHA pathway). We believe that the DHA pathway is of little importance for the glycerol dissimilation because of the low enzyme activities in this pathway and because the mutant, despite wild-type levels for these enzymes, is unable to utilize glycerol for growth. Thus, the main route for dissimilation seems to be the G3P pathway.

Glycerol production occurs via synthesis of triose phosphate in the glycolytic pathway followed by reduction of dihydroxyacetone phosphate by a nicotinamide adenine dinucleotide-dependent G3P dehydrogenase and subsequent dephosphorylation of G3P via a phosphatase. The specific activities of these enzymes showed an approximately two-fold increase in cell-free extracts of salt stressed, glycerol producing cells (A. Nilsson unpublished). The molecular level at which this activation takes place is not yet known. It is however of interest that the specific activity of the NAD-dependent dehydrogenase is enhanced by relatively small changes in salt concentration which may be of importance for glycerol production in osmotically dehydrated cells. To understand better the kinetic regulation of the glycerol metabolism we are now purifying two enzymes of the pathways for biochemical studies.

REFERENCES

Adler, L. (1978). Properties of alkaline phosphatase of the halotolerant
    yeast *Debaryomyces hansenii*. Biochem. biophys. Acta, $\underline{522}$,
    113-121.

Adler, L. & Gustafsson, L. (1980). Polyhydric alcohol production and
    intracellular amino acid pool in relation to halotolerance
    of the yeast *Debaryomyces hansenii*. Arch. Microbiol., $\underline{124}$,
    123-130.

Adler, L., Falk, K-E., Norkrans, B. & Ångström, J. (1981). Polyol and
    water in the salt-tolerant yeast *Debaryomyces hansenii* as
    studied by ^{1}H nuclear magnetic resonance. FEMS Microbiol.
    Lett., $\underline{11}$, 269-271.

Adler, L., Blomberg, A. & Nilsson, A. (1985). Glycerol metabolism and
    osmoregulation in the salt-tolerant yeast *Debaryomyces
    hansenii*. J. Bact., $\underline{162}$, 300-306.

Brown, A.D. (1976). Microbial water stress. Bact. Rev., $\underline{40}$, 803-846.

Brown, A.D. (1978). Compatible solutes and extreme water stress in
    eucaryotic micro-organisms. Adv. microbial. Physiol., $\underline{17}$,
    181-242.

Crowe, L.M., Mouradian, R., Crowe, J.H., Jackson, S.A. & Womersley, C.
    (1984). Effects of carbohydrates on membrane stability at
    low water activities. Biochem. biophys. Acta, $\underline{769}$, 141-150.

Gustafsson, L. & Norkrans, B. (1976). On the mechanism of salt
    tolerance. Production of glycerol and heat during growth of
    Debaryomyces hansenii. Arch. Microbiol., $\underline{110}$, 177-183.

Harris, R.F. (1981). Effect of water potential on microbial growth and
    activity. $\underline{In}$ Water potential relations in soil microbiology,
    Soil Science Society of America special publication
    number 9, eds. J.F. Parr, W.R. Gardner & L.F. Elliott,
    pp. 23-95. Madison, Wisconsin: Soil Science Society of
    America.

Hobot, J.A. & Jennings, D.H. (1981). Growth of *Debaryomyces hansenii*
    and *Saccharomyces cerevisiae* in relation to pH and salinity.
    Exp. Mycol., $\underline{5}$, 217-228.

Le Rudulier, D., Strom, A.R., Dandekar, A.M., Smith, L.T. &
    Valentine, R.C. (1984). Molecular biology of osmoregulation.
    Science, N.Y., $\underline{224}$, 1064-1068.

Meredith, S.A. & Romano, A.H. (1977). Uptake and phosphorylation of
    2-deoxy-D-glucose by wild type and respiratory-deficient
    bakers' yeast. Biochem. biophys. Acta, $\underline{497}$, 745-759.

Myers, J.S. & Jakoby, W.B. (1973). Effect of polyhydric alcohols on
    kinetic parameters of enzymes. Biochem. biophys. Res.
    Commun., $\underline{51}$, 631-636.

Norkrans, B. (1966). Studies on marine occurring yeasts: growth related
    to pH, NaCl concentration and temperature. Arch. Mikrobiol.,
    $\underline{54}$, 374-392.

Norkrans, B. & Kylin, A. (1969). Regulation of the potassium to sodium
    ratio and of the osmotic potential in relation to salt
    tolerance in yeasts. J. Bact., $\underline{100}$, 836-845.

Ruwart, M.J. & Suelter, C.H. (1971). Activation of yeast pyruvate kinase
    by natural and artificial cryoprotectants. J. biol. Chem.,
    $\underline{246}$, 5990-5993.

Yancey, P.H., Clark, M.E., Hand, S.C., Bowlus, R.D. & Somero, G.N. (1982). Living with water stress: evolution of osmolyte systems. Science, N.Y., <u>217</u>, 1214-1222.

10 BIOCHEMICAL INDICATORS OF THE ROLE OF FUNGI AND
    THRAUSTOCHYTRIDS IN MANGROVE DETRITAL SYSTEMS

R.H. Findlay

J.W. Fell

N.K. Coleman

J.R. Vestal

INTRODUCTION
        Interest in mangrove fungi is, to a large extent, based on
the economic importance of the mangrove community (mangal) throughout
the tropics (Saenger *et al.* 1983). Distribution and taxonomy of higher
fungi have been extensively studied by Kohlmeyer and others (Kohlmeyer &
Kohlmeyer 1979), while the occurrence of the fungal-like
thraustochytrids has been the subject of a series of papers by Ulken
(e.g. Ulken 1984). The role of fungi in the mangal is less well known,
although it is presumably similar to that of other forest and swamp
ecosystems. The interest in our laboratory has been on the microbial
role in leaf decomposition. The conversion of the leaf carbon to
microbial protein is important as a basis of an extensive marine and
estuarine food web (Odum *et al.* 1982). Our early studies documented the
sequence of fungal infestation during leaf decomposition (Fell & Master
1973) and demonstrated the potential role of *Phytophthora* spp. in
nitrogen enrichment (Fell & Master 1980).

Our present studies examine: 1) biochemical changes during leaf
decomposition as an indication of changes in microbial community
structure; 2) enzyme activities to determine the function of the
microbial community. The premise on which we based our biochemical
studies is that compounds that occur in all organisms can be used to
estimate microbial community biomass, while other biochemical
constituents, those restricted in distribution, can be used as signature
compounds to partition the microbial community (White 1983). The
compounds that we have used, for both the measurement of total community
biomass and the partitioning of the microbial community, were
phospholipid fatty acids (PFA) which were assayed as fatty acid methyl
esters (FAME). Phospholipids are a component of every cellular
membrane, are not utilized as storage products and are rapidly degraded
at cell death (White 1983). Therefore examination of the phospholipid
bound fatty acids ensures that only living and recently dead biomass
are analyzed.

Specific fatty acids have been identified as signature compounds of
bacterial populations. Perry *et al.* (1979) identified a group of fatty
acids as signature compounds for bacteria in marine sediments. These
include iso- and anteiso-branched chains, cyclopropyl 17:0 and 19:0,
10 methylpalmitic acid, cis-vaccenic acid (18:1w7) and two branched
chain monounsaturated acids (i15:1w8 and i17:1w8). Gillan *et al.* (1983)

characterized four "chemotypes" of marine bacteria, each with different fatty acid signatures. Fatty acids have also been utilized to differentiate prokaryotic and eukaryotic organisms (White 1983). Prokaryotes, with the exception of cyanobacteria, do not contain polyunsaturated fatty acids which are usually present in the polar lipids of eukaryotic cells. Erwin (1973) noted several important trends among the polyunsaturated fatty acids of eukaryotic microorganisms. First, 20:5w3 and 22:6w3 tend to be major fatty acids in the phospholipids of marine algae and marine Metazoa. Secondly, photosynthetic green algae, higher plants and higher fungi do not contain long-chain, highly unsaturated fatty acids greater than 18 carbons. Their polyunsaturated phospholipids contain only 18:2w6 and 18:3w3. The third trend was that the phospholipid of animal-like Protozoa is highly enriched in w6 polyunsaturated acids.

There is a vast literature on fatty acids of fungi which has been reviewed by Arnold (1981) and Wassef (1977). Polyunsaturated fatty acids common in higher fungi have been reported in organic detritus of *Spartina*, *Posidonia* and *Avicennia* spp. (Schultz & Quinn 1973; Wannigama *et al.* 1981; Nichols *et al.* 1982) and in microcosm studies of fouling communities where fungi were stimulated (White *et al.* 1980). However, there has been very little use of these fatty acids, 18:2w6 and 18:3w3, as fungal signature compounds owing to their presence in plant lipid. Among the thraustochytrids Ellenbogen *et al.* (1969) reported the presence of the fatty acid 22:5w6, which has the potential as a signature compound. A review of the literature indicates this fatty acid is restricted among microorganisms to parasitic protozoans (Erwin 1973), heterotrophic *Euglena gracilis* Klebs (Erwin 1973), *Ochromonas danica* Prings. (Erwin 1973) and several marine phytoplanktors (Ackman *et al.* 1968; Chuecas & Riley 1969).

The activity of various enzymes, the rate of DNA, RNA, and lipid synthesis, oxygen utilization and carbon dioxide production form a continuum of measures of metabolic activity similar to the biochemical constituents used as biomass measures. Activities common to all organisms will reflect the metabolic activity of the community while activities restricted in distribution will offer insight into the metabolic activities of a subset of the microbial assemblage. The assays that we have employed cover such a continuum. Flourescein diacetate (FDA) is hydrolyzed to flourescein by a wide variety of intra- and extra-cellular hydrolytic enzymes from both prokaryotic and eukaryotic microorganisms. The rate of ^{14}C-acetate incorporation into lipid will measure the metabolic activity of those microorganisms that can utilize acetate. At the substrate levels and incubation time used, this measure will probably be biased towards the bacteria (White *et al.* 1980). The production of glucose from carboxy-methylcellulose measures the metabolic activity of the portion of the microbial community active in cellulose decomposition. In various decomposition studies with detritus from species of *Carex*, *Spartina*, *Pinus* and *Quercus*, measurements of microbial activity have shown a positive correlation with microbial biomass. Measurements of hydrolytic enzyme activity have correlated with substrate weight loss (Morrison *et al.* 1977;

Bobbie *et al.* 1978; Federle & Vestal 1980; White 1983; Tenore *et al.* 1984).

This chapter describes a study to determine if fatty acid and enzyme analyses can be utilized to monitor microbial community structure and activity in decomposing mangrove leaves. Thraustochytrids are used as a model component to follow their introduction and activity within the system. Thraustochytrids were selected owing to their ubiquitous nature in marine environments and their presumed importance in detrital systems.

MATERIALS AND METHODS

There were two phases to the study: 1) a laboratory examination of PFA and of selected enzyme activities of thraustochytrids; 2) a field study to examine PFA and enzyme activities during different stages of mangrove leaf decomposition.

Laboratory studies

The PFA of pure cultures of thraustochytrids was examined using the techniques outlined by Bobbie & White (1980); while general hydrolytic enzyme activity was measured by the cleavage of fluorescein diacetate (FDA) to fluorescein (Schnurer & Rosswall 1982). Two strains of thraustochytrids isolated from mangrove leaves and one strain of *Thraustochytrium multirudimentale* Goldstein obtained from G. Bahnweg were used in this study.

Field studies

Leaves were selected in the field to represent the various stages from living to advanced decay as estimated by colour (green, yellow, brown and black). Living (green) and senescent (yellow) leaves were picked from trees. When senescent leaves fall naturally into the water, they float for 1-2 days prior to sinking. Therefore we collected yellow floating leaves and yellow, brown, and black leaves from the water-sediment interface. These leaves represent a two month or longer decomposition process (Fell & Master 1980). Approximately 25% of the leaf was extracted for lipid analysis. Fifteen replicate disks (0.6 cm diam) were removed from the remainder of the leaf with five leaf disks being assayed for hydrolytic enzyme activity, cellulase activity and the rate of ^{14}C-acetate incorporation into lipid. Four replicate leaves were used for each of the decomposition stages. Phospholipid fatty acids and FDA were assayed as above. The rate of ^{14}C-acetate incorporation into lipid was determined as reported by McKinley *et al.* (1983). Cellulase activity was measured by the method of Petterson & Porath (1966). The leaf disks were leached in filter sterilized (0.45 μm) sea water for 2 h prior to analysis to eliminate interference from coloured leachates. Thraustochytrids were isolated from the leaves using standard techniques for lower fungi (Fell & Master 1975).

RESULTS AND DISCUSSION
Pure cultures
    Table 10.1 shows the fatty acid composition (as weight per cent, calculated as µg FAME/µg Total FAME x 100) of five strains of thraustochytrids. The major fatty acids are 16:0, 22:6w3 and 22:5w6 and the minor fatty acids are 20:4w6 and 20:5w3. *T. roseum* Goldstein also contained substantial amounts of 18:1w9. The phospholipid fatty acid 22:5w6 is proposed as a signature compound for the thraustochytrids, because it occurred in all the strains examined as a major fatty acid and because it is in limited distribution in other marine organisms. The average amount of PFA bound 22:5w6 recovered from the three strains analyzed in this study was 70 µg gram dry weight, a value that will be used when calculating thraustochytrid biomass. The only other marine microorganisms reported to contain this fatty acid are among the Bacillariophyceae and the Chrysophyta (Ackman *et al.* 1968; Chuecas & Riley 1969). Seven of twenty species of these algae contained this fatty acid among their total saponifiable lipid with typical weight per cent values ranging from 1.0% to trace amounts. The Bacillariophyceae typically contain large amounts of 16:3w4 (4.0 to 12.5 weight per cent) and 20:5w3 (7.5 to 30.0 weight per cent) and show ratios of 16:3w4/22:5w6 and 20:5w3/22:5w6 that range from 54:1 to 4:1 and 250:1 to 15:1 respectively. The Chrysophyta also show high levels of 20:5w3 in their total saponifiable lipid (4.0 to 31.0 weight per cent) and a ratio of 20:5w3/22:5w6 that ranges from 120:1 to 4:1. There are conflicting reports in the literature as to whether this group contains high levels of 16:3w4 or 18:4w3 (Ackman *et al.* 1968; Cheucas & Riley 1969; Wood 1974). In contrast, the average ratio of 20:5w3/22:5w6 for the strains of thraustochytrids examined was 1:4. These ratios are an important consideration when evaluating the field data.

Enzyme and uptake studies demonstrated that all strains were FDA positive, a result of the presence of general hydrolytic enzymes, however none of the strains studied showed cellulase activity, nor were they capable of incorporating significant levels of ^{14}C–acetate.

    Field study
    Changes during senescence. Table 10.2 shows the nM FAME/ g dry wt of the FAME recovered from the phospholipid of 6 decomposition stages of mangrove leaves. The major PFA of green mangrove leaves were 16:0, 18:1w9, 18:2w6 and 18:3w3 with substantial amounts of 16:1w13t, 17:0, and 18:0. A comparison of green leaves to yellow senescent leaves showed that total PFA decreased by approximately 75% while the leaves were still on the trees. Two monenoic fatty acids, 16:1w13t and 18:1w9 showed the largest decreases with only 6.5% and 5.2% remaining in the senescent leaves. Saturated fatty acids decreased by an average of 65%. The two polyunsaturated fatty acids, 18:2w6 and 18:3w3, showed an average decrease of 80%. These decreases were the most variable. In two of the leaves these fatty acids decreased by 96.5% while in the other two leaves examined they decreased by 65%. The fatty acids which originate in the mangrove leaf can be removed from the pool of PFA by

Table 10.1 Weight per cent of fatty acid methyl esters for thraustochytrids

FAME[1]	Isolate T-2[2]	Isolate T-4[2]	*T. multirudimentale*[2] Goldstein	*T. aureum*[3] Goldstein	*T. roseum*[3] Goldstein
14:0	0.92	1.87	2.27	2.3	0.7
15:0	0.26	0.25	3.53	4.2	0.9
16:1ω7	0.09	0.11	2.16	0.8	1.3
16:0	48.83	64.38	39.70	27.1	29.3
17:0	0.06	nd	nd	1.4	nd
18:3ω6	0.02	0.03	2.99	TR	0.7
18:4ω3	0.03	TR	nd	nd	nd
18:2ω6	TR	TR	5.89	2.6	3.2
18:3ω3	0.02	nd	TR	TR	TR
18:1ω9	0.08	0.16	12.37	5.7	34.6
18:0	0.06	1.33	4.40	1.3	4.6
20:4ω6	1.23	1.10	3.89	4.8	1.8
20:5ω3	2.67	0.04	6.67	5.9	6.1
30:3ω6	0.05	0.15	nd	nd	TR
20:4ω3	0.19	nd	0.35	nd	nd
20:2ω6	TR	0.03	nd	nd	TR
20:0	TR	nd	nd	nd	nd
20:5ω6	8.93	10.94	7.39	9.5	6.6
22:6ω3	36.13	18.17	11.24	34.1	10.8
22:4ω6	TR	0.03	TR	0.8	TR
22:5ω3	0.15	0.28	0.31	nd	TR
22:0	0.05	0.10	TR	nd	nd

nd = not detected.

TR = trace.

[1] fatty acids are named by: the number of carbons ':'; the number of double bonds 'ω'; the position of the first double bond from the aliphatic or omega end of the molecule.

[2] this study.

[3] Ellenbogen *et al.* 1969.

one of several mechanisms: the phospholipid can be reabsorbed by the
plant; the phosphate can be cleaved by either plant or microbial
phosphatase to yield a diacyle glyceride; a fatty acid may be cleaved
from the phospholipid by lipase to produce a free fatty acid: polyenoic
fatty acids may be chemically oxidized. Assuming the first three
mechanisms removed all fatty acids at a similar rate and unsaturated
fatty acids oxidized at a rate proportional to the number of double
bonds present then the rate of decrease of 16:1w13t should predict the
minimum rate of decrease of the mangrove polyenoic fatty acids. Using
the average decrease for 16:1w13t (94.5%) it is predicted that 365 ng g
dry wt. of 18:2w6 and 429 ng g dry wt. of 18:3w3 should remain in the
senescent leaves. The mean values for both of these fatty acids
(1028 and 1311 ng g dry wt) are far in excess of this expected value.
This indicates the presence of significant fungal biomass in the
senescent leaves. The appearance of the bacterial fatty acids a 15:0,
cy 17:0, cy 19:0 and 19:0 in the senescent leaves indicates that bacteria
are also present.

Microbial biomass may be estimated from the amount of fatty acids
recovered. We stress that assumptions are required to make these
estimates and that these estimates should be considered 'order of
magnitude' comparisons. Fatty acid profiles and the percentage of the
hyphal dry weight attributed to fatty acids are available (Wassef 1977)
for three genera commonly isolated from decaying mangrove leaves (Fell
& Master 1980). Reported amounts are variable: *Aspergillus* 0.3 mg,
Penicillium 2 mg and *Fusarium* 12 mg 18:3w3 g dry wt of hyphae.
Arbitrarily using the figure available for *Penicillium*, the average
amount of fungal hyphae in the leaves was 442 µg of fungal hyphae/gm dry
wt of leaf. Leaf A (Figure 10.1, D.P. 1) showed the greatest
concentration of fungal biomass (1320 µg) with little or no fungal
biomass in leaf B. Similarly, using 5.7 mg of 18:1w7 g dry wt of *Vibrio
cholerae* Pacini (Findlay *et al.* unpublished data) the leaves averaged
26 µg of bacteria g dry wt with a high of 64 µg in leaf B and little or
no bacterial biomass in leaf A. Since a 15:0, cy 17:0 and cy 19:0, 19:0
and 18:w7 all appear to be markers for different bacterial chemotypes
(Gillan *et al.* 1983) estimates of total bacterial biomass should be
raised by a factor of four. It is apparent from these data that
microbes were associated with mangrove leaves before leaf fall and the
community may be bacterial and/or fungal.

Changes during decomposition. Based on this study, and several of our
unpublished studies, we have found that thraustochytrids can be
isolated from the leaf within 24-48 h after leaf fall and continued to
be associated with the leaf throughout the entire decomposition process.

Biomass. As found with senescent leaves, the occurrence of the
signature fatty acid infers the presence of microorganisms while the
biomass of those organisms can be estimated by the nanograms of fatty
acid recovered. To assess the role of microorganisms in the
decomposition process the senescent leaves, yellow floating leaves and

Table 10.2　Nanograms of fatty acid methyl esters per gram dry weight of mangrove leaf

FAME[1]	LEAF CATEGORIES					
	1 Green on tree	2 Senescent on tree	3 Yellow floating	4 Yellow sunken	5 Brown sunken	6 Black sunken
14:0	113 ± 80	108 ± 55	142 ± 219	46 ± 14	75 ± 63	136 ± 185
a 15:0		13 ± 26	129 ± 243	14 ± 19	12 ± 14	165 ± 273
15:0	161 ± 67	143 ± 46	162 ± 236	36 ± 14	36 ± 29	197 ± 358
16:1w13t	2372 ± 1750	155 ± 78	34 ± 25	36 ± 20	32 ± 28	40 ± 24
16:0	33662 ± 17287	12968 ± 2740	2638 ± 301	2493 ± 759	1677 ± 2218	1395 ± 1698
a 17:0	689 ± 187	80 ± 62	144 ± 264	29 ± 34	8 ± 10	258 ± 481
17:0	nd	11 ± 22	77 ± 154	10 ± 12	14 ± 17	158 ± 256
cy 17:0	1641 ± 520	846 ± 169	266 ± 220	183 ± 91	98 ± 138	293 ± 405
18:3w6	nd	nd	nd	nd	nd	nd
18:4w3	nd	nd	nd	nd	nd	nd
18:2w6	5590 ± 1459	1028 ± 1024	346 ± 380	552 ± 365	8 ± 16	36 ± 28
18:3w3	6572 ± 5417	1311 ± 1427	349 ± 340	1076 ± 352	8 ± 16	10 ± 16
18:1w9	20237 ± 7981	1045 ± 468	648 ± 370	829 ± 457	28 ± 19	363 ± 473
18:1w7	650 ± 296	149 ± 77	108 ± 89	102 ± 100	6 ± 12	301 ± 370
18:0	1502 ± 705	720 ± 214	405 ± 224	224 ± 121	110 ± 126	547 ± 921
cy 19:0	nd	16 ± ±9	46 ± 92	3 ± 5	nd	119 ± 184
19:0	nd	91 ± 169	116 ± 231	8 ± 10	nd	233 ± 453
20:4w6	nd	nd	nd	7 ± 12	24 ± 48	46 ± 87
20:5w3	nd	nd	nd	2 ± 3	nd	5 ± 11
20:0	Tr	257 ± 271	336 ± 620	32 ± 26	6 ± 10	15 ± 14
22:5w6	nd	nd	3 ± 5	4 ± 3	1 ± 3	34 ± 47
22:6w3	nd	nd	8 ± 8	6 ± 6	3 ± 4	73 ± 12
TOTAL FAME	72851 ± 28145	18642 ± 3116	6045 ± 2580	5735 ± 2153	2142 ± 2567	4590 ± 6632
FDA Activity[2]	29 ± 3		37 ± 3	26 ± 4	11 ± 2	11 ± 2
Cellulase activity[3]	632 ± 72		1044 ± 11	1100 ± 189	432 ± 72	211 ± 50
14C-acetate incorporation[4]	1048 ±		2095 ± 762	1619 ± 95	4571 ± 1143	19047 ± 3238

nd = not detected.

1 FAME given as ng g dry wt of leaf extracted and designated as: the number of carbons ':'; the number of double bonds 'w'; the position of the first double bond from the aliphatic or omega end of the molecule. Unsaturated fatty acids are all cis except when denoted by 't' which indicates trans double bonds. The prefixes 'i' and 'a' indicate iso- and anteiso-branching while 'cy' indicates a cyclopropyl ring. X ± S.D. N=4.

2 nmoles of fluorescein h 2.8 cm² leaf. X ± S.D. N=4.

3 nmoles of glucose h 2.8 cm² of leaf. X ± S.D. N=4.

4 DMP 2.8 cm² of leaf. X ± S.D. N=4.

yellow, brown and black sunken leaves were compared. Analysis of
variance on these categories of leaves followed by a Student-Newman-
Kuels (SNK) test for significantly different means (p = 0.05) showed that
senescent leaves contained significantly higher levels of total PFA than
categories 3 through to 6 (Table 10.2). A test for linear trends show
a significant (p < 0.01) trend of decreasing total FAME. Similarly, the
levels of 16:1w13t, 16:0 and 18:0 were significantly higher in the

Figure 10.1 Fatty acid methyl ester profiles of mangrove leaves. Each
bar graph represents the fatty acid profile of a single mangrove leaf.
Decomposition Period 1 (D.P. 1) is yellow senescent leaves and D.P. 2
is yellow floating leaves. Fatty acids 1 through to 20 are as follows:
14:0, a 15:0, 15:0, 16:1w7, 16:1w13t, 16:0, a 17:0, cy 17:9, 17:0,
18:2w6, 18:3w3, 18:1w9, 18:1w7, 18:0, cy 19:0, 19:0, 20:0, 22:5w6,
22:6w3, and 22:4w6.

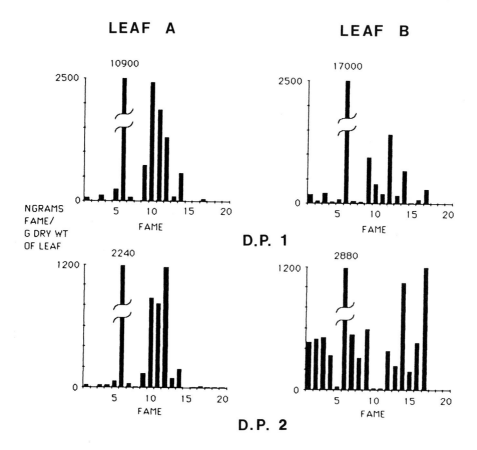

senescent leaves and there was a significant trend $(0.05 > p > 0.01)$ towards decreased amounts of these fatty acids as the leaves decayed. The amount and types of FAME recovered from a decaying mangrove leaf are influenced by two factors: the amount, if any, of original mangrove PFA remaining; the biomass and community structure of the microbial assemblage present on/in the leaf. The decrease in total FAME and in 16:1w13t, 16:0 and 18:0 that occurred during the decay were likely due to loss of mangrove PFA.

The polyunsaturated fatty acids 22:5w6 and 22:6w3 were not present in the senescent leaves but were found in each category of leaf recovered from the water. This indicates that marine organisms including thraustochytrids colonized the leaves soon after they entered the water. As the ratio of 20:5w3/22:5w6 averaged 1:5 in the few samples in which 20:5w3 was present and neither 16:3w4, nor 18:4w3, were detected, we have concluded that phytoplankton did not contribute significantly to the 22:5w6 recovered from the decaying mangrove leaves. The bacterial fatty acids first observed in the senescent leaves were present throughout the degradation process.

Although microorganisms colonized and were involved in the degradation of the mangrove leaves, the variable nature of the community and the small sample size used in this experiment were such that no significant changes in the nanogram amounts of marker fatty acids for fungi, bacteria or thraustochytrids could be detected. Using the values derived from the literature and developed in this chapter it is possible to compare the biomass of these components at specific stages of decomposition. Fungal biomass, as previously calculated from the amount of 18:3w3 recovered as dry weight of *Penicillium* hyphae averaged 129 µg g dry wt of leaf extracted in the yellow floating leaves, peaked at 538 µg g dry wt of leaf in yellow sunken leaves, decreased to approximately 5 µg g dry wt of leaf in the brown and black leaves. Bacterial biomass, calculated from the amount of 18:1w7 recovered as dry weight of *Vibrio cholerae* averaged 17 µg g dry wt of leaf in the yellow floating leaves, 18 µg g dry wt of leaf in the yellow sunken leaves, 1 µg in the brown leaves and increased to 52 µg in the black sunken leaves. As in the senescent leaves at least four chemotypes of bacteria appear to be present and estimates of total bacterial biomass should be increased to 68, 72, 4 and 204 µg g dry wt of leaf. Thraustochytrid biomass averaged 37 µg g dry wt of yellow floating leaf, 47 µg g dry wt of leaf in yellow sunken leaves, decreases to 15 µg g dry wt of leaf in brown leaves and increased to 430 µg g dry wt of leaf in the black sunken leaves.

Community structure. The calculation of weight per cent (ng FAME/ng Total FAME x 100, data not shown) for each of the fatty acids present in the FAME profiles allows investigation of changes in the microbial community structure. There was a significant trend towards an increase in the weight per cent of 14:0, a 15:0, 16:1w7, cy 17:0, 18:1w7, cy 19:0, 22:5w6 and 22:6w3 during the decomposition process. In addition, the weight per cent of a 15:0, 16:1w7, cy 17:0, 18:1w7, cy 19:0, 22:5w6 and

22:6w3 were significantly higher (ANOVA, SNK, P = 0.05) in the black leaves. This indicates that bacteria and thraustochytrids increased in importance within the community as the leaves decayed and became a significantly larger proportion of the microbial community in the black leaves. The weight per cent of the polyenoic fatty acid 18:3w3 was significantly higher in the sunken yellow leaves than in any other decaying leaves. This indicates that fungi comprised a significantly greater proportion of the microbial community at this time. The weight per cent of 16:0 showed a significant trend of decreasing relative abundance with senescent leaves being significantly different from black leaves.

Microbial activity. The pattern of activity for both general hydrolytic enzymes, measured as FDA hydrolysis, and cellulase activity were similar (Table 10.2) with substantial levels of activity in green leaves, the highest activity in yellow leaves and a decrease in activity during decomposition. Analysis of variance followed by the SNK (p = 0.05) test for significant means showed yellow leaves to be significantly different in activity from black leaves. The uptake and incorporation of ^{14}C acetate into lipid showed the opposite trend with little activity in green or yellow leaves and a significant (ANOVA, SNK, p = 0.05) increase in black leaves. The significant increase in relative abundance of fungi in yellow leaves in conjunction with peak hydrolytic enzyme activity suggests that fungi may play a central role early in the decomposition process. Similarly, the significant increase in the rate of labelled acetate incorporation into lipid in black leaves is at a time when bacteria constitute a significantly greater proportion of the microbial community which suggests that these organisms are responsible for this measured increase in metabolic activity.

Variance between leaves. Figure 10.1 shows the FAME profiles of 4 individual leaves. The first pair represent senescent leaves. Leaf A was enriched in 18:2w6 and 18:3w3, while leaf B was enriched in 14:0, a 15:0, 16:1w7, cy 17:0, cy 19:0 and 20:0. The second pair of leaves are floating yellow leaves. A similar pattern was observed with leaf A being enriched in 18:2w6, 18:3w3 and 18:1w9 while leaf B is enriched in 14:0, a 15:0, 15:0, 16:1w7, a 17:0, cy 17:0, 17:0, 18:0, cy 19:0 and 20:0. In addition 22:5w6 and 22:6w3 are also present. The figure further illustrates the variability in the microbial community present early in the decomposition process. The microbial community of the 'A' leaves was dominated by fungal biomass with no biochemical evidence of significant bacterial biomass. The FAME profiles from the 'B' leaves were dominated by bacterial signature compounds and the amounts of 18:2w6 and 18:3w3 were below the levels predicted by calculation of the remaining mangrove detrital PFA. The other leaves in each treatment (not illustrated) showed intermediate patterns with markers for both fungi and bacteria. This variability in community composition helps to explain the variance in the data. The variance of a data set can be expressed in terms of components of variance (standard deviation/mean x 100). Components of variance of homogeneous samples (pure cultures and thoroughly mixed sediments) assayed as in this study are typically 10 per cent. Components of variance for natural marine sediments, which are considered to be dominated by bacterial biomass, typically range from 25 to 50 per cent. The component of variance for the nanogram amounts of the signature lipids recovered from the leaves early in the decomposition process for fungi, bacteria and thraustochytrids were in excess of 100 per cent. This was due to the presence of these organisms on some leaves and their absence on others. Experiments in progress will test if the original C/N ratio, per cent water content of the senescent leaves, or some other variable, determines which organisms colonize the leaves early in the decomposition process.

REFERENCES

Ackman, R.G., Tocher, C.S. & McLachlan, J. (1968). Marine phytoplankter fatty acids. J. Fish. Res. Bd. Canada, 25, 1603-1620.

Arnold, W.N. (1981). Lipids. In Yeast Cell Envelopes: biochemistry, biophysics, and ultrastructure, ed. W.N. Arnold, pp. 97-114. Boca Raton: CRC Press.

Bobbie, R.J., Morrison, S.J. & White, D.C. (1978). Effects of substrate biodegradability on the mass and activity of the associated estuarine microbiota. Appl. environ. Microbiol., 35, 179-184.

Bobbie, R.J. & White, D.C. (1980). Characterization of benthic microbial community structure by high resolution gas chromatography of fatty acid methyl esters. Appl. environ. Microbiol., 39, 1212-1222.

Chuecas, L. & Riley, J.P. (1969). Component fatty acids of the total lipids of some marine phytoplankton. J. mar. biol. Ass. U.K., 49, 97-116.

Ellenbogen, B.B., Aaronson, S., Goldstein, S. & Belsky, M. (1969). Polyunsaturated fatty acids of aquatic fungi: possible phylogenetic significance. Comp. Biochem. Physiol., 29, 805-811.

Erwin, J.A. (1973). Comparative biochemistry of fatty acids in eukaryotic microorganisms. In Lipids and Biomembranes of Eukaryotic Microorganisms, ed. J.A. Erwin, pp. 41-143. New York: Academic Press.

Federle, T.W. & Vestal, J.R. (1980). Microbial colonization and decomposition of *Carex* litter in an Arctic lake. Appl. environ. Microbiol., 39, 888-893.

Fell, J.W. & Master, I.M. (1973). Fungi associated with the decay of mangrove (*Rhizophora mangle* L.) leaves in South Florida. In Estuarine Microbial Ecology, eds. L.H. Stevenson & R.R. Colwell, pp. 455-466. Columbia: University of South Carolina Press.

Fell, J.W. & Master, I.M. (1975). Phycomycetes (*Phytophthora* spp. nov. and *Pythium* sp. nov.) associated with degrading mangrove (*Rhizophora mangle*) leaves. Can. J. Bot., 53, 2908-2922.

Fell, J.W. & Master, I.M. (1980). The association and potential role of fungi in mangrove detrital systems. Botanica mar., 23, 257-263.

Gillan, F.T., Johns, R.J., Verheyen, T.V., Nichols, P.D., Esdaile, R.J. & Bavor, H.J. (1983). Monounsaturated fatty acids as specific bacterial markers in marine sediments. In Advances in Organic Geiochemistry 1981, ed. M. Bjoroy, pp. 198-206. New York: Wiley.

Kohlmeyer, J. & Kohlmeyer, E. (1979). Marine Mycology, The Higher Fungi. New York: Academic Press.

McKinley, V.L., Federle, T.W. & Vestal, J.R. (1983). Effects of petroleum hydrocarbons on plant litter microbiota in an Arctic lake. Appl. environ. Microbiol., 43, 129-135.

Morrison, S.J., King, J.D., Bobbie, R.J., Bechtold, R.E. & White, D.C. (1977). Evidence of microflora succession on allochthonous plant litter in Apalachicola Bay, Florida, U.S.A. Mar. Biol., 41, 229-240.

Nichols, P.D., Klumpp, D.W. & Johns, R.B. (1982). Lipid components of the seagrasses *Posidonia australis* and *Heterozostera tasmanica* as indicators of carbon source. Phytochem., 21, 1613-1621.

Odum, W.E., McIvor, C.C. & Smith, T.J. (1982). The ecology of the mangroves of South Florida: a community profile. U.S. Dept. of the Interior FWS/OBS-81/24.

Perry, G.J., Volkman, J.K., Johns, R.B. & Bavor Jr., H.J. (1979). Fatty acids of bacterial origin in contemporary marine sediments. Geochim. Cosochim. Acta, 43, 1715-1725.

Petterson, G. & Porath, J. (1966). A cellulolytic enzyme from *Penicillium notatum*. Meth. Enzym., 8, 603-607.

Saenger, P.E., Hegerl, J. & Davie, J.D.S. (1983). Global Status of Mangrove Ecosystems. Commission on Ecology Papers 3. International Union for Conservation of Nature and Natural Resources.

Schnurer, J. & Rosswall, T. (1982). Fluorescein diacetate hydrolysis as a measure of total microbial activity in soil and litter. Appl. environ. Microbiol., 43, 1256-1261.

Tenore, K.R., Hanson, R.B., McClain, J., Maccubbin, A.E. & Hodson, R.E. (1984). Changes in decomposition and nutritional value to a benthic deposit feeder of decomposing detritus pools. Bull. mar. Sci., 35, 299-311.

Ulken, A. (1984). The fungi of the mangal ecosystem. In Hydrobiology of the Mangal, eds. F.D. Por & I. Dor, pp. 27-33. The Hague: W. Junk.

Wannigama, G.P., Volkman, J.K., Gillan, F.T., Nichols, P.D. & Johns, R.B. (1981). A comparison of lipid components of the fresh and dead leaves and pneumatophores of the mangrove *Avicennia marina*. Phytochem., 20, 659-666.

Wassef, M.K. (1977). Fungal lipids. In Advances in Lipid Research, Vol. 15, eds. R. Paoletti & L. Kritchevsky, pp. 159-230. New York: Academic Press.

White, D.C. (1983). Analysis of microorganisms in terms of quantity and activity in natural environments. In Microbes in their natural environment, eds. J.H. Slater, R. Whittenbury & J.W.T. Wimpenny, Society for General Microbiology Symposium, 34, 37-66.

White, D.C., Bobbie, R.J., Nickels, J.S., Fazio, S.D. & Davis, W.M. (1980). Nonselective biochemical methods for the determination of fungal mass and community structure in estuarine detritial miacroflora. Botanica mar., <u>23</u>, 239-250.

Wood, B.J.B. (1974). Fatty acids and saponifiable lipids. <u>In</u> Algal Physiology and Biochemistry, ed. W.D.P. Stewart, pp. 236-265. Berkeley: University of California Press.

11 BIOLOGY AND PHYLOGENY OF THE LABYRINTHULALES AND THRAUSTOCHYTRIALES

S.T. Moss

INTRODUCTION

The Labyrinthulales and the Thraustochytriales are orders of obligately marine eukaryotes which may be readily isolated from estuarine, coastal and oceanic habitats worldwide. Although they are nearly ubiquitous on isolation plates for marine fungi they have rarely been found by direct microscopical examination of substrates in the natural environment. There are 39 validly described species of which only eight belong to the Labyrinthulales. Since description of the first labyrinthulids, *Labyrinthula macrocystis* Cienk. and *L. vitellina* Cienk., by Cienkowski in 1867 (Cienkowski 1867) and the type species of *Thraustochytrium*, *T. proliferum* Sparrow in 1936 (Sparrow 1936), species assigned to these orders have been the subjects of many research publications (e.g. Porter 1969, 1974; Perkins 1970, 1972, 1974; Gaertner 1972; Darley *et al.* 1973; Alderman *et al.* 1974; Porter & Kochert 1978; Bahnweg 1979; Chamberlain 1980; Moss 1980, 1985; Nakatsuji & Bell 1980; Ulken *et al.* 1985). However, in spite of these studies neither order has been satisfactorily assigned to any higher taxon and their affinity with the Mycota remains uncertain. Correspondingly their inclusion in a volume devoted to the biology of marine fungi necessitates explanation. Following the initial publications by Cienkowski and Sparrow, the study of these marine organisms has been virtually the exclusive domain of the marine mycologist. This results from their similar gross morphology, reproductive behaviour, ecology and methods of isolation, culture and examination to those of either the slime moulds, in the case of the Labyrinthulales, or the zoosporic fungi for the Thraustochytriales. Owing to these superficial fungal similarities and their study predominantly by mycologists the Thraustochytriales and Labyrinthulales constitute the subjects of this and several other chapters of this book.

THRAUSTOCHYTRIALES

The Thraustochytriales is an order of obligately marine achlorophyllous organisms characterized by monocentric thalli which are attached to their substrata by branched, radiating epi- or endo-biotic rhizoid-like extensions of the thallus, the ectoplasmic net (Figure 11.1). Asexual reproduction is normally by conversion of the vegetative thallus to a zoosporangium within which are formed many laterally biflagellate, heterokont zoospores (Figures 11.4,5). The anteriorly directed flagellum possesses a bilateral row of mastigonemes while the posterior flagellum is of the whiplash type and is the shorter

of the two. Release of zoospores is by the partial or complete
disintegration of the sporangial wall. Following a swarm period
zoospores settle, encyst and form daughter thalli (Figures 11.7a-c). A
second type of asexual reproductive propagule, the aplanospore, has been
recorded for species from all genera but only in the monotypic genus
Aplanochytrium is the aplanospore the only spore type known (Bahnweg &
Sparrow 1972). No sexual reproductive stage has been found in any
species of the Thraustochytriales.

The Thraustochytriales contains seven genera with 31 species, namely:
Thraustochytrium Sparrow (1936) emend. Johnson et Sparrow (1961) with
sixteen species; *Japonochytrium* Kobayashi et Ookubo (1953), monotypic;
Schizochytrium Goldstein et Belsky (1964), two species; *Althornia* Jones
et Alderman (1971), monotypic; *Aplanochytrium* Bahnweg et Sparrow (1972),
monotypic; *Labyrinthuloides* Perkins (1973 a), four species; *Ulkenia*
Gaertner (1977), with six species. These genera are separated primarily
on sporangial morphology and the mechanism of spore release. The
vegetative thalli of thraustochytrids are uninucleate, contain single
dictyosomes, centrioles associated with shallow nuclear pockets and are
bounded by discrete cell walls (Figure 11.1a). The ectoplasmic net
emerges from each thallus through a discontinuity in the cell wall
(Figure 11.1b). Although the organization of the vegetative thallus
superficially resembles that of the chytrids, this similarity is not
supported at the ultrastructural level. The initial ultrastructural
study by Goldstein *et al*. (1964) stimulated more detailed investigations
by several workers on a range of genera and species (Jones & Alderman
1971; Kazama 1972 a,b, 1974 a,b, 1975; Perkins 1972, 1973 a,b, 1976;
Darley *et al*. 1973; Alderman *et al*. 1974; Harrison & Jones 1974 a,b,c;
Chamberlain 1980; Moss 1980, 1985; Raghu Kumar 1982 a,b). Three
characters have received particular attention and have been the prime
reasons for the present uncertain phylogenetic affinities of the
Thraustochytriales, these are: the structure of the cell wall; the
possession of an ectoplasmic net; the presence of a structure
associated with the origin of the ectoplasmic net from the thallus,
termed the sagenogenetosome. The vegetative thallus of each
thraustochytrid examined possessed a cell wall composed of a few
(Figure 11.2a) to many (Figure 11.2d) compact layers of closely
adpressed scales. Only where the wall is disrupted can the individual
scales be resolved (Figure 11.2b). Each scale is 2-3 nm thick,
circular with a diameter of 0.5-1 µm and lacks ornamentations. The

Figures 11.1a-c Thraustochytriales.
Figure 11.1a *Thraustochytrium* sp., vegetative thallus showing the
single nucleus (N), mitochondria (M), dictyosome (D), centrioles (C)
and cell wall (W). T.E.M., bar = 1 µm.
Figure 11.1b *Thraustochytrium aggregatum*, vegetative thallus showing
emergence of the ectoplasmic net (EN) from within the thallus wall (W).
T.E.M., bar = 1 µm.
Figure 11.1c *Ulkenia visurgensis*, vegetative thalli (VT) and a
sporangium containing aplanospores (A) with ectoplasmic nets (EN).
S.E.M., bar = 10 µm.

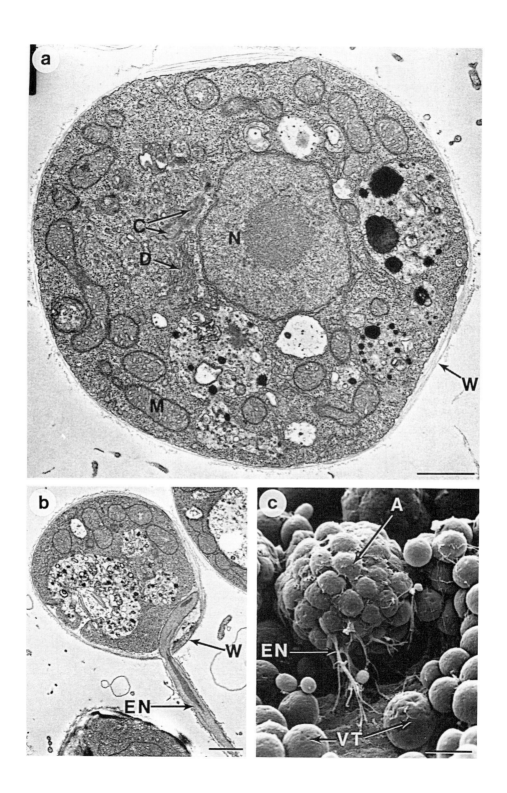

scales are formed in the dictyosome cisternae throughout thallus development (Figures 11.2c, 4b) and are deposited on to the wall either directly or indirectly in dictyosome-derived vesicles. The ectoplasmic net lacks a cell wall and is delimited only by a plasma membrane which is continuous with that of the thallus (Figure 11.3c). Internally the net lacks organelles but does contain membranous inclusions within an amorphous matrix (Figures 11.1b, 3c). The cytoplasm of the thallus is separated from the lumen of the ectoplasmic net by an electron-opaque aggregate termed the sagenogenetosome. Although the complexity of the sagenogenetosome varies between species its basic construction remains similar. Convergent on the sagenogenetosome are lamellae of the thallus endoplasmic reticulum (Figure 11.3a) which are frequently in continuity with a region of convoluted smooth endoplasmic reticulum surrounding a ribosome-free region of cytoplasm, the paranuclear body, which Kazama (1980) speculated to be a microbody. Paranuclear bodies are characteristic of many thraustochytrids and are frequently closely associated with the nucleus (Moss 1980). Within the sagenogenetosome the lamellae of endoplasmic reticulum are constricted, anastomose with each other (Figures 11.3a,b) and become directly or indirectly continuous with the membrane system within the ectoplasmic net system (Figure 11.3c). The presence of a wall composed of scales, a sagenogenetosome and an ectoplasmic net are now considered diagnostic for the Thraustochytriales and assignment of any organism to this order necessitates ultrastructural confirmation of their presence. Even in retrospect it is not possible to resolve with any degree of certainty any one of these characters at the light microscope level. However it must be emphasized that speciation and also assignment of species to genera are based on characters which may be readily observed with the light microscope.

Thraustochytrium, the type genus of the Thraustochytriales, is characterized by monocentric thalli which discharge their zoospores by either the partial (e.g. *T. roseum* Goldstein, *T. globosum* Kobayashi et Ookubo, *T. striatum* Schneider) or the complete (e.g. *T. aggregatum* Ulken, *T. antarcticum* Bahnweg et Sparrow) disintegration of the sporangial wall. The mechanism of zoosporogenesis is one of the principal criteria used for the speciation of this genus, although it is also used at the generic level in the order. Two basic types of

Figures 11.2a-d Thraustochytriales, transmission electron micrographs.
Figure 11.2a *Thraustochytrium* sp., thin multilamellate cell wall (W) bounded externally by a layer of mucilage (arrowed). Bar = 250 nm.
Figure 11.2b *Thraustochytrium* sp., disrupted cell wall showing the several layers of wall scales (S). Bar = 250 nm.
Figure 11.2c *Ulkenia visurgensis*, dictyosome adjacent to the nucleus (N) of a young vegetative thallus. Note the wall scales (S) in the distal cisternae. Bar = 250 nm.
Figure 11.2d *Ulkenia visurgensis*, multinucleate sporangium prior to cytokinesis. Note the thick cell wall (W), nuclei (N), each with an associated dictyosome (D), lipid bodies (L) and a central vacuole (V). Bar = 2 μm.

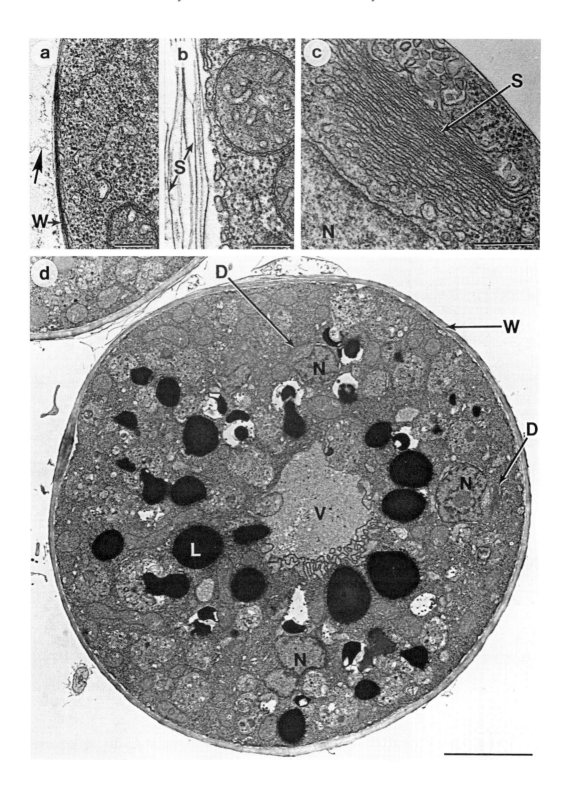

cleavage are recognized within the Thraustochytriales, namely,
progressive cleavage in which karyokinesis is completed prior to
cytokinesis and successive bipartition in which each nuclear division is
followed immediately by cytoplasmic cleavage. This latter type of
cleavage gives rise to the tetrads of cells which characterize species
of *Schizochytrium*. Cytokinesis in the Thraustochytriales involves
either invagination of the plasma membrane alone (Kazama 1975; Moss
1980, 1985) or the fusion of plasma membrane invaginations with
cytoplasmically derived cleavage vesicles (Moss 1985). In aplanospore
formation of *Ulkenia visurgensis* (Ulken) Gaertner wall scales from the
parent thallus are invaginated with the plasma membrane during
cytokinesis (Moss 1980), hence the aplanospores possess cell walls
throughout their development (Figure 11.3d). The presence of a cell
wall composed of scales has also been reported on the zoospores of some
species of the Thraustochytriales (Jones & Alderman 1971; Kazama
1974 a; Raghu Kumar 1982 b) whereas in other species a wall is lacking
(Darley *et al.* 1973; Kazama 1980). None of the zoospores of the
Thraustochytrium species which I have examined possessed a cell wall
although wall scales were found associated with the site of flagella
insertion (Figures 11.7a,b). Only after retraction of the flagella into
the zoospore vacuole followed by the rounding-up of the settled zoospore
were wall scales deposited on to the plasma membrane of the young
thallus (Figure 11.7c). In addition to the two basic types of cleavage
described, other cytokinetic characters are considered of taxonomic
significance. Within the genus *Thraustochytrium* three types of
sporangia can be identified - nonproliferous, monoproliferous and
multiproliferous (Figure 11.6). In nonproliferous forms (Figure 11.4a)
the entire sporoplasm cleaves to form spores (e.g. *Thraustochytrium
globosum*, *T. roseum*, *T. arudimentale* Artemchuk), whereas in the
monoproliferous forms (e.g. *T. kinnei* Gaertner, *T. motivum* Goldstein,
T. proliferum) a region of multinucleate sporoplasm at the base of the
sporangium remains uncleaved to form a proliferation body around which
a wall is deposited (Figure 11.5). Following zoospore release the
proliferation body enlarges to form a secondary sporangium. In
multiproliferous forms (e.g. *T. multirudimentale* Goldstein,

Figures 11.3a-d Thraustochytriales, transmission electron micrographs.
Figure 11.3a *Ulkenia visurgensis*, sagenogenetosome (SA) and associated
lamellae of the endoplasmic reticulum (ER). Bar = 500 nm.
Figure 11.3b *Thraustochytrium aggregatum*, transverse section of a
sagenogenetosome showing the associated anastomosed elements of the
endoplasmic reticulum. Bar = 500 nm.
Figure 11.3c *Thraustochytrium aggregatum*, sagenogenetosome (SA) at the
base of the ectoplasmic net (EN). The net is delimited by an
evagination of the thallus plasma membrane (PM). Note the continuity
(arrowed) of the endomembrane system of the ectoplasmic net with that of
the sagenogenetosome. Bar = 500 nm.
Figure 11.3d *Ulkenia visurgensis*, sporangium with aplanospores (A).
Each aplanospore is uninucleate, has a thin cell wall and an ectoplasmic
net (EN). The sporangial wall (W) is thin compared to that of the
precleavage sporangium (arrowed). Bar = 2 μm.

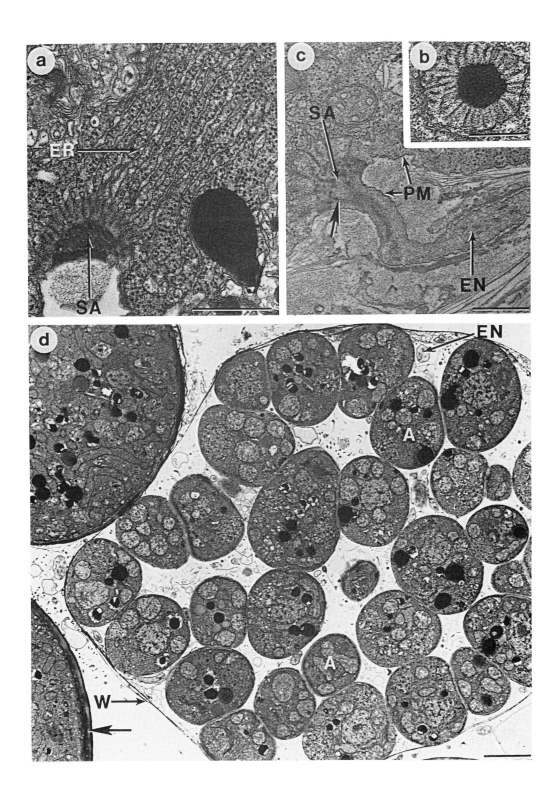

T. kerguelensis Bahnweg et Sparrow, *T. rossii* Bahnweg et Sparrow) two or more proliferation bodies are cleaved from the sporoplasm, each of which gives rise to a secondary sporangium. The ontogeny of the secondary sporangia of the Thraustochytriales differs from that of the Saprolegniales. In the Saprolegniales the secondary sporangia are initiated only after zoospore release and form by resumed growth of the subtending hypha through the basal septum, they do not develop from a delimited region of sporoplasm.

The single species of *Japonochytrium*, *J. marinum* Kobayashi et Ookubo, differs from the nonproliferous species of *Thraustochytrium* only by the presence of a subsporangial dilation, the apophysis, of the ectoplasmic net and the release of zoospores through an apical pore in the sporangial wall. Each of the genera *Althornia*, *Aplanochytrium*, *Schizochytrium* and *Ulkenia* can be readily distinguished from other genera of thraustochytrids. *Althornia crouchii* Jones et Alderman is the only species of the Thraustochytriales which is free-floating and lacks an ectoplasmic net and sagenogenetosome. However the wall is composed of scales and the mode of zoosporulation and zoospore type conform to those characteristic of the Thraustochytriales. *Aplanochytrium kerguelensis* Bahnweg et Sparrow differs from other thraustochytrids in that the aplanospores, which form by successive bipartition, are the only spore type known. Cleavage of the vegetative thallus in species of *Schizochytrium* is also by successive bipartition. This produces the regular tetrads of vegetative cells characteristic of the genus (Figures 11.6, 7d); the vegetative cells eventually undergo progressive cleavage to form zoospores. Both the vegetative thalli and young sporangia of *Ulkenia* species resemble those of *Thraustochytrium*. However, species of *Ulkenia* are characterized by the release of the uncleaved sporoplast from the sporangial wall as either a uninucleate (*U. amoeboidea* (Bahnweg et Sparrow) Gaertner, *U. minuta* Raghu Kumar) or multinucleate (*U. visurgensis*, *U. profunda* Gaertner, *U. sarkariana* Gaertner, *U. radiata* Gaertner) amoeboid cell which then cleaves to produce typical thraustochytriaceous zoospores (Gaertner 1977; Raghu Kumar 1982 a,b).

The seventh genus of the Thraustochytriales, *Labyrinthuloides*, was described by Perkins (1973 a) and owing to the motility of the cuneiform vegetative cells was placed in the Labyrinthulales. However, species of *Labyrinthuloides* are morphologically dissimilar to those of *Labyrinthula* and their assignment to the Thraustochytriales seems a better disposition. Unlike the ectoplasmic net of *Labyrinthula* species, that

Figures 11.4a-c *Thraustochytrium* sp., nonproliferous zoosporangium, transmission electron micrographs.
Figure 11.4a Zoosporangium prior to spore release. Bar = 2 μm.
Figure 11.4b Zoospore initial showing a dictyosome (D) with wall scales (S). Note the lack of a cell wall. Bar = 1 μm.
Figure 11.4c Zoospore with laterally inserted flagella (F). Note the transverse sections of flagella (F) and mastigonemes (MA) in the intersporal space. Bar = 1 μm.

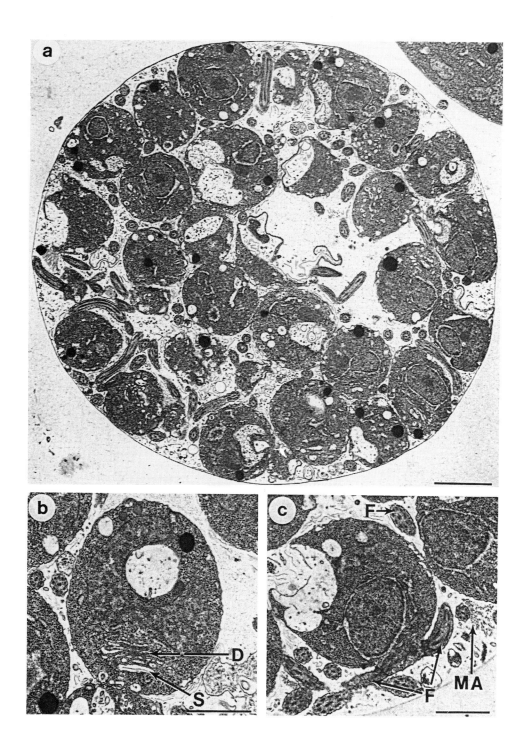

of species of *Labyrinthuloides* does not enrobe the cell but radiates
from it in a manner characteristic of the thraustochytrids. Although
zoosporulation has been observed in species of *Labyrinthuloides* the
normal reproductive unit is the aplanospore which is formed by
successive bipartition of the thallus (Perkins 1973 a; Quick 1974). In
both its production of aplanospores and cleavage by successive
bipartition the genus *Labyrinthuloides* is very similar to *Aplanochytrium*
kerguelensis. Additional similarities between these two genera have
been identified at the biochemical level. Ulken *et al.* (1985) found the
cell wall composition of species from both genera to be similar with

Figure 11.5 *Thraustochytrium* sp., monoproliferous zoosporangium with
zoospore initials (Z) and basal, walled proliferation body.
Transmission electron micrograph, bar = 2μm.

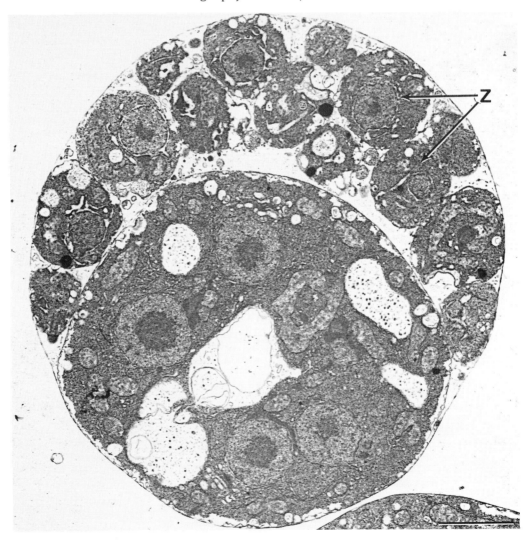

fucose as the major sugar, whereas in other genera of the thraustochytrids examined galactose was the major sugar (Darley *et al.* 1973; Bahnweg & Jäckle Chapter 12). These similarities not only support inclusion of the genus *Labyrinthuloides* in the Thraustochytriales but indicate that the genera *Labyrinthuloides* and *Aplanochytrium* may be congeneric. For a more comprehensive description of the genera and species of the Thraustochytriales the reader is referred to Karling's monograph on the biflagellate phycomycetes (Karling 1981).

Figure 11.6 Diagrammatic representation of the basic sporangial types in the zoosporic Thraustochytriales.

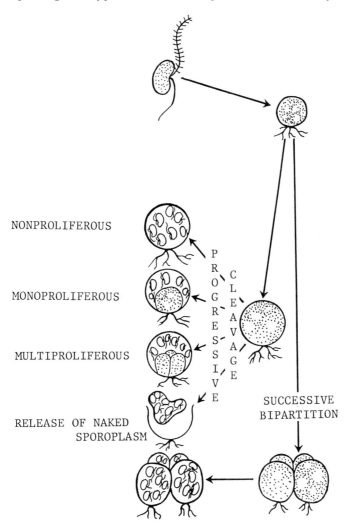

LABYRINTHULALES
        Following description of the first labyrinthulids by
Cienkowski in 1867 these primarily marine organisms have been placed
variously in the Protozoa, algae, Mycota and Protista. This diversity
of taxonomic affiliations reflects both the unusual morphology and life
style of the labyrinthulids and our incomplete knowledge of their
biology. The order contains the single genus *Labyrinthula* with eight
species. Other genera have been assigned to the Labyrinthulales but are
now considered either misplaced (*Labyrinthorhiza* Chadefaud, 1956;
Labyrinthomyxa marina Mackin et Ray, 1966 (= *Dermocystidium marinum*
Mackin, Owen et Collier, 1950)) or synonymous (*Chlamydomyxa* Archer,
1875; *Labyrinthomyxa* Duboscq, 1921; *Pseudoplasmodium* Molisch, 1926;
Labyrinthodictyon Valkanov, 1972) with *Labyrinthula*.

Species of *Labyrinthula* are characterized by spindle-shaped, < 30 μm
long, vegetative cells (Figure 11.8a) which move at rates of up to
200 μm min $^{-1}$ through a network of hyaline ectoplasmic net elements
(Young 1943; Porter 1969; Stey 1969; Bartsch 1971). The ectoplasmic
nets of adjacent *Labyrinthula* cells and daughter cells formed following
binary fission of spindle-cells fuse to produce a colony of independent
cells within a common net system. The net elements extend radially and
spindle-cells migrate, frequently in unilateral chains, towards the net
apices. Spindle-cells move independently, may reverse direction and
may move in different directions within the same net element. The
spindle-cells are uninucleate, frequently possess two large vacuoles,
one or two dictyosomes, mitochondria, peripherally positioned lipid
bodies but lack centrioles associated with the interphase nucleus
(Figure 11.9). Although species of the Labyrinthulales are
superficially unlike any species of the Thraustochytriales the two
orders possess fine structural characters common to both and which auger
for their close phylogeny. These include a wall composed of scales, an
ectoplasmic net and sagenogenetosomes.

The spindle-cell of species of *Labyrinthula* is bounded by a wall
constructed of a single layer of unornamented scales. Each scale is
2-4 nm thick, circular with a diameter of 0.5-0.75 μm and closely
adpressed to the cell plasma membrane. Adjacent scales overlap each
other by 10-150 nm but do not fuse (Figure 11.8d). The unilamellate
wall is most apparent at the sites of the sagenogenetosomes where the

Figure 11.7b *Thraustochytrium* sp., recently encysted zoospore (left
side of micrograph) showing remains of retracted flagella (F) and
mastigonemes (MA) in the vacuole. Note the wall scales (S) associated
with the region where flagella were inserted. Following retraction of
the flagella the encysted zoospore becomes spherical (arrowed).
Bar = 1 μm.
Figure 11.7c *Thraustochytrium* sp., wall region of a recently encysted
zoospore showing a single layer of scales (S). Bar = 250 nm.
Figure 11.7d *Schizochytrium aggregatum*, tetrad of vegetative cells
formed by successive bipartition. All four cells are retained within
the original wall (W) of the parent thallus. Bar = 1 μm.

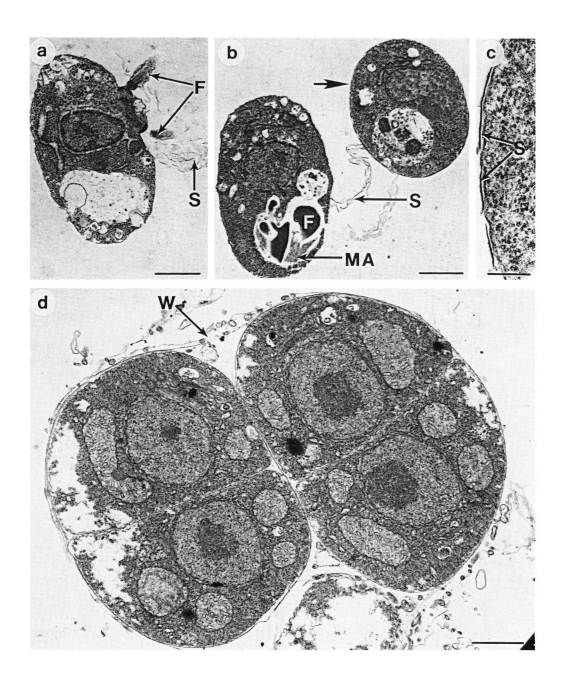

Figures 11.7a–d Thraustochytriales, transmission electron micrographs.
Figure 11.7a *Thraustochytrium* sp., released zoospore with laterally
inserted flagella (F). Note that the wall scales (S) are associated
only with the region of flagella insertion. Bar = 1 μm.

plasma membrane is invaginated (Figures 11.8c,d). As in the
Thraustochytriales, the scales are formed in the dictyosome cisternae
(Figures 11.9a,b) but only appear to be deposited extracellularly in the
plane of cleavage between daughter cells (Figure 11.9c). Several
sagenogenetosomes (= bothrosome, Porter 1969) are distributed over the
surface of the spindle-cell. Each sagenogenetosome comprises an
electron-opaque aggregate associated with a single lamella of the

Figures 11.8a-d *Labyrinthula* sp.

Figure 11.8a Spindle-cells (SC) within the ectoplasmic net (EN).
S.E.M., bar = 10 µm.

Figure 11.8b Sagenogenetosome (SA) initial and associated lamella of
the endoplasmic reticulum (ER) adjacent to the plasma membrane (PM) of
the cleavage plane (arrowed) between daughter cells. T.E.M.,
bar = 500 nm.

Figure 11.8c Sagenogenetosome (SA) and ectoplasmic net (EN) initial
formed by evagination of the plasma membrane (PM). T.E.M.,
bar = 500 nm.

Figure 11.8d Ectoplasmic net (EN) penetrated through the unilamellate
cell wall of the spindle cell. Note the characteristic overlap
(arrowed) of the wall scales (S). T.E.M., bar = 500 nm.

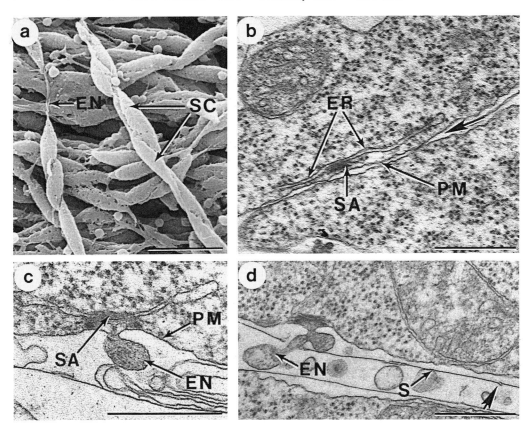

endoplasmic reticulum. The sagenogenetosomes appear to form *de novo*, most commonly in the plane of cleavage (Figure 11.8b) and prior to development of the associated ectoplasmic net. The ectoplasmic net forms by evagination of the plasma membrane adjacent to the sagenogenetosome initial (Figure 11.8c) and penetrates through the unilamellate wall (Figure 11.8d). A branch from the lamella of endoplasmic reticulum extends through the sagenogenetosome to maintain continuity with the membrane components within the net system (Figure 11.8c). The ectoplasmic net lacks a cell wall and contains no organelles. The several ectoplasmic nets formed by each spindle-cell fuse with each other to enrobe the spindle-cell within the net system (Schwab-Stey & Schwab 1974 a) and nets of adjacent spindle-cells fuse to form a common ectoplasmic net system. Reproduction of the spindle-cells is normally by binary fission (Figure 11.9c) which follows mitosis, the only vegetative stage at which centriolar equivalents, termed protocentrioles, have been observed (Perkins 1970; Porter 1972). The cleavage plane is initially transverse but becomes oblique as the daughter cells elongate and scales are deposited to complete the wall of each daughter cell. In culture, spindle-cells frequently aggregate to form sori. Hollande & Enjumet (1955) documented the production of heterokont, stigmate zoospores from such sori in *L. algeriensis* Hollande et Enjumet, and Perkins & Amon (1969) reported meiosis during zoosporulation, although the stage at which syngamy occurs in the life cycle has not been identified.

The spindle-cells of all species of *Labyrinthula* are characterized by movement through the ectoplasmic net system although the cells lack cilia, flagella, podia and any other visible means of propulsion. However, Nakatsuji & Bell (1980) proposed an actomyosin system with myosin in the net and actin on the surface of the spindle-cells.

Speciation of the genus *Labyrinthula* is based primarily on the dimensions of the spindle-cell (Pokorny 1967) although the validity of all eight species described is uncertain.

ECOLOGY

Most information on the occurrence and numbers of labyrinthulids and thraustochytrids in nature is based on either pine pollen baiting techniques (Gaertner 1968) or, more rarely, dilution plate methods using a modified Vishniac's medium (Fuller *et al.* 1964). There are very few reports of direct observations (Riemann & Schrage 1983).

The thraustochytrids have been isolated from a wide range of marine habitats worldwide. The numbers of individual propagules are greatest in coastal and estuarine sediments (Gaertner 1967 a,b; Riemann & Schrage 1983) where Ulken (1983) found $40-50 \times 10^3$ propagules per litre of sediment but they have also been found associated with algae (Miller & Jones 1983), marine angiosperms, the mantle cavities of molluscs (Alderman & Jones 1971), organic detritus (Amon 1978), saline soils (Booth 1971; Artemchuk 1972) and in the water column (Gaertner & Raghu

Kumar 1980). Species of the Thraustochytriales have also been isolated
from inland saline lakes (Amon 1978) but none from fresh water.
Thraustochytrids are considered euryhaline (Bremer 1974; Jones &
Harrison 1976), obligately aerobic (Goldstein & Belsky 1964) with
temperature optima between 25 and 30°C, although for species isolated
from Antarctic waters Bahnweg (1979) indicated an optimum of only 4°C.
With few exceptions the mode of nutrition is considered saprotrophic
with the normally epibiotic thalli attached to their substrates by
epi- or endo-biotic ectoplasmic nets. Perkins (1972, 1973 b) speculated
that the ectoplasmic net aids in the transport of enzymes to the
substrate and the products of extracellular digestion back to the
thallus. Miller & Jones (1983) found that thraustochytrids were
responsible for the early stages in the degradation of moribund seaweed,
although Goldstein & Belsky (1964) had noted the inability of
thraustochytrids to utilize living algal tissues. The only record of a
plant parasitic thraustochytrid is that of a species of *Schizochytrium*
found on the diatom *Thalassionema nitzschioides* (Grun.) Hustedt
(Gaertner 1979). Quick (1974) described a close association between
Labyrinthuloides schizochytrops Quick and the sea grass *Halodule
wrightii* (Ascherson) Ascherson but the nutritional status of this
relationship is not clear.

Since 1980 there have been several reports of animal pathogenic species
of the Thraustochytriales. Polglase (1980, 1981) reported *Ulkenia
amoeboidea* Gaertner in the octopus *Eledone cirrhosa* (Lamarck) and
thraustochytrids associated with lesions in rainbow trout, Jones & O'Dor
(1983) found thraustochytrids associated with gill lesions of squid and
McLean & Porter (1982) identified a thraustochytrid causing the Yellow
Spot disease of a nudibranch. These associations are considered in more
detail elsewhere in this book (see Porter, Chapter 13; Polglase *et al.*,
Chapter 14; Alderman & Polglase, Chapter 17).

Species of *Labyrinthula* are cosmopolitan in coastal waters and are most
readily isolated from species of the marine angiosperms *Zostera* and
Spartina, although they can also be isolated from coastal sediments,
algae and organic detritus. Some of their more obscure recorded
habitats include the root hairs of trees grown in soils of low salinity
(Aschner 1958), inland saline soils (Amon 1978), the spore cases of
vesicular-arbuscular mycorrhizal fungi in sand dunes (Koske 1981),
diatoms (Chamberlain 1985) and in one report parasitic on a fresh water

Figures 11.9a-c *Labyrinthula* sp., transmission electron micrographs.
Figure 11.9a Dictyosome adjacent to the nucleus (N) of a spindle-cell
showing the presence of wall scales (S) in the distal cisternae.
Bar = 1 µm.
Figure 11.9b Higher magnification of the dictyosome cisterna arrowed in
Fig. 9a within which is a wall scale (S). Bar = 100 nm.
Figure 11.9c Recently divided spindle-cell with daughter cells still
retained within the parental wall (W). Each daughter cell contains a
single nucleus (N), lipid bodies (L), a single dictyosome (D),
mitochondria (M) and several sagenogenetosomes (SA). Bar = 1 µm.

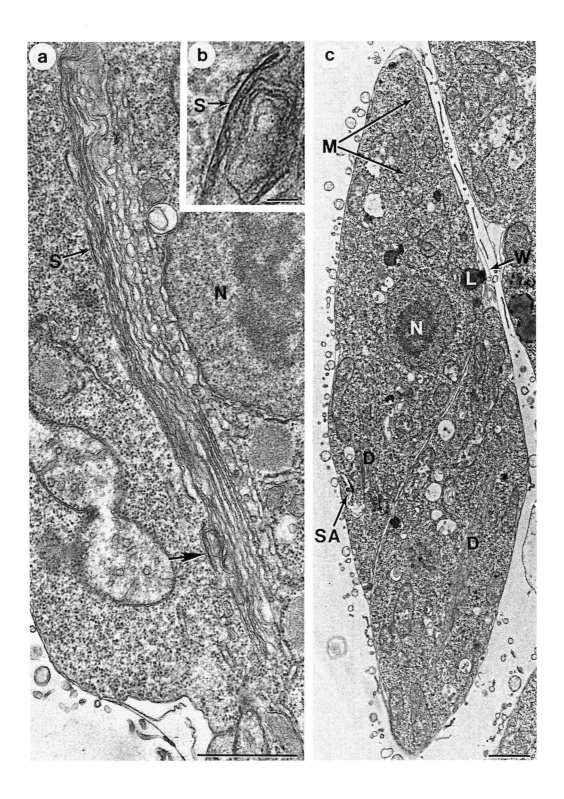

alga (Zopf 1892). Renn (1936) reported species of *Labyrinthula* to cause the "wasting" disease of *Zostera marina* L. However, although spindle-cells of *Labyrinthula* species are ubiquitous in the dead cells of *Z. marina* leaves their role as pathogens has not been proven.

PHYLOGENY
        Previous to the fine structural studies of species of the Thraustochytriales which revealed their characteristic possession of multilamellate walls composed of dictyosome-derived scales and ectoplasmic nets, these organisms were accepted as members of the zoosporic fungi. Their reproduction by heterokont, laterally biflagellate planospores and the chytrid-like morphology of the thallus were considered of sufficient phyletic significance for their assignment to the Saprolegniales, Oomycetes (Sparrow 1936, 1943, 1973; Dick 1973; Gaertner 1977) and the Heterokontimycotina (Dick 1976). Conversely, species of the Labyrinthulales have been a taxonomic enigma since their original description by Cienkowski. They have been variously assigned to the Rhizopodea (Calkins 1934; Honigberg *et al.* 1964; Schwab-Stey & Schwab 1974 b), Mycetozoa (Bessey 1950; Pokorny 1967; Olive 1970), Myxothallophyta (Fitzpatrick 1930); Chrysophyceae and Haptophyceae of the algae (Hollande & Enjumet 1955), Mycota (Whittaker 1969), Protista (Olive 1975; Whittaker & Margulis 1978; Corliss 1984) and Protoctista (Margulis 1981). These several affiliations have resulted from the many different interpretations of the trophic spindle-cells and associated net elements of the labyrinthulids (e.g. net plasmodium, Cienkowski 1867; podia, Zopf 1892, Schwab-Stey & Schwab 1974 b; slime-tracks, Hohl 1966, Porter 1969).

Ultrastructural studies of both the labyrinthulids and thraustochytrids have contributed information of considerable taxonomic significance which not only supports their close phyletic affinity but separates them from other taxa. The presence of a cell wall composed of scales and, in the thraustochytrids, a chemical composition high in protein and containing either galactose or, in species of *Labyrinthuloides* and *Aplanochytrium*, fucose (Ulken *et al.* 1985; Bahnweg & Jäckle Chapter 12) as the major wall polysaccharide separates these organisms from the zoosporic fungi. In the Mycota the walls are characteristically fibrillar (Aronson 1965) and in the heterokont fungi the principal wall polysaccharides are β-1, 3-, β-1, 4- or β-1, 6- glucans (Vaziri-Tehrani & Dick 1980). No information is available on the chemical composition of the cell wall in species of *Labyrinthula*. Cell walls composed of scales have been described in the Chrysophyceae, Haptophyceae and Prasinophyceae of the algae (Belcher & Swale 1967, 1971; Dodge 1973; Hearth *et al.* 1975), the amoebo-flagellate stage of the mycetozoan *Ceratiomyxella tahitiensis* Olive et Stoianovitch (Furtado & Olive 1971, 1972) and the protists *Diplophrys* spp. and *Sorodiplophrys stercorea* (Cienk.) Olive et Dykstra (Dykstra 1976; Dykstra & Porter 1984). In all these organisms wall scales are either chemically dissimilar to those of the thraustochytrids or have not been characterized and sagenogenetosomes are absent. However in species of *Diplophrys* and in *S. stercorea* ectoplasmic nets lacking organelles are present and the

cells exhibit a gliding movement similar to that of species of
Labyrinthuloides (Dykstra & Porter 1984).

The taxonomic affinities of the Labyrinthulales and Thraustochytriales
remain unresolved. The sagenogenetosome appears unique to species of
the labyrinthulids and thraustochytrids and combined with some of their
other ultrastructural and biochemical characters support their
separation from all other taxa. Their assignment to the Kingdom
Protista Haeckel (1866) seems the least conjectural position. However,
within the Protista the diversity of organisms is large with many
artificially separated taxa. It is unfortunate that the different
nomenclatural systems used for protists has resulted in a number of
synonymous names for the higher-level groupings of the labyrinthulids
and thraustochytrids, e.g. Labyrinthulina (Olive 1975), Labyrinthomorpha
(Levine *et al*. 1980), Labyrinthulomycota (Margulis 1981), with an equal
diversity of names at the ordinal and familial levels.

Many of the characters used in the delimitation of genera and species of
the thraustochytrids and labyrinthulids are unsatisfactory and the
number of morphological characters considered of taxonomic importance is
small. There is a need for a less conservative approach to the taxonomy
of these organisms and investigations similar to the biochemical studies
of Bahnweg & Jäckle (Chapter 12) and the ribosomal RNA studies of Porter
& Kochert (1978) may aid in both a more natural classification of the
thraustochytrids and labyrinthulids and the elucidation of their
phyletic affinities.

ACKNOWLEDGEMENTS
   I wish to acknowledge the support received for this research
from NERC grant GR3/3978.

REFERENCES
Alderman, D.J., Harrison, J.L., Bremer, G.B. & Jones, E.B.G. (1974).
      Taxonomic revisions in the marine biflagellate fungi:
      ultrastructural evidence. Mar. Biol., 25, 345-357.
Alderman, D.J. & Jones, E.B.G. (1971). Shell disease of oysters. Fishery
      Investigations, London, Ser. II, no. 8, 1-19.
Amon, J.P. (1978). Thraustochytrids and labyrinthulids of terrestrial,
      aquatic and hypersaline environments of the Great Salt Lake,
      USA. Mycologia, 70, 1299-1301.
Archer, W. (1875). On *Chlamydomyxa labyrinthuloides*. Q. Jl microsc.
      Sci., 15, 107-130.
Aronson, J.M. (1965). The cell wall. In The Fungi: An Advanced Treatise,
      Vol. IVB, eds. G.C. Ainsworth & A.S. Sussman, pp. 49-76.
      New York: Academic Press.
Artemchuk, N.J. (1972). The fungi of the White sea. III. New
      Phycomycetes, discovered in the Great Salma Strait of the
      Kandalakshial Bay. Veröff. Inst. Meeresforsch. Bremerh., 13,
      231-237.
Aschner, M. (1958). Isolation of *Labyrinthula macrocystis* from soil.
      Bull. Res. Coun. Israel Sect. D., 174-179.

Bahnweg, G. (1979). Studies on the physiology of Thraustochytriales. I. Growth requirements and nitrogen nutrition of *Thraustochytrium* spp., *Schizochytrium* sp., *Japonochytrium* sp., *Ulkenia* spp., and *Labyrinthuloides* spp. Veröff. Inst. Meeresforsch. Bremerh., <u>17</u>, 245-268.

Bahnweg, G. & Sparrow, F.K. (1972). *Aplanochytrium kerguelensis* gen. nov. spec. nov., a new phycomycete from subantarctic marine waters. Arch. Mikrobiol., <u>81</u>, 45-49.

Bartsch, G. (1971). Cytologische beobachtungen an *Labyrinthula coenocystis* Schmoller bie verschiedenen kulturbedingungen. Z. allg. Mikrobiol., <u>11</u>, 79-90.

Belcher, J.H. & Swale, M.F. (1967). *Chromulina placentula* sp. nov. (Chrysophyceae), a freshwater nanoplankton flagellate. Br. phycol. Bull., <u>3</u>, 257-267.

Belcher, J.H. & Swale, M.F. (1971). The microanatomy of *Phaeaster pascheri* Scherffel (Chrysophyceae). Br. phycol. Bull., <u>6</u>, 157-169.

Bessey, E.A. (1950). Morphology and Taxonomy of Fungi. Philadelphia: Blakiston.

Booth, T. (1971). Distribution of certain soil inhabiting chytrid and chytridiaceous species related to some physical and chemical factors. Can. J. Bot., <u>49</u>, 1743-1755.

Bremer, G.B. (1974). The ecology of marine lower fungi. <u>In</u> Recent Advances in Aquatic Mycology, ed. E.B. Gareth Jones, pp. 313-333. London: Elek Science.

Calkins, G.N. (1934). The Biology of Protozoa. Philadelphia: Lea & Febiger.

Chadefaud, M. (1956). Sur un *Labyrinthula* de Roscoff. C.r. hebd. Séanc. Acad. Sci., Paris, <u>243</u>, 1794-1797.

Chamberlain, A.H.L. (1980). Cytochemical and ultrastructural studies on the cell walls of *Thraustochytrium* spp. Botanica mar., <u>23</u>, 669-677.

Chamberlain, A.H.L. (1985). Labyrinthulids and thraustochytrids as parasites of benthic diatoms. Abstract, The Fourth International Marine Mycology Symposium, p. 28. Portsmouth Polytechnic, U.K.

Cienkowski, L. (1867). Über den bau und die entwicklung der Labyrinthuleen. Arch. mikrosk. Anat. EntwMech., <u>3</u>, 274-310.

Corliss, J.O. (1984). The kingdom Protista and its 45 phyla. Biosystems, <u>17</u>, 87-126.

Darley, W.M., Porter, D. & Fuller, M.S. (1973). Cell wall composition and synthesis via Golgi-directed scale formation in the marine eucaryote, *Schizochytrium*, with a note on *Thraustochytrium* sp. Arch. Mikrobiol., <u>90</u>, 89-106.

Dick, M.W. (1973). Saprolegniales. <u>In</u> The Fungi: An Advanced Treatise, Vol. IVB, eds. G.C. Ainsworth, F.K. Sparrow & A.L. Sussman, pp. 113-144. New York: Academic Press.

Dick, M.W. (1976). The ecology of aquatic phycomycetes. <u>In</u> Recent Advances in Aquatic Mycology, ed. E.B. Gareth Jones, pp. 513-542. London: Elek Science.

Dodge, J.D. (1973). The Fine Structure of Algal Cells. New York: Academic Press.

Duboscq, O. (1921). *Labyrinthomyxa sauvageani* n.g. n.sp. Protomyxan
    parasite de *Laminaria lejolisii* Sauvageau. C.r. Séanc. Soc.
    Biol., 84, 27-33.
Dykstra, M.J. (1976). Wall and membrane biogenesis in the unusual
    labyrinthulid-like organism *Sorodiplophrys stercorea*.
    Protoplasma, 87, 329-346.
Dykstra, M.J. & Porter, D. (1984). *Diplophrys marina*, a new scale-
    forming protist with labyrinthulid affinities. Mycologia,
    76, 626-632.
Fitzpatrick, H.M. (1930). The Lower Fungi: Phycomycetes. New York:
    McGraw-Hill.
Fuller, M.S., Fowles, B.E. & McLaughlin, D.J. (1964). Isolation and pure
    culture study of marine phycomycetes. Mycologia, 56,
    745-756.
Furtado, J.S. & Olive, L.S. (1971). Ultrastructure of the protostelid
    Ceratiomyxella tahitiensis, including scale formation. Nova
    Hedwigia, 21, 537-576.
Furtado, J.S. & Olive, L.S. (1972). Scale formation in a primitive
    mycetozoan. Trans. Am. microscop. Soc., 91, 594-596.
Gaertner, A. (1967 a). Niedere, mit pollen köderbare Pilze in der
    südlichen Nordsee. Veröff. Inst. Meeresforsch. Bremerh., 10,
    159-165.
Gaertner, A. (1967 b). Ökologische untersuchungen an einem marinen
    Pilz aus der Unigebung von Helgoland. Helgolander. Wiss.
    Meeresunters., 15, 181-192.
Gaertner, A. (1968). Eine methode des quantitativen nachweises niederer,
    mit pollen köderbarer pilze im meerwasser und im sediment.
    Veröff. Inst. Meeresforsch. Bremerh. Suppl., 3, 75-92.
Gaertner, A. (1972). Characters used in the classification of
    thraustochytriaceous fungi. Veröff. Inst. Meeresforsch.
    Bremerh., 13, 183-194.
Gaertner, A. (1977). Revision of the Thraustochytriaceae (lower marine
    fungi). I. *Ulkenia* nov. gen., with description of three new
    species. Veröff. Inst. Meeresforsch. Bremerh., 16, 139-157.
Gaertner, A. (1979). Some fungal parasites found in the diatom
    populations of the Rosfjord area (South Norway) during March
    1979. Veröff. Inst. Meeresforsch. Bremerh., 18, 29-33.
Gaertner, A. & Raghu Kumar, S. (1980). Ecology of the Thraustochytrids
    (lower marine fungi) in the Fladen Ground and other parts of
    the North Sea I. Meteor Forsch.-Ergebnisse, 22, 165-185.
Goldstein, S. & Belsky, M. (1964). Axenic culture of a new marine
    phycomycete possessing an unusual type of asexual
    reproduction. Am. J. Bot., 51, 72-78.
Goldstein, S., Moriber, L. & Hershinov, B. (1964). Ultrastructure of
    Thraustochytrium aureum, a biflagellate marine Phycomycete.
    Mycologia, 56, 897-904.
Haeckel, E. (1866). General Morphologie der Organismen. Berlin:
    G. Reimer.
Harrison, J.L. & Jones, E.B.G. (1974 a). Ultrastructural aspects of the
    marine fungus *Japonochytrium* sp. Arch. Mikrobiol., 96,
    305-317.

Harrison, J.L. & Jones, E.B.G. (1974 b). Ultrastructural observations on the formation of zoospores in *Thraustochytrium kinnei* Gaertner. Trans. mycol. Soc. Japan, 15, 273-288.

Harrison, J.L. & Jones, E.B.G. (1974 c). Zoospore discharge in *Thraustochytrium striatum*. Trans. Brit. mycol. Soc., 62, 283-288.

Hearth, W., Kuppel, A., Franke, W.W. & Brown, R.M. (1975). The ultrastructure of the scale cellulose from *Pleurochrysis scherffelii* under various experimental conditions. Cytobiologie, 10, 268-284.

Hohl, H.R. (1966). The fine structure of slimeways in *Labyrinthula*. J. Protozool., 13, 41-43.

Hollande, A. & Enjumet, M. (1955). Sur l'évolution et la systématique des Labyrinthulidae; Étude de *Labyrinthula algeriensis* nov. sp. Annls Sci. nat. (Zool.), 11 sér., 17, 357-368.

Honigberg, B.M., Balamath, W., Bovee, E.C., Corliss, J.O., Gojdics, M., Hall, R.P., Kudo, R.R., Levine, N.D., Loeblich, A.R., Weiser, J. & Wenrich, D.H. (1964). A revised classification of the Phylum Protozoa. J. Protozool., 11, 7-20.

Johnson, T.W. & Sparrow, F.K. (1961). Fungi of Oceans and Estuaries, New York, Cramer.

Jones, E.B.G. & Alderman, D.J. (1971). *Althornia crouchii* gen. et sp. nov., a marine biflagellate fungus. Nova Hedwigia, 21, 381-399.

Jones, E.B.G. & Harrison, J.L. (1976). Physiology of marine phycomycetes. In Recent Advances in Aquatic Mycology, ed. E.B. Gareth Jones, pp. 261-278. London: Elek Science.

Jones, G.M. & O'Dor, R.K. (1983). Ultrastructural observations on a thraustochytrid fungus parasitic in the gills of squid (*Ilex illecebrosus* LeSueur). J. Parasit., 69, 903-911.

Karling, J.S. (1981). Predominantly Holocarpic and Eucarpic Simple Biflagellate Phycomycetes. Vaduz: J. Cramer.

Kazama, F.Y. (1972 a). Ultrastructure of *Thraustochytrium* sp. zoospores. I. kinetosome. Arch. Mikrobiol., 83, 179-188.

Kazama, F.Y. (1972 b). Ultrastructure of *Thraustochytrium* sp. zoospores. II. Striated inclusions. J. Ultrastruct. Res., 41, 60-66.

Kazama, F.Y. (1974 a). Ultrastructure of *Thraustochytrium* sp. zoospores. IV. External morphology with notes on the zoospores of *Schizochytrium* sp. Mycologia, 66, 272-280.

Kazama, F.Y. (1974 b). The ultrastructure of nuclear division in *Thraustochytrium* sp. Protoplasma, 82, 155-175.

Kazama, F.Y. (1975). Cytoplasmic cleavage during zoosporogenesis in *Thraustochytrium* sp.: ultrastructure and the effects of colchicine and D_2O. J. Cell Sci., 17, 155-170.

Kazama, F.Y. (1980). The zoospore of *Schizochytrium aggregatum*. Can. J. Bot., 58, 2434-2446.

Kobayashi, Y. & Ookubo, M. (1953). Studies on the marine Phycomycetes. Bull. natn. Sci. Mus., Tokyo, 33, 53-65.

Koske, R.E. (1981). *Labyrinthula* inside the spores of a vesicular-arbuscular mycorrhizal fungus. Mycologia, 73, 1175-1180.

Levine, N.D., Corliss, J.O., Cox, F.E.G., Deroux, G., Grain, J., Honigberg, B.M., Leedale, G.F., Loeblich, A.R., III,

Lom, J., Lynn, D.H., Merinfeld, E.G., Page, F.C., Poljansky, G., Sprague, V., Vávra, J. & Wallace, F.G. (1980). A newly revised classification of the Protozoa. J. Protozool., 27, 37-58.

McLean, N. & Porter, D. (1982). The yellow-spot disease of *Tritonia diomedea* Bergh, 1894 (Mollusca: Gastropoda: Nudibranchia): encapsulation of the thraustochytriaceous parasite by host amoebocytes. J. Parasit., 68, 243-252.

Mackin, J.G., Owen, H.M. & Collier, A. (1950). Preliminary note on the occurrence of a new protistan parasite, *Dermocystidium marinum* n.sp. in *Crassostrea virginica* (Gmelin). Science, 111, 328-329.

Mackin, J.G. & Ray, S.M. (1966). The taxonomic relationship of *Dermocystidium marinum* Mackin, Owen and Collier. J. Invert. Pathol., 8, 544-545.

Margulis, L. (1981). Symbiosis in Cell Evolution. San Francisco: Freeman.

Miller, J.D. & Jones, E.B.G. (1983). Observations on the association of thraustochytrid marine fungi with decaying seaweed. Botanica mar., 26, 345-351.

Molisch, H. (1926). *Pseudoplasmodium aurantiacum* n.g. et n.sp. eine neue Acrasiee aus Japan. Sci. Rep. Tohoku Univ., Ser. 4, 1, 119-134.

Moss, S.T. (1980). Ultrastructure of the endomembrane-sagenogenetosome-ectoplasmic net complex in *Ulkenia visurgensis* (Thraustochytriales). Botanica mar., 23, 73-94.

Moss, S.T. (1985). An ultrastructural study of taxonomically significant characters of the Thraustochytriales and the Labyrinthulales. J. Linn. Soc. (Bot), 91, 329-357.

Nakatsuji, N. & Bell, E. (1980). Control by calcium of the contractility of *Labyrinthula* slimeways and of the translocation of *Labyrinthula* cells. Cell Motility, 1, 17-29.

Olive, L.S. (1970). The Mycetozoa: a revised classification. Bot. Rev., 36, 59-87.

Olive, L.S. (1975). The Mycetozoans. New York: Academic Press.

Perkins, F.O. (1970). Formation of centriole and centriole-like structures during meiosis and mitosis in *Labyrinthula* sp. (Rhizopodea, Labyrinthulida) an electron microscope study. J. Cell Sci., 6, 629-653.

Perkins, F.O. (1972). The ultrastructure of holdfasts, "rhizoids", and "slime tracks" in thraustochytriaceous fungi and *Labyrinthula* spp. Arch. Mikrobiol., 84, 95-118.

Perkins, F.O. (1973 a). A new species of marine labyrinthulid *Labyrinthuloides yorkensis* gen. nov. spec. nov. - cytology and fine structure. Arch. Mikrobiol., 90, 1-17.

Perkins, F.O. (1973 b). Observations of thraustochytriaceous (Phycomycetes) and labyrinthulid (Rhizopodea) ectoplasmic nets on natural and artificial substrates - an electron microscope study. Can. J. Bot., 51, 485-491.

Perkins, F.O. (1974). Phylogenetic considerations of the problematic thraustochytriaceous-labyrinthulid-*Dermocystidium* complex based on observations of fine structure. Veröff. Inst. Meeresforsch. Bremerh., Suppl. 5, 45-63.

Perkins, F.O. (1976). Fine structure of lower marine and estuarine
    fungi. In Recent Advances in Aquatic Mycology, ed.
    E.B. Gareth Jones, pp. 279-312. London: Elek Science.
Perkins, F.O. & Amon, J.P. (1969). Zoosporulation in *Labyrinthula* sp.,
    an electron microscope study. J. Protozool., 16, 235-257.
Pokorny, K.S. (1967). *Labyrinthula*. J. Protozool., 14, 697-708.
Polglase, J.L. (1980). A preliminary report on the Thraustochytrid(s)
    and Labyrinthulid(s) associated with a pathological
    condition in the lesser octopus *Eledone cirrhosa*. Botanica
    mar., 23, 699-706.
Polglase, J.L. (1981). Thraustochytrids as potential pathogens of marine
    animals. Bull. Br. mycol. Soc., 16 (Suppl. 1), 5.
Porter, D. (1969). Ultrastructure of *Labyrinthula*. Protoplasma, 67,
    1-19.
Porter, D. (1972). Cell division in the marine slime mold, *Labyrinthula*
    sp. and the role of the bothrosome in extracellular membrane
    production. Protoplasma, 74, 427-448.
Porter, D. (1974). Phylogenetic considerations of the
    Thraustochytriaceae and Labyrinthulaceae. Veröff. Inst.
    Meeresforsch. Bremerh., Suppl. 5, 19-44.
Porter, D. & Kochert, G. (1978). Ribosomal RNA molecular weights and the
    phylogeny of *Labyrinthula*. Expl. Mycol., 2, 346-351.
Quick, J.A. (1974). *Labyrinthuloides schizochytrops* n. sp., a new
    marine *Labyrinthula* with spheroid "spindle" cells. Trans.
    Am. Microsc. Soc., 93, 344-365.
Raghu Kumar, S. (1982 a). Fine structure of the thraustochytrid *Ulkenia*
    amoeboidea. I. Vegetative thallus and formation of the
    amoeboid stage. Can. J. Bot., 60, 1092-1102.
Raghu Kumar, S. (1982 b). Fine structure of the thraustochytrid *Ulkenia*
    amoeboidea. II. The amoeboid stage and formation of
    zoospores. Can. J. Bot., 60, 1103-1114.
Renn, C.E. (1936). The persistence of the eel-grass disease and
    parasite on the American Atlantic coast. Nature, Lond.,
    138, 507-508.
Riemann, F. & Schrage, M. (1983). On a mass occurrence of a
    thraustochytrioid protist (Fungi Rhizopodan Protozoa) in an
    Antarctic Anaerobic Marine sediment. Veröff. Inst.
    Meeresforsch. Bremerh., 19, 191-202.
Schwab-Stey, H. & Schwab, D. (1974 a). Hypothetical origin of the
    particular membrane relationships in *Labyrinthula*.
    Protoplasma, 81, 125-130.
Schwab-Stey, H. & Schwab, D. (1974 b). Scanning electron microscopic
    investigation on the motile stage of *Labyrinthula*
    coenocystis Schmoller. Cytobiol., 8, 383-394.
Sparrow, F.K. (1936). Biological observations on the marine fungi of
    Woods Hole waters. Biol. Bull. mar. biol. Lab., Woods Hole,
    70, 236-263.
Sparrow, F.K. (1943). The Aquatic Phycomycetes Exclusive of the
    Saprolegniaceae and *Pythium*. Ann Arbor: University of
    Michigan Press.
Sparrow, F.K. (1973). Mastigomycotina (Zoosporic Fungi). In The Fungi:
    An Advanced Treatise, Vol. IVB, eds. G.C. Ainsworth,

F.K. Sparrow & A.L. Sussman, pp. 61-73. New York: Academic Press.

Stey, H. (1969). Elektronenmikroskopische untersuchung an *Labyrinthula coenocystis*. Z. Zellforsch. mikrosk. Anat., 102, 387-418.

Ulken, A. (1983). Phycomyceten im Watt des Jadebusens. Veröff. Inst. Meeresforsch. Bremerh., 19, 177-183.

Ulken, A., Jäckle, I. & Bahnweg, G. (1985). Morphology, nutrition and taxonomy of an *Aplanochytrium* sp. from the Sargasso Sea. Mar. Biol., 85, 89-95.

Valkanov, A. (1972). Untersuchungen über die struktur und den entwicklungszyklus von *Labyrinthodictyon magnificum* Valk. Arch. Protistenk., 114, 426-443.

Vaziri-Tehrani, B. & Dick, M.W. (1980). Neutral and amino sugars from the cell walls of Oomycetes. Biochem. Systemat. Ecol., 8, 105-108.

Whittaker, R.H. (1969). New concepts of kingdoms of organisms. Science, 163, 150-159.

Whittaker, R.H. & Margulis, L. (1978). Protist classification and the kingdoms of organisms. Biosystems, 10, 3-18.

Young, E.L. (1943). Studies on *Labyrinthula*. The etiologic agent of the wasting disease of eel-grass. Am. J. Bot., 30, 586-593.

Zopf, W. (1892). Zur kenntnis der Labyrinthuleen, einer familie der Mycetozoen. Beitr. Physiol. Morph. nied. Organ., 2, 36-48.

G. Bahnweg

I. Jäckle

Thraustochytriales and Labyrinthulales are marine protists
which present considerable difficulties with regard to their taxonomy,
even at higher levels of classification. Species of *Labyrinthula*
Cienkowski (1867) parasitize a large variety of marine algae and marine
vascular plants. Despite extensive studies, description of a general
life cycle is still provisional and possibly incomplete (Olive 1975).
Labyrinthula has been considered a member of the Mycetozoa (Zopf 1892),
of the slime moulds (Bonner 1967), of the Rhizopodea (Honigberg *et al.*
1964) and also of the Chrysophyta (Hollande & Enjumet 1955).
Thraustochytriales are marine saprophytes first observed on decaying
algae by Sparrow (1936). Sporangia of thraustochytrids typically
possess an ectoplasmic net (Perkins 1973 a). Reproduction is asexual by
laterally biflagellate zoospores formed by progressive cleavage.
Bipartitioning of sporangia or protoplasts and amoeboid stages may occur
in some species. *Althornia* Alderman & Jones (1971) lacks an ectoplasmic
net, *Aplanochytrium* Bahnweg & Sparrow (1972) reproduces by aplanospores.
The number of morphological and developmental characters used in
classification of thraustochytrids is small and was reviewed in detail
by Gaertner (1972).

Electron microscopic studies on Labyrinthulales and Thraustochytriales
revealed structures shedding more light on their relationship. Perkins
(1972, 1973 a) noted the principal identity of the fine structure of
organelle-free ectoplasmic nets and special net-forming structures,
sagenogenetosomes, in both thraustochytrids and labyrinthulids.
Sagenogenetosomes are thus far discovered in *Labyrinthula* (Perkins 1972,
1973 a), *Labyrinthuloides* (Perkins 1973 b), *Thraustochytrium*,
Schizochytrium (Perkins 1972), *Japonochytrium*, *Aplanochytrium* (Alderman
et al. 1974) and *Ulkenia* (Moss 1980). Unlike any fungal cell wall the
thraustochytrid wall was found to be multilamellate consisting of
compressed layers of Golgi-derived scales (Darley *et al.* 1973; Perkins
1973 a; Alderman *et al.* 1974). Perkins (1973 a) found a thin wall-like
structure in *Labyrinthula* composed of only a single layer. Moss (1983)
demonstrated this to be a single layer of scales. Other details of the
ultrastructure including zoospore structure and flagellation of
Labyrinthulales and Thraustochytriales were reviewed by Perkins (1974,
1976). Evaluation of the results of those studies led to the conclusion
that thraustochytrids and labyrinthulids should not be included in the
Protozoa, Oomycetes or any group of fungi (Porter 1974; Perkins 1974,
1976). This view was corroborated by results of cell wall composition

studies of some thraustochytrids and by comparison of ribosomal RNA
molecular weights of thraustochytrids and labyrinthulids (Porter 1974).

Physiological and biochemical characters have been employed in the
classification of organisms lacking significant morphological and
developmental features, e.g. yeasts and bacteria. Based on nutritional
characters of 27 isolates of thraustochytrids representing 16 species of
5 genera (Bahnweg 1979 a,b), similarity coefficients were calculated of
all pairs of isolates and subjected to cluster analysis. Results for a
product-moment-correlation coefficient and subsequent average linkage
cluster analysis (Schäffer 1974) are presented in Figure 12.1. Other
coefficients as well as cluster analysis with complete or single linkage
resulted in almost identical arrangements of isolates. Although
phenetic similarities displayed in Figure 12.1 do not necessarily
indicate relatedness they serve to illustrate enormous heterogeneity in
the thraustochytrids. Bahnweg (unpublished) studied 37 selected
nutritional characters of more than 200 isolates of thraustochytrids
including species studied previously (Bahnweg 1979 a,b) and additional
ones, e.g. *Schizochytrium minutum*, *Ulkenia profunda*, *Thraustochytrium
pachydermum*, *Labyrinthuloides* spp. and *Aplanochytrium* spp. The data
agreed well with those obtained previously (Bahnweg 1979 a,b)
corroborating the view that substrate spectra may be suitable for a more
complete characterization of thraustochytrids.

Composition of cell wall carbohydrates has been considered a useful tool
in the classification of fungi (Bartnicki-Garcia 1968). Darley *et al.*
(1973) found that cell walls of *Schizochytrium aggregatum* and
Thraustochytrium sp. were characterized by high protein and low
polysaccharide contents, the latter being composed predominantly of
galactose and xylose. Cell wall analysis of a large number of isolates
of thraustochytrids and *Dermocystidium* sp. Goldstein & Moriber (1966) by
methods reported recently (Ulken *et al.* 1985) revealed considerable
diversity in cell wall composition. Polysaccharides accounted for
approximately 30% of the cell wall dry weight. Values as low as 5-7%
and as high as 60% were observed in some species. Galactose was the
major sugar in wall carbohydrates of *Thraustochytrium*, *Schizochytrium*,
Japonochytrium and *Ulkenia*. In addition, considerable amounts of xylose
and rhamnose (*T. rossii*) were detected (Table 12.1). Some species of
Thraustochytrium and *Ulkenia* contained a hexose tentatively identified
as 3-0-methyl-galactose by electrophoretic mobility (Lindberg & Swan
1960). Gross composition of cell walls of *Labyrinthuloides* and
Aplanochytrium was similar to thraustochytrids except for that of the
major sugar of the cell wall polysaccharides which was fucose, a hexose
not detected in any of the other thraustochytrids (Table 12.1). Sugars
of wall polysaccharides of thraustochytrids are similar to wall and
capsular carbohydrates of marine algae, where galactose and xylose are
major constituents, while rhamnose, mannose and 3,6-anhydro-galactose,
traces of which were detected by us in *T. pachydermum*, *T. aureum* and
T. aggregatum, were minor components (Percival 1978).
Polysaccharide-sulphate esters are widely distributed in walls of marine
red, green and brown algae (Percival 1978; McCandless & Craigie 1979).
Chamberlain (1980) detected sulphate in cell walls of *Thraustochytrium*.

Figure 12.1 Cluster analysis (average linkage) of thraustochytrids
(origin of isolates in Bahnweg 1979 a) based on a similarity matrix
calculated with a product-moment correlation coefficient.

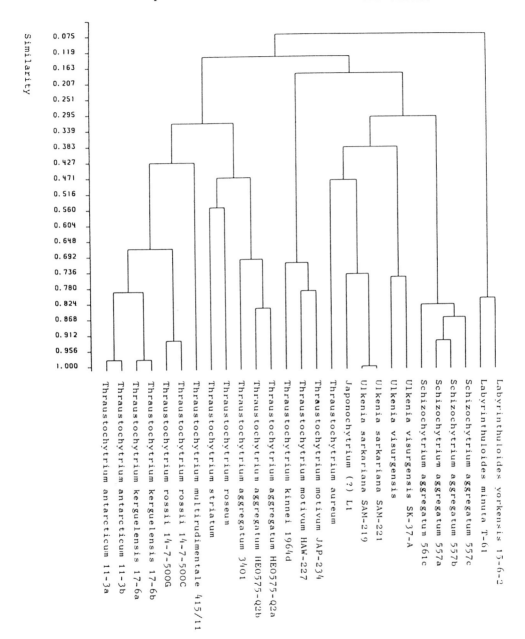

Table 12.1 Sugars in cell walls of thraustochytrids (% of polysaccharides) (Gal = galactose, Glu = glucose, Man = mannose, 3-0-M-Gal = 3-0-methyl-galactose, Hex = unidentified hexose, Xyl = xylose, Fuc = fucose, Rham = rhamnose)

Isolate[1]	Gal	Glu	Man	3-0-M-Gal	Hex	Xyl	Fuc	Rham	Sugar acid
T. antarcticum 11-3a	60-70	3-4	1-1.5	-	-	20-22	-	6	4-8
T. kinnei 1694d	70-90	2-4	4-8	-	-	8-10	-	2-4	9
T. motivum HAW-227	68-75	3	3-4	-	-	10-15	-	6	trace
T. motivum JAP-234	62-75	1.5-3	1.5-5	-	-	12-18	-	5	trace
T. aureum	38-47	9-12	-	14-18	-	20-25	-	-	7.5
T. kerguelensis 17-6a	45-60	5	2.5	-	-	30-35	-	5-10	5
T. multirudimentale 415/11	50-70	-	8-16	-	-	16-26	-	4	0.4
T. rossii 14-7-500G	28-36	1.2	-	-	-	20-32	-	32-36	6
T. striatum	60	-	-	10-12	-	19-26	-	-	2-5
T. roseum	50-65	-	-	10-15	-	20-25	-	-	1-2
T. pachydermum AII9	63-76	3-6	1-6	0.6	-	6-10	-	-	6
T. aggregatum HE0575-Q2a	70-90	2-6	-	-	-	5-10	-	5-10	1-2
S. aggregatum 557c	72-86	1-2	0.2	0.5	-	12	-	-	1-3
J. marinum (?) L1	60-72	2	2	-	-	20-30	-	-	1
U. visurgensis	61-77	1.5-3	0.3	10	-	10-15	-	-	0.5
U. visurgensis SK-37-A	74	1.5-2	0.5-2	2-7	-	15-22	-	-	0.3
U. sarkariana SAM-219	60-80	6	3	6	-	12-15	-	-	0.5
U. sarkariana SAM-221	65-80	5-8	-	5-8	-	16	-	-	0.5
L. minuta T-61	10-15	-	-	-	-	3-5	60-75	-	15-20
L. yorkensis 15-6-2	10-15	-	-	-	-	3.5	70-80	-	5
D. sp. LIS-37	12.5	6-12	12-18	-	12.5	6-9	6	12.5	56
D. sp. WH-49	11-22	14-17	14-22	-	10-12	5.8	5	11	17-20
A. sp. HE0379-DÜFKK1	17	-	trace	-	-	23	60	trace	6
A. sp. HE0379-JH11	29	-	trace	-	-	20	45	trace	-
A. sp. HE0379-NFW6	41	-	trace	-	-	9	50	-	18
A. sp. HE0379-NFW12	18	-	trace	-	-	8	65	-	20
A. sp. N5994	9	-	-	-	-	15	56	-	16
A. sp. N4367	14	4	-	-	-	8	59	-	12
A. sp. 2400/57	23	8	trace	-	-	14	44	-	12

[1] T. = Thraustochytrium, S. = Schizochytrium, J. = Japonochytrium, U. = Ulkenia, L. = Labyrinthuloides, D. = Dermocystidium Goldstein et Moriber, A. = Aplanochytrium.

Up to 10% of the cell wall dry matter of *Aplanochytrium* spp. studied by us was made up of sulphate. Sulphate was also found in cell walls of halobacteria (Steber & Schleifer 1975), however, and its significance may be ecologic rather than taxonomic.

DNA base composition (guanin + cytosin content or G+C value) is a useful

Table 12.2   DNA base composition of thraustochytrids

Isolate	Mol % G+C
Thraustochytrium kinnei Gaertner 1694d	48.2
T. kinnei 1709c	48.5
T. aureum Goldstein	61.3
T. aureum 3552b	58.3
T. multirudimentale Goldstein 415/11	53.4
T. multirudimentale (Kumar)	60.5
T. striatum Schneider	62.6
T. roseum Goldstein	63.4
T. aggregatum Ulken 3401	56.8
T. aggregatum HEO575-Q2a	56.1
S. aggregatum Goldstein et Belsky 561a	62.2
S. aggregatum 557c	61.7
Ulkenia visurgensis (Ulken) Gaertner	49.5
U. visurgensis (Schneider)	52.6
U. sarkariana Gaertner SAM-219	50.5
U. sarkariana SAM-221	50.8
U. profunda Gaertner (Gaertner)	48.2
Thraustochytrium antarcticum Bahnweg et Sparrow 11-3a	38.8
T. kerguelensis Bahnweg et Sparrow 17-6a	36.4
T. rossii Bahnweg et Sparrow 14-7-500G	34.8
T. pachydermum Scholz AII9	30.1
Labyrinthuloides yorkensis Perkins T-20	35.5
L. yorkensis 15-6-2	34.6
L. minuta (Watson et Raper) Perkins T-61	32.7
Aplanochytrium sp. HEO379-JH11	39.6
A. sp. HEO379-NFW6	40.6
A. sp. HEO379-NFW12	39.8
A. sp. N5994	39.3
A. sp. 2400-57	37.6
Dermocystidium sp. WH-49	42.3

Table 12.3 DNA/DNA hybridizations among thraustochytrids (low G+C group)

Isolates	11-3a	17-6a	*T. rossii*	WH-49	T-20	NFW12	N5994	2400/57
T. pachydermum AII9	53±6	61±12	52±8	32±7		40±15	22±6	42±20
T. antarcticum 11-3a	100	76±16	88±19	27±10	31±11	62±26	79±12	46±9
T. kerguelensis 17-6a		100	72±18	33±8		58±18		
T. rossii 14-7-500G			100	33±12				
Dermocystidium WH-49				100	44±26	45±5	13±11	5±2
L. yorkensis T-20					100	22±17		23±19
Aplanochytrium NFW12						100		5±16
Aplanochytrium N5994							100	66±12
Aplanochytrium 2400/57								100

Table 12.4 DNA/DNA hybridizations among thraustochytrids (high G+C group)

Isolates	Q2b	3401	561a	557c	*T. aureum*	*U. vis.*	*U. vis.*(S)	219	221
T. aggregatum Q2a	96±10	80±11	62±8	56±8	10±3				
T. aggregatum Q2b	100	86±15							
T. aggregatum 3401		100		50±5					
S. aggregatum 561a			100	92±12	6±10				
S. aggregatum 557c				100	15±7				
T. aureum					100	10±4	15±4	20±8	25±6
U. visurgensis (Schneider)						100	96±6	30±17	32±12
U. visurgensis							100	35±10	32±8
U. sarkariana SAM-219								100	92±12
U. sarkariana SAM-221									100

character in taxonomy. Eukaryotic protists (Protozoa, algae, fungi) display a diversity of G+C values almost comparable to the prokaryotes (Mandel 1970; Storck & Alexopoulos 1970; Shapiro 1970). G+C values of thraustochytrids are listed in Table 12.2. It is surprising to find two distinct groups separated by a difference in G+C value. The first includes species of *Thraustochytrium*, *Ulkenia* and *Schizochytrium* with G+C values between 48% and 63%, the second includes the antarctic thraustochytrids *Thraustochytrium rossii*, *T. kerguelensis* and *T. antarcticum*, *Labyrinthuloides*, *Aplanochytrium*, *T. pachydermum* and *Dermocystidium sensu* Goldstein & Moriber with G+C values between 30% and 42%.

DNA/DNA hybridization has been applied by us to a number of thraustochytrids. The method employed was based on spectrophotometric determination of initial renaturation rates (DeLey *et al.* 1970). Eukaryotic DNA contains organelle and repetitive DNA in addition to unique DNA. The procedure of DeLey *et al.* (1970) was therefore altered such that first Wetmur-Davidson plots were derived from DNA renaturation rates (Wetmur & Davidson 1968) to determine the time span after which most of the repetitive sequences had renatured. This was the case after 10-15 minutes and our studies were therefore based on comparison of renaturation rates between 15 and 25 minutes. Results are presented in Tables 12.3 and 12.4. Different isolates of the same species have homologies over 80% (*Thraustochytrium aggregatum*, *Schizochytrium aggregatum*, *Ulkenia visurgensis* and *U. sarkariana*). Antarctic thraustochytrids display homology values of 72% to 88% between different species (Table 12.3). Different isolates of *Aplanochytrium* and *Labyrinthuloides*, on the other hand, have low homology values indicating the presence in the oceans of a diverse flora of such organisms practically inseparable from one another on the basis of morphology. Only a few studies have applied the spectrophotometric DNA/DNA hybridization technique to the taxonomy of fungi. Martini & Phaff (1973) and Kurtzman (1984) determined homologies of yeasts. Jahnke (1984) studied homologies of Basidiomycotina and compared results of spectrophotometric and radioactive labelling (filter) techniques. The significance of homology values obtained by a spectrophotometric technique was tentatively estimated as follows: 40-60% = distantly related, 70-80% = closely related, 90-100% = very closely related or identical. DNA/DNA hybridization methods were found to be useful only at the species level. Construction of a phylogenetic tree requires the study of more conservative molecular structures. Sequences of 5 S ribosomal RNA of *Ulkenia visurgensis* and *Schizochytrium aggregatum* determined by MacKay & Doolittle (1982) were included by Huysmans *et al.* (1983) in the construction of a similarity phenogram. The dissimilarity between the thraustochytrids and any other group of eukaryotic protists was surprisingly high. Hori *et al.* (1985) used the available 5 S ribosomal RNA data to reconstruct a phylogenetic tree and found the thraustochytrids to be a very old group which evolved before the fungi, Protozoa, brown algae and green plants.

ACKNOWLEDGEMENTS
The authors thank A. Gaertner, A. Ulken, S.R. Kumar

(Bremerhaven), E.B.G. Jones (Portsmouth), J. Schneider (Kiel), D. Porter (Athens, Georgia) and F.O. Perkins (Gloucester Point, Virginia) for contributing cultures to this study, part of which was supported by the Deutsche Forschungsgemeinschaft.

REFERENCES

Alderman, D.J. (1974). Ultrastructure of stages of *Hyalochlorella marina* Poyton. Veröff. Inst. Meeresforsch. Bremerh., Suppl. 5, 251-261.

Alderman, D.J. & Jones, E.B.G. (1971). Physiological requirements of two marine phycomycetes, *Althornia crouchii* and *Ostracoblabe implexa*. Trans. Br. mycol. Soc., 57, 213-225.

Alderman, D.J., Harrison, J.L., Bremer, G.B. & Jones, E.B.G. (1974). Taxonomic revisions in the marine biflagellate fungi: the ultrastructural evidence. Mar. Biol., 25, 345-357.

Bahnweg, G. (1979 a). Studies on the physiology of Thraustochytriales. I. Growth requirements and nitrogen nutrition of *Thraustochytrium* spp., *Schizochytrium* sp., *Japonochytrium* sp., *Ulkenia* spp., and *Labyrinthuloides* spp. Veröff. Inst. Meeresforsch. Bremerh., 17, 245-268.

Bahnweg, G. (1979 b). Studies on the physiology of Thraustochytriales. II. Carbon nutrition of *Thraustochytrium* spp., *Schizochytrium* sp., *Japonochytrium* sp., *Ulkenia* spp., and *Labyrinthuloides* spp. Veröff. Inst. Meeresforsch. Bremerh., 17, 269-273.

Bahnweg, G. & Sparrow, F.K. (1972). *Aplanochytrium kerguelensis* gen. nov. spec. nov., a new phycomycete from subantarctic marine waters. Arch. Mikrobiol., 81, 45-49.

Bartnicki-Garcia, S. (1968). Cell wall chemistry, morphogenesis, and taxonomy of fungi. A. Rev. Microbiol., 22, 87-108.

Bonner, J.T. (1967). The cellular slime molds. Princeton: Princeton University Press.

Chamberlain, A.H.L. (1980). Cytochemical and ultrastructural studies on the cell walls of *Thraustochytrium* spp. Botanica mar., 23, 669-677.

Cienkowski, L. (1867). Über den Bau und die Entwicklung der Labyrinthuleen. Arch. mikrosk. Anat. EntwMech., 3, 274-310.

Darley, W.M., Porter, D. & Fuller, M.S. (1973). Cell wall composition and synthesis via Golgi-directed scale formation in the marine eucaryote, *Schizochytrium*, with a note on *Thraustochytrium* sp. Arch. Mikrobiol., 90, 89-106.

DeLey, J., Cattoir, H. & Reynaerts, A. (1970). The quantitative measurement of DNA hybridization from renaturation rates. Eur. J. Biochem., 12, 133-142.

Gaertner, A. (1972). Characters used in the classification of thraustochytriaceous fungi. Veröff. Inst. Meeresforsch. Bremerh., 13, 183-194.

Goldstein, S. & Moriber, L. (1966). Biology of a problematic marine fungus, *Dermocystidium* sp. I. Development and cytology. Arch. Mikrobiol., 53, 1-11.

Hollande, A. & Enjumet, M. (1955). Sur l'évolution et la systématique des Labyrinthulidae; étude de *Labyrinthula algeriensis* nov. sp. Annls. Sci. nat. Zool. Biol. Anim., 17, 357-368.

Honigberg, B.M., Balamuth, W., Bovee, E.C., Corliss, J.O., Gojdics, M., Hall, R.P., Kudo, R.R., Levine, N.D., Loeblich, A.R., Weiser, J. & Wenrich, D.H. (1964). A revised classification of the phylum protozoa. J. Protozool., 11, 7-20.

Hori, H., Lim, B.L. & Osawa, S. (1985). Evolution of green plants as deduced from 5 S rRNA sequences. Proc. natn. Acad. Sci. USA, 82, 820-823.

Huysmans, E., Dams, E., Vandenberghe, A. & DeWachter, R. (1983). The nucleotide sequences of the 5 S rRNAs of four mushrooms and their use in studying the phylogenetic position of basidiomycetes among the eukaryotes. Nucleic Acids Res., 11, 2871-2880.

Jahnke, K.D. (1984). Artabgrenzung durch DNA-Analyse bei einigen Vertretern der Strophariaceae (Basidiomycetes). Biblthca Mycol., 96. Vaduz: J. Cramer.

Kurtzman, C.P. (1984). DNA base sequence complementarity and the definition of fungal taxa. Microbiol. Sci., 1, 44-48.

Lindberg, B. & Swan, B. (1960). Paper electrophoresis of carbohydrates in germanate buffer. Acta Chem. Scand., 14, 1043-1050.

MacKay, R.M. & Doolittle, W.F. (1982). Two thraustochytrid 5 S ribosomal RNAs. Nucleic Acids Res., 10, 8307-8310.

Mandel, M. (1970). DNA base composition of eucaryotic protists. In Handbook of Biochemistry, eds. H.A. Sober, R.A. Harte & E.K. Sober, pp. H75-H79. Cleveland, Ohio: The Chemical Rubber Company.

Martini, A. & Phaff, H.J. (1973). The optical determination of DNA-DNA homologies in yeasts. Ann. Microbiol., 23, 59-68.

McCandless, E.L. & Craigie, J.S. (1979). Sulfated polysaccharides in red and brown algae. A. Rev. Pl. Physiol., 30, 41-53.

Moss, S.T. (1980). Ultrastructure of the endomembrane-sagenogenetosome-ectoplasmic net complex in *Ulkenia visurgensis* (Thraustochytriales). Botanica mar., 23, 73-94.

Moss, S.T. (1983). Ultrastructure of the Labyrinthulales and Thraustochytriales. Its taxonomic significance. Paper presented at the 5th Meeting Int. Soc. Evol. Protistol. Banyuls-Sur-Mer, France, June 6-9.

Olive, L.S. (1975). The Mycetozoans. New York: Academic Press.

Percival, E. (1978). Sulfated polysaccharides metabolized by the marine Chlorophyceae - a review. In Carbohydrate Sulfates, ed. R.G. Schweiger, pp. 203-212. Washington, D.C.: Amer. Chem. Soc.

Perkins, F.O. (1972). The ultrastructure of holdfasts, "rhizoids", and "slime tracks" in thraustochytriaceous fungi and *Labyrinthula* spp. Arch. Mikrobiol., 84, 95-118.

Perkins, F.O. (1973 a). Observations of thraustochytriaceous (Phycomycetes) and labyrinthulid (Rhizopodea) ectoplasmic nets on natural and artificial substrates. Can. J. Bot., 51, 485-491.

Perkins, F.O. (1973 b). A new species of marine labyrinthulid
    Labyrinthuloides yorkensis gen. nov. spec. nov. - cytology
    and fine structure. Arch. Mikrobiol., 90, 1-17.
Perkins, F.O. (1974). Phylogenetic considerations of the problematic
    thraustochytriaceous-labyrinthulid-*Dermocystidium* complex
    based on observations of fine structure. Veröff. Inst.
    Meeresforsch. Bremerh., Suppl. 5, 45-63.
Perkins, F.O. (1976). Fine structure of lower marine and estuarine
    fungi. In Recent Advances in Aquatic Mycology, ed.
    E.B. Gareth Jones, pp. 279-312. London: Elek Science.
Porter, D. (1974). Phylogenetic considerations of the
    Thraustochytriaceae and Labyrinthulaceae. Veröff. Inst.
    Meeresforsch. Bremerh., Suppl. 5, 19-44.
Schäffer, J.B. (1974). Beschreibung und Benutzeranleitung des
    Wishartschen Clusteranalyse-Pakets. GSF-Bericht MD 97.
    München: Ges. Strahlen- u. Umweltforschung mbH.
Shapiro, H.S. (1970). Distribution of purines and pyrimidines in
    deoxyribonucleic acids. In Handbook of Biochemistry, eds.
    H.A. Sober, R.A. Harte & E.K. Sober, pp. H80-H92.
    Cleveland, Ohio: The Chemical Rubber Company.
Sparrow, F.K. (1936). Biological observations on the marine fungi of
    Woods Hole waters. Biol. Bull., 70, 236-263.
Steber, J. & Schleifer, K.H. (1975). *Halococcus morrhuae*: a sulfated
    heteropolysaccharide as the structural component of the
    bacterial cell wall. Arch. Microbiol., 105, 173-177.
Storck, R. & Alexopoulos, J. (1970). Deoxyribonucleic acids of fungi.
    Bact. Rev., 34, 126-154.
Ulken, A., Jäckle, I. & Bahnweg, G. (1985). Morphology, nutrition and
    taxonomy of an *Aplanochytrium* sp. from the Sargasso Sea.
    Mar. Biol., 85, 89-95.
Wetmur, J.G. & Davidson, N. (1968). Kinetics of renaturation of DNA. J.
    Mol. Biol., 31, 349-370.
Zopf, W. (1892). Zur Kenntnis der Labyrinthuleen, einer Familie der
    Mycetozoen. Beitr. Physiol. Morphol. nied. Organ., 2, 36-48.

13 MYCOSES OF MARINE ORGANISMS: AN OVERVIEW OF PATHOGENIC
    FUNGI

D. Porter

INTRODUCTION

There exists a wide range of different marine fungi which
interact with marine algae, vascular plants, invertebrates and fish
(Johnson & Sparrow 1961; Jones 1976 a; Kohlmeyer & Kohlmeyer 1979;
Alderman 1982). These interactions may be classified as symbioses (in
the broad sense) and include commensalism, mutualism and parasitism.
The following discussion will be restricted mainly to parasitic marine
fungi.

Among the parasitic marine fungi many are known to be devastating
pathogens. However, even today, our knowledge of most of these
interactions, including those which result in severe epizootics and
epiphytotics, is limited. Information about pathogenic marine fungi is,
for the most part, in the form of descriptions of the disease syndromes
and/or morphological descriptions of the suspected fungal agents. In
some cases it has been difficult to demonstrate clearly whether an
implicated fungus is, in fact, the primary cause of a particular
disorder or merely a secondary invader. The relationship between
species of *Labyrinthula* and the wasting disease of eelgrass (*Zostera
marina* L.) is a prime example of the problems in determining the
causative agent for a particular disease syndrome (Pokorny 1967;
Rasmussen 1977).

Continued research in basic marine mycology is an essential resource for
the pathologist concerned with diseases of marine algae, plants and
animals, not only in their natural environment but also in the more
artificial habitats used for the controlled cultivation of marine
organisms. Mariculture has revealed serious fungal pathogens both of
algae and invertebrates. If left unchecked such pathogens can cause
significant losses of cultured organisms. An awareness of the potential
of certain marine fungi to become serious pathogens in crowded
mariculture operations is best achieved by identification and
understanding of the biology of possible pathogens in their natural
habitat. Thus the early observations of *Lagenidium callinectes* Couch,
an oomycete parasite of the eggs of the blue crab (Couch 1942;
Rogers-Talbert 1948), contributed important background information when
the fungus was subsequently found to be a widespread pathogen of the
eggs and larvae of cultured penaeid shrimps (Lightner & Fontaine 1973),
lobsters (Nilson *et al.* 1976) and Dungeness crabs (Armstrong *et al.*
1976). Another example of the disease potential of marine fungi is

found in Japan where species of the red alga *Porphyra* have been grown
for more than 200 years as a food stuff, nori, with present day annual
yields of several hundred thousand tonnes. In certain localities the
fungus pathogen *Pythium porphyrae* Takahashi et Sasaki, which causes the
red wasting disease of *Porphyra* spp., has been reported to destroy an
entire crop within 2-3 weeks (Andrews 1976).

Evolution of fungi in the marine environment has resulted in a wide
diversity of fungal morphologies, substrate utilization, and
associations with other organisms. The demonstration of this diversity
has been the backbone of marine mycology for decades and will
undoubtedly continue in the future. Parasitic or potentially parasitic
marine fungi are found in the Ascomycotina, Deuteromycotina, the
Oomycetes, Chytridiomycetes, Hyphochytridiomycetes and
Plasmodiophoromycetes of the Mastigomycotina, and recently even in the
thraustochytrids.

ASCOMYCOTINA
Among the more than 30 species of parasitic marine
Ascomycotina only a few are known to attack sea grasses or salt marsh
halophytes. One of these is *Buergenerula spartinae* Kohlmeyer et
Gessner, a weak parasite of the culms of *Spartina alterniflora* Loisel.
on which it produces no macroscopic symptoms (Kohlmeyer & Gessner 1976).
Even more unusual are ascomycete parasites of marine animals, of which
the best documented example is *Trichomaris invadens* Hibbits, Hughes et
Sparks, a parasite of the Alaskan tanner crab, *Chionoecetes bairdi*
Rathbun (Hibbits *et al.* 1981). *T. invadens* infects primarily the
epidermal and connective tissues of the crustacean (Sparks 1982) and
produces a mat of black ascocarps on the surface of the carapace and
legs (Porter 1982).

Most of the reported parasitic marine Ascomycotina are associated with
macroalgae (Kohlmeyer 1979), however all these appear to be highly
coevolved interrelationships; none is known to cause epidemics. A
broad range of morphological and taxonomic diversities are to be found
in the ascomycete parasites of algae. *Spathulospora* spp. infect single
cells of species of the southern hemisphere red, filamentous alga,
Ballia. They form small crustose thalli on the algal surface which
supports the ascocarps (Kohlmeyer & Kohlmeyer 1975). *Chadefaudia* spp.
infect species of various red algae on which they form immersed or
superficial ascocarps with little macroscopic morphological alteration
to the host plant (Kohlmeyer 1973; Stegenga & Kemperman 1984). Species
of *Haloguignardia* induce galls in the stipes of *Sargassum* spp., but with
no other apparent adverse effect on the brown algal host (Kohlmeyer
1971). Possibly the best studied ascomycete algal parasite association
is that of *Phycomelaina laminariae* (Rostrup) Kohlm. on *Laminaria
saccharina* (L.) Lam. and other species of *Laminaria*. Hyphae of
P. laminariae invade the meristoderm of the stipe and form blackened
stromatic patches in which spermogonia and ascocarps form (Schatz 1983).
P. laminariae appears to predispose the kelp stipes to invasion by
saprobic fungi, ultimately resulting in the breakage and loss of the

distal blades (Schatz 1984).

Several marine Ascomycotina form obligate associations with macroalgae. These associations appear to be mutualistic and have been termed mycophycobioses (Kohlmeyer & Kohlmeyer 1972). The brown algae, *Ascophyllum nodosum* (L.) LeJolis and *Pelvetia canaliculata* (L.) Dcne. et Thuret are uniformly associated with *Mycosphaerella ascophylli* Cotton (Webber 1967). The red algae, *Apophlaea lyallii* Hooker et Harvey and *A. sinclairii* Harvey, endemic to New Zealand, are similarly associated with *M. apophlaeae* Kohlm. (Kohlmeyer & Hawkes 1983). These fungi are systemic and grow intercellularly throughout the algae. The only macroscopic expressions of the fungi are their seasonal fruiting bodies. An atypical vitamin requirement of *M. ascophylli* for biotin suggests a close nutritional interrelationship with *A. nodosum* (Fries 1979).

The ascomycete *Turgidosculum ulvae* (Reed) Kohlm. et Kohlm. has only been reported as growing within the sac-like thallus of the green alga, *Blidingia minima* var. *vexata* Norris collected along the west coast of North America (Reed 1902; Schatz 1980). The fungus fills the central cavity of the alga and grows between the algal cells. Ascocarps and pycnidia are also readily formed within the alga. The association results in the plant becoming black, leathery and somewhat gnarled. Kohlmeyer (1974) suggested that this association is another case of mycophycobiosis, but it appears that the association is not obligate as there are also normal uninfected green plants of *B. minima* var. *vexata* in the field (Reed 1902; Porter unpublished). In fact the fungus is able periodically to penetrate and parasitize cells of the alga (Porter unpublished). Thus, *T. ulvae* should be regarded as a weak parasite of the host alga. In spite of this, the fungus may confer survival benefit to the alga. Intertidal invertebrate grazers, such as limpets, will consume the green, uninfected alga and leave the black, leathery, infected alga undisturbed (Cubit 1974). It would be of interest to determine whether the apparently unattractive nature of the infected alga is a result of its texture or the presence of toxic secondary metabolites produced by the fungus.

DEUTEROMYCOTINA
Few marine Deuteromycotina are known to have a parasitic mode of nutrition. Most are clearly saprobic and involved in the breakdown of wood and other dead plant and algal materials (Jones 1976 b). In fact, Kohlmeyer & Kohlmeyer (1979) listed only one parasitic deuteromycete species, *Spaceloma cecidii* Kohlm., and described it as a hyperparasite on the galls caused by *Haloguignardia* spp. in *Sargassum* and *Cystoseira*. *Cytospora rhizophorae* Kohlm. et Kohlm. is a pycnidial fungus which is parasitic on roots and seedlings of *Rhizophora* spp. (mangrove) and causes dieback of young plants (Kohlmeyer & Kohlmeyer 1979).

Although not strictly a marine fungus, *Fusarium solani* (Mart.) Sacc., is a highly successful opportunistic parasite of various marine crustaceans in culture situations where it has caused epidemic destruction of shrimp larvae and young adults (Lightner 1981; Hose *et al.* 1984). Other species of Deuteromycotina reported to be parasitic on marine animals are discussed in chapter 17 of this volume by Alderman and Polglase. *Ostracoblabe implexa* Bornet et Flahault, the causative agent of oyster shell disease (Alderman & Jones 1971), may be tentatively assigned to the Deuteromycotina; it has regularly septate hyphae but no known reproductive spores.

Blodgettia bornetii Wright is a chlamydospore forming deuteromycete which is only found growing in the cell walls of the green algae, *Cladophora fuliginosa* Kütz. and *Siphonocladus rigida* Howe (Kohlmeyer & Kohlmeyer 1979). Although the relationship is often described as a mutualistic one, recent observations (TeStrake & Aldrich 1984) suggest that different fungi may be associated with the two algae and that the association may not be strictly obligate. Further studies are needed to clarify these relationships.

MASTIGOMYCOTINA
Oomycetes
Of all the marine fungi, it is the Oomycetes that exhibit the greatest diversity of parasitic species. Adapted to an aquatic habitat by the production of zoospores, these fungi range in morphology from small, globose, monocentric, endobiotic parasites such as *Ectrogella perforans* Petersen which can cause epidemic infections of species of *Licmophora* and other benthic pennate diatoms (Sparrow 1960) to mycelial, eucarpic destructive pathogens such as *Pythium porphyrae* on species of *Porphyra* (Kazama 1979). The spectrum of host organisms parasitized by species of the Oomycetes includes diatoms, filamentous and foliose red, brown and green algae, nematodes, and the eggs, larvae and young adults of various marine crustaceans. A few are even hyperparasites on other marine fungi. In addition, species of Oomycetes have recently been implicated as the primary agents in an ulcerative mycosis in menhaden fish (Noga & Dykstra 1986).

Of the six marine species of *Olpidiopsis* (Lagenidiales), a genus of endobiotic, holocarpic, monocentric parasites, four are found in filamentous algae and two are hyperparasites of other marine fungi (Karling 1981). *Olpidiopsis andreei* (Lagerheim) Karling has a broad host range, having been described from green (*Acrisiphonia* spp.), brown (*Ectocarpus* spp.) and red (*Ceramium* sp.) algae (Sparrow 1969). *Olpidiopsis antithamnionis* Whittick et South, on the other hand, appears to be host specific in the red alga, *Antithamnion floccosum* (O.F. Müll.) Kleen (Whittick & South 1972). *Olpidiopsis globosa* Anastasiou et Churchland parasitizes the saprobic deuteromycete, *Papulaspora halima* Anastasiou, causing greatly enlarged swellings in the hyphal tips of the host.

The three marine species of *Lagenidium* are parasitic on the eggs and/or larvae of various crustaceans (Alderman 1982), or they are associated (saprobically?) with algal material (Fuller *et al.* 1964). *Lagenidium* has proven to be a serious disease in crustacean mariculture where extremely high mortalities can occur in just a few days of exposure to zoospores (Lightner 1981). Zoospore cysts on the surface of the eggs or larval cuticle germinate to produce a ramifying mycelium which rapidly replaces the host tissue and subsequently forms a new population of zoospores by the conversion of the fungal thallus into zoosporangia (Lightner 1981).

Lagenisma coscinodisci Drebes is a virulent, endobiotic parasite of species of the planktonic, centric diatoms, *Coscinodiscus* and *Palmeria*. *Lagenisma coscinodisci* is morphologically similar to species of *Lagenidium* and it has been the subject of a series of excellent developmental studies by Schnepf & coworkers in 1978 (cited in Karling 1981). This species has a particularly unusual form of sexual reproduction in which single celled oogonia and antheridia develop from encysted zoomeiospores (Schnepf *et al.* 1978). In natural populations *L. coscinodisci* is associated with the rapid decline of host diatom cell numbers (Gotelli 1971; Grahame 1976) and is thus one of the few marine fungi to be directly implicated in the control of phytoplankton blooms.

Petersenia, Sirolpidium and *Haliphthoros* are genera of predominantly marine species in the Lagenidiales that are characterized by their production of lobed or branched endobiotic thalli which tend to fragment at maturity into several sporangial units (Karling 1981). Only *Petersenia* has fresh water species. The recently described *Petersenia palmariae* van der Meer et Pueschel is a parasite of the foliar red alga, *Palmaria mollis* (Setchell et Gardner) van der Meer et Bird, a close relative of the commercially harvested dulse (van der Meer & Pueschel 1985). The fungus forms translucent blisters on the algal fronds and is able to cause the rapid demise of sporelings, particularly at temperatures below 15°C. A similar disease of the red alga *Chondrus crispus* Stackh., an important source of carrageenan, caused by *Petersenia pollagaster* (Petersen) Sparrow is described by Molina in chapter 15 of this volume. *Sirolpidium* has species which are parasites of filamentous algae, and one which parasitizes the larvae of clams and oysters (Vishniac 1955). With the report of a *Sirolpidium* sp. causing a serious disease in cultured larvae of penaeid shrimps (Lightner 1977), mycologists should be aware of this otherwise obscure marine oomycete genus. *Haliphthoros milfordensis* Vishniac forms a branched mycelium of relatively stout hyphae. It is a facultative, although virulent, parasite of the eggs, larvae and juvenile stages of various crustaceans, particularly in culture systems (Fisher *et al.* 1975; Hatai *et al.* 1980).

Littoral marine nematodes are parasitized and killed by several species of endobiotic lagenidiaceous fungi (Newell *et al.* 1977). These include *Myzocytium vermicola* (Zopf) Fischer emend. Newell, Cefalu et Fell, a *Lagenidium*-like fungus, *Haptoglossa heterospora* Drechsler and *Gonimochaete latitubus* Newell, Cefalu et Fell. The latter two species are described as lagenidioid fungi as they produce only aplanospores, but in a manner similar to their zoosporic relatives.

With few exceptions, the parasitic marine members of the Saprolegniales are holocarpic, endobiotic species but like other Saprolegniales, produce dimorphic zoospores. One exception is *Leptolegnia baltica* Höhnk et Vallin which caused a massive mortality of the marine copepod, *Eurytemora hirundoides* Nordquist, in the Baltic Sea. *L. baltica* produces a branched mycelium with extramatrical filamentous sporangia (Höhnk & Vallin 1953). Several endobiotic, holocarpic members of the Saprolegniales are parasites of marine algae. Morphologically, the simplest of these is the relatively common diatom parasite, *Ectrogella perforans*, mentioned above. The infection of the pennate diatom, *Licmophora hyalina* Agardh, and the development of the endobiotic sporangium has been described in detail (Raghu Kumar 1980 a,b). *Eurychasma dicksoni* (Wright) Magnus is another common algal parasite. It has been reported from more than 20 species of filamentous red and brown algae in north temperate waters (Jenneborg 1977). This endobiotic parasite enlarges and ruptures the infected host cell to form the mature sporangium within which the primary zoospores encyst. After the release of secondary zoospores, the sporangium contains a network of primary cyst walls (Sparrow 1969). *Eurychasma tumifaciens* (Magnus) Sparrow is a parasite of the nodal cells of species of the red alga *Ceramium* where it causes localized abnormal growth of the host (Sparrow 1936).

The genus *Atkinsiella*, with two marine species, *A. dubia* (Atkins) Vishniac and *A. hamanaensis* Bian et Egusa, is an unusual saprolegnialean genus with a thallus of very broad, bulbous, unsegmented lobes that may be 1-2 mm in diameter (Fuller *et al.* 1964). The protoplast collapses into the centre of these lobes to cleave into primary zoospores which then encyst within the thallus. Secondary zoospores are released through stout discharge tubes (Sparrow 1973). These fungi are endobiotic parasites of primarily the eggs of various marine crustaceans (Atkins 1954 b; Bian & Egusa 1980). *Atkinsiella* spp. do not, at present, appear to pose a serious disease threat to natural populations or cultured crustaceans because they seem only to be able to infect eggs detached from ovigerous females and held in sea water (Lightner 1981).

Leptolegniella marina (Atkins) Dick is a little known parasite of marine crustacean eggs (Atkins 1954 a; Johnson & Pinschmidt 1963). Its branched thallus appears to be holocarpic (Dick 1971).

The Peronosporales are mainly terrestrial plant parasites but with a relatively few marine species in *Pythium* and *Phytophthora*. Only the genus *Pythium* has parasitic marine species. *P. porphyrae*, which may be identical to *P. marinum* Sparrow (Kazama 1979), is a devastating pathogen of *Porphyra* spp., both in natural populations (Fuller *et al.* 1966) and in mariculture operations (Kazama 1979). The circular lesions caused by species of *Pythium* in *Porphyra* spp. are a result of the direct penetration of the algal cells by the hyphae (Kazama & Fuller 1970). A second algal disease caused by a species of *Pythium* has been reported. *P. undulatum* var. *littorale* Höhnk causes necrotic lesions in the blades of *Fucus distichus* L. in natural populations in British Columbia (Thompson 1981). The disease is not severe, however, as the alga is apparently able to produce a hypersensitive response which results in

the localized autolysis of algal cells surrounding the infected tissue.
An abscission zone of these algal cells causes the diseased tissue to
fall out of the alga (Thompson 1981). *Pythium thalassium* Atkins is a
possible parasite of crab eggs (Atkins 1955). However it has been
rejected from the genus (Plaats-Niterink 1981) and is more likely a
lagenideaceous fungus.

Chytridiomycetes

Relatively few species of Chytridiomycetes have been
described from the marine environment. Most of these are monocentric,
epibiotic chytrids from moribund filamentous algae and intertidal or
estuarine sediments baited with pollen (Sparrow 1960). There are a few
which are primary parasites, however, such as *Rozella marina* Sparrow
(Johnson & Sparrow 1961) which is a parasite of the relatively common
Chytridium polysiphoniae Cohn. *R. marina* invades the sporangium of
C. polysiphoniae, causing hypertrophy and then filling the host cell
with its own sporangium (Sparrow 1936). This parasite has since gone
unreported, probably because it is necessary to be familiar with the
host morphology and development in order to detect the presence of such
an endobiotic parasite of another chytrid.

Hyphochytridiomycetes

Several species of parasitic marine Hyphochytridiomycetes
have been reported. These include three species of *Anisolpidium* which
form endobiotic, monocentric thalli in cells of several species of
filamentous brown algae including species of *Ectocarpus*, *Pylaiella* and
Sphacellaria (Johnson & Sparrow 1961). *Hyphochytrium peniliae* Artemchuk
et Zelzinskaia, a polycentric hyphochytrid, was described as causing a
severe mycosis of the planktonic cladoceran, *Penilia avirostris* Dana by
growing throughout the host body (Artemchuk & Zelzinskaia 1969).
However, their illustrations and description leave some doubt as to the
inclusion of this fungus in the genus *Hyphochytrium*.

Plasmodiophoromycetes

The members of the Plasmodiophoromycetes are all
endoparasites of flowering plants, algae and oomycetous fungi. The
marine species are nearly all parasites of seagrasses and are known only
by their plasmodial and cyst stages (Karling 1968). The sori of cysts
occur within individual or chains of pea-sized galls induced in the host
tissue. Three marine species of *Plasmodiophora* have been described.
P. diplantherae (Ferd. et Winge) Cook, the most common species, produces
internodal galls on *Halodule* spp. (Hartog 1965). *P. bicaudata*
J. Feldmann produces cysts with polar bristles in internodal galls on
Zostera nana Roth. and *P. halophilae* Ferd. et Winge forms individual
galls on the petioles of *Halophila* spp. *Phagomyxa algarum* Karling is an
endoparasite of several filamentous brown algae (Karling 1944). It does
not induce host cell enlargement and cysts have not been observed, but
it has a trophic plasmodial phase which suggests affinities with the
Plasmodiophoromycetes.

THRAUSTOCHYTRIALES
        A consideration of the thraustochytrids stretches the limits
of our discussion of the diversity of parasitic marine fungi. Although
studied by mycologists, these zoosporic protista are probably more
closely related to chrysophycean algae (Perkins 1974; Porter 1974).
They are generally considered to be saprobic, however, thraustochytrids
have rarely been observed *in situ*. Because of their superficial
morphological similarity to the true chytrids and their common
occurrence in coastal waters, mycologists anticipated finding them as
epibiotic parasites of marine phytoplankton, but this habitat has been
reported only once, for *Schizochytrium* sp. (Gaertner 1979). Recently,
however, certain thraustochytrids have been reported as parasites of
octopus (Polglase 1980), nudibranch (McLean & Porter 1982) and squid
(Jones & O'Dor 1983). It is interesting that each of these cases
involves a shell-less mollusc. Are tissues of these invertebrates more
prone to thraustochytrid invasion or are there thraustochytrid parasites
of other organisms that have not yet been detected?

LABYRINTHULALES
        Like the thraustochytrids, the marine slime moulds in the
genus *Labyrinthula* have more historic than phylogenetic connection to
the true fungi. They are sometimes included with the thraustochytrids
in a single division (Olive 1975), but unlike the thraustochytrids,
Labyrinthula spp. have obvious parasitic capabilities. Colonies of
Labyrinthula spp. are able to digest bacteria, yeast, diatoms and other
microorganisms (Pokorny 1967; Porter 1972). *Labyrinthula* spp. has long
been associated with the wasting disease of *Zostera marina* L. (Renn
1936; Short *et al.* 1986) and other sea grasses (Armiger 1964).
Although *Labyrinthula* cells are able to penetrate the cell walls of the
sea grasses (Perkins 1973), it is still considered, by some, to be a
secondary invader (Rasmussen 1977).

CONCLUSIONS
        Parasitic marine fungi are found in diverse taxonomic
groups and in various associative relationships with their host
organisms. It is likely that with further exploration of the marine
environment by trained mycologists more fungal diseases of algae, sea
grasses and invertebrates will be revealed. In addition, as more marine
organisms are brought into controlled monoculture, fungal epidemics are
likely to become more frequent.

Future research in marine mycology should include a spectrum of
investigational areas including: 1, continued exploration of marine
fungal diversity; 2, clear determination of the fungal agents of
diseases of marine organisms; 3, elucidation of the structural and
chemical interactions between fungi and their hosts; 4, description of
the effects of pathogenesis at the population level; 5, determination of
effective control measures to combat fungal pathogens of marine
organisms.

REFERENCES

Alderman, D.J. (1982). Fungal diseases of aquatic animals. In Microbial Diseases of Fish, ed. R.J. Roberts, Society for General Microbiology, Special Publication, 9, 189-242. London: Academic Press.

Alderman, D.J. & Jones, E.B.G. (1971). Shell disease of oysters. Fishery Investigations, London, Ser. II, no. 8, 1-19.

Andrews, J.H. (1976). The pathology of marine algae. Biol. Rev., 51, 211-253.

Armiger, L.C. (1964). An occurrence of *Labyrinthula* in New Zealand *Zostera*. N.Z. Jl. Bot., 2, 3-9.

Armstrong, D.A., Buchanan, D.V. & Caldwell, R.S. (1976). A mycosis caused by *Lagenidium* sp. in laboratory reared larvae of the Dungeness crab, *Cancer magister*, and possible chemical treatments. J. Invert. Path., 28, 329-336.

Artemchuk, N.J. & Zelzinskaia, L.M. (1969). The marine fungus *Hyphochytrium peniliae* n. sp. infecting the zooplanktonic crustacean *Penilia avirostris* (Dana). Mik. Fitopat., 3, 356-359.

Atkins, D. (1954 a). Further notes on a marine member of the Saprolegniaceae, *Leptolegnia marina* n. sp. infecting certain invertebrates. J. mar. biol. Ass. U.K., 33, 613-625.

Atkins, D. (1954 b). A marine fungus *Plectospira dubia* n. sp. (Saprolegniaceae) infecting crustacean eggs and small Crustacea. J. mar. biol. Ass. U.K., 33, 721-732.

Atkins, D. (1955). *Pythium thalassium* n. sp. infecting the egg mass of the pea crab, *Pinnotheres pisum*. Trans. Br. mycol. Soc., 38, 31-46.

Bian, B.Z. & Egusa, S. (1980). *Atkinsiella hamanaensis* sp. nov. isolated from cultivated ova of the mangrove crab, *Scylla serrata* (Forsskål). J. Fish Dis., 3, 373-385.

Couch, J.N. (1942). A new fungus on crab eggs. J. Elisha Mitchell scient. Soc., 58, 158-162.

Cubit, J.D. (1974). Interactions of Seasonally Changing Physical Factors and Grazing Affecting High Intertidal Communities on a Rocky Shore. Ph.D. thesis, University of Oregon, U.S.A.

Dick, M.W. (1971). Leptolegniellaceae fam. nov. Trans. Br. mycol. Soc., 57, 417-425.

Fisher, W.S., Nilson, E.H. & Schleser, R.A. (1975). Effect of the fungus *Haliphthoros milfordensis* on the juvenile stages of the American lobster *Homarus americanus*. J. Invert. Path., 26, 41-45.

Fries, N. (1979). Physiological characteristics of *Mycosphaerella ascophylli*, a fungal endophyte of the marine brown alga *Ascophyllum nodosum*. Physiologia Pl., 45, 117-121.

Fuller, M.S., Fowles, B.E. & McLaughlin, D.J. (1964). Isolation and pure culture study of marine phycomycetes. Mycologia, 56, 745-756.

Fuller, M.S., Lewis, B. & Cook, P. (1966). Occurrence of *Pythium* sp. on the marine alga *Porphyra*. Mycologia, 58, 313-318.

Gaertner, A. (1979). Some fungal parasites found in the diatom populations of the Rosfjord area (south Norway) during March 1979. Veröff. Inst. Meeresforsch. Bremerh., 18, 29-33.

Gotelli, D. (1971). *Lagenisma coscinodisci*, a parasite of the marine
    diatom *Coscinodiscus*, occurring in the Puget Sound,
    Washington. Mycologia, 63, 171-174.
Grahame, E.S. (1976). The occurrence of *Lagenisma coscinodisci* in
    Palmeria hardmaniana from Kingston harbour, Jamaica. Br.
    phycol. J., 11, 57-61.
Hartog, C. den, (1965). Some notes on the distribution of *Plasmodiophora
    diplantherae*, a parasitic fungus on species of *Halodule*.
    Persoonia, 4, 15-18.
Hatai, K., Bian, B.Z., Baticados, C.A. & Egusa, S. (1980). Studies on
    the fungal diseases in crustaceans. II. *Haliphthoros
    philippinensis* sp. nov. isolated from cultivated larvae of
    the jumbo tiger prawn (*Penaeus monodon*). Trans. mycol. Soc.
    Jap., 21, 47-55.
Hibbits, J., Hughes, G.C. & Sparks, A.K. (1981). *Trichomaris invadens*
    gen. et sp. nov., an ascomycete parasite of the tanner crab
    Chionoecetes bairdi Rathbun (Crustacea, Brachyura). Can. J.
    Bot., 59, 2121-2128.
Höhnk, W. & Vallin, S. (1953). Epidemisches absterben von *Eurytemora* in
    Bothnischen Meerbusen, verursacht durch *Leptolegnia baltica*
    nov. spec. Veröff. Inst. Meeresforsch. Bremerh., 2, 215-223.
Hose, J.E., Lightner, D.V., Redman, R.M. & Danald, D.A. (1984).
    Observations on the pathogenesis of the imperfect fungus,
    Fusarium solani, in the Californian brown shrimp, *Penaeus
    californiensis*. J. Invert. Path., 44, 292-303.
Jenneborg, L.-H. (1977). *Eurychasma*-infection of marine algae. Changes
    in algal morphology and taxonomic consequenses. Botanica
    mar., 20, 499-507.
Johnson, T.W. Jr. & Pinschmidt, W.C. Jr. (1963). *Leptolegnia marina*
    Atkins in blue crab ova. Nova Hedwigia, 5, 413-418.
Johnson, T.W. Jr. & Sparrow, F.K. Jr. (1961). Fungi in Oceans and
    Estuaries. Weinheim: J. Cramer.
Jones, E.B.G. (ed). (1976 a). Recent Advances in Aquatic Mycology.
    London: Elek Science.
Jones, E.B.G. (1976 b). Lignicolous and algicolous fungi. In Recent
    Advances in Aquatic Mycology, ed. E.B. Gareth Jones,
    pp. 1-49. London: Elek Science.
Jones, G.M. & O'Dor, R.K. (1983). Ultrastructural observations on a
    thraustochytrid fungus parasitic in the gills of squid
    (*Illex illecebrosus* Le Sueur). J. Parasit., 69, 903-911.
Karling, J.S. (1944). *Phagomyxa algarum* n. g., n. sp. an unusual
    parasite with plasmodiophoalean and proteomyxan
    characteristics. Am. J. Bot., 31, 38-52.
Karling, J.S. (1968). The Plasmodiophorales, 2nd ed. Vaduz: J. Cramer.
Karling, J.S. (1981). Predominantly Holocarpic and Eucarpic Simple
    Biflagellate Phycomycetes. Vaduz: J. Cramer.
Kazama, F.Y. (1979). *Pythium* 'red rot disease' of *Porphyra*. Experientia,
    35, 443-444.
Kazama, F.Y. & Fuller, M.S. (1970). Ultrastructure of *Porphyra perforata*
    infected with *Pythium marinum*, a marine fungus. Can. J.
    Bot., 48, 2103-2107.
Kohlmeyer, J. (1971). Fungi from the Sargasso Sea. Mar. Biol., 8,
    344-350.

Kohlmeyer, J. (1973). Fungi from marine algae. Botanica mar., 16, 201–215.

Kohlmeyer, J. (1974). Higher fungi as parasites and symbionts of algae. Veröff. Inst. Meeresforsch. Bremerh., Suppl. 5, 339–356.

Kohlmeyer, J. (1979). Marine fungal pathogens among Ascomycetes and Deuteromycetes. Experientia, 35, 437–439.

Kohlmeyer, J. & Gessner, R.V. (1976). *Buergenerula spartinae* sp. nov., an ascomycete from salt marsh cordgrass, *Spartina alterniflora*. Can. J. Bot., 54, 1759–1766.

Kohlmeyer, J. & Hawkes, M.W. (1983). A suspected case of mycophycobiosis between *Mycosphaerella apophlaeae* (Ascomycetes) and *Apophlaea* spp. (Rhodophyta). J. Phycol., 19, 257–260.

Kohlmeyer, J. & Kohlmeyer, E. (1972). Is *Ascophyllum nodosum* lichenized? Botanica mar., 15, 109–112.

Kohlmeyer, J. & Kohlmeyer, E. (1975). Biology and geographical distribution of *Spathulospora* species. Mycologia, 67, 629–637.

Kohlmeyer, J. & Kohlmeyer, E. (1979). Marine Mycology: The Higher Fungi. New York: Academic Press.

Lightner, D.V. (1977). Larval mycosis of shrimps. In Disease Diagnosis and Control in North American Marine Aquaculture, ed. C.J. Sindermann, pp. 36–41. New York: Elsevier.

Lightner, D.V. (1981). Fungal diseases of marine Crustacea. In Pathogenesis of Invertebrate Microbial Diseases, ed. E.W. Davidson, pp. 451–484. Totowa, N.J.: Allanheld, Osmun.

Lightner, D.V. & Fontaine, C.T. (1973). A new fungal disease of the white shrimp *Penaeus setiferus*. J. Invert. Path., 22, 94–99.

McLean, N. & Porter, D. (1982). The yellow spot disease of *Tritonia diomedea* Bergh (Mollusca, Gastropoda, Nudibranchia); encapsulation of the thraustochytriaceous parasite by host amoebocytes. J. Parasit., 68, 243–252.

Newell, S.Y., Cefalu, R. & Fell, J.W. (1977). *Myzocytium*, *Haptoglossa*, and *Gonimochaete* (Fungi) in littoral marine nematodes. Bull. Mar. Sci., 27, 177–207.

Nilson, E.H., Fisher, W.S. & Shleser, R.A. (1976). A new mycosis of larval lobster (*Homarus americanus*). J. Invert. Path., 27, 177–183.

Noga, E.J. & Dykstra, M.J. (1986). Oomycete fungi associated with ulcerative mycosis in menhaden, *Brevoortia tyrannus* (Latrobe). J. Fish Dis., 9, 47–53.

Olive, L.S. (1975). The Mycetozoans. New York: Academic Press.

Perkins, F.O. (1973). Observations of thraustochytriaceous (Phycomycetes) and labyrinthulid (Rhizopodea) ectoplasmic nets on natural and artificial substrates – an electron microscope study. Can. J. Bot., 51, 485–491.

Perkins, F.O. (1974). Phylogenetic considerations of the problematic thraustochytriaceous-labyrinthulid-*Dermocystidium* complex based on observations of fine structure. Veröff. Inst. Meeresforsch. Bremerh., Suppl. 5, 45–63.

Plaats-Niterink, A.J. van der (1981). Monograph of the Genus *Pythium*. Studies in Mycology, 21. Baarn: Centraalbureau voor Schimmelcultures.

Pokorny, K.S. (1967). *Labyrinthula*. J. Protozool., 14, 697–708.

Polglase, J.L. (1980). A preliminary report on the thraustochytrid(s) and labyrinthulid(s) associated with a pathological condition in the lesser octopus *Eledone cirrhosa*. Botanica mar., 23, 699-706.

Porter, D. (1972). Cell division in the marine slime mold, *Labyrinthula* sp. and the role of the bothrosome in extracellular membrane production. Protoplasma, 74, 427-448.

Porter, D. (1974). Phylogenetic considerations of the Thraustochytriaceae and Labyrinthulaceae. Veröff. Inst. Meeresforsch. Bremerh., Suppl. 5, 19-44.

Porter, D. (1982). The appendaged ascospores of *Trichomaris invadens* (Halosphaeriaceae), a marine ascomycete parasite of the tanner crab, *Chionoecetes bairdi*. Mycologia, 74, 361-373.

Raghu Kumar, C. (1980 a). An ultrastructural study of the marine diatom *Licmophora hyalina* and its parasite *Ectrogella perforans*. I. Infection of host cells. Can. J. Bot., 58, 1280-1290.

Raghu Kumar, C. (1980 b). An ultrastructural study of the marine diatom *Licmophora hyalina* and its parasite *Ectrogella perforans*. II. Development of the fungus in its host. Can. J. Bot., 58, 2557-2574.

Rasmussen, E. (1977). The wasting disease of eelgrass (*Zostera marina*) and its effects on environmental factors and fauna. In Seagrass Ecosystems: a Scientific Perspective, ed. C.P. McRoy & C. Helfferich, pp. 1-51. New York: Marcel Dekker.

Reed, M. (1902). Two new ascomycetous fungi parasitic on marine algae. Univ. Calif. Publ. Bot., 1, 141-164.

Renn, C.E. (1936). The wasting disease of *Zostera marina*. I. A phytological investigation of the diseased plant. Biol. Bull., 70, 148-158.

Rogers-Talbert, R. (1948). The fungus *Lagenidium callinectes* Couch (1942) on eggs of the blue crab in Chesapeake Bay. Biol. Bull. mar. biol. Lab., Woods Hole, 95, 214-228.

Schatz, S. (1980). Taxonomic revision of two pyrenomycetes associated with littoral-marine green algae. Mycologia, 72, 110-117.

Schatz, S. (1983). The developmental morphology and life history of *Phycomelaina laminariae*. Mycologia, 75, 762-772.

Schatz, S. (1984). Degradation of *Laminaria saccharina* by saprobic fungi. Mycologia, 76, 426-432.

Schnepf, E., Deichgräber, G. & Drebes, G. (1978). Development and ultrastructure of the marine, parasitic oomycete, *Lagenisma coscinodisci* (Lagenidiales): sexual reproduction. Can. J. Bot., 56, 1315-1325.

Short, F.T., Mathieson, A.C. & Nelson, J.I. (1986). Recurrence of the eelgrass wasting disease at the border of New Hampshire and Maine, U.S.A. Mar. Ecol. Prog. Ser., in press.

Sparks, A.K. (1982). Observations on the histopathology and probable progression of the disease caused by *Trichomaris invadens*, an invasive ascomycete, in the tanner crab, *Chionoecetes bairdi*. J. Invert. Path., 40, 242-254.

Sparrow, F.K. (1936). Biological observations on the marine fungi of Woods Hole waters. Biol. Bull. mar. biol. Lab., Woods Hole, 70, 236-263.

Sparrow, F.K. (1960). Aquatic Phycomycetes, 2nd ed. Ann Arbor, Michigan: University of Michigan Press.

Sparrow, F.K. (1969). Zoosporic marine fungi from the Pacific Northwest (U.S.A.). Arch. Mikrobiol., *66*, 129–146.

Sparrow, F.K. (1973). The peculiar marine phycomycete *Atkinsiella dubia* from crab eggs. Arch. Mikrobiol., *93*, 137–144.

Stegenga, H. & Kemperman, Th.C.M. (1984). A note on the genus *Chadefaudia* G. Feldmann (Ascomycetes) in Southern Africa, with the description of a new species. Botanica mar., *27*, 443–447.

TeStrake, D. & Aldrich, H.C. (1984). Ultrastructure of two associations involving marine fungi and green algae. Botanica mar., *27*, 515–519.

Thompson, T.A. (1981). Some Aspects on the Taxonomy, Ecology and Histology of *Pythium* Species Associated with *Fucus distichus* in Estuaries and Marine Habitats of British Columbia. M.S. thesis, University of British Columbia, Canada.

van der Meer, J.P. & Pueschel, C.M. (1985). *Petersenia palmariae* n. sp. (Oomycetes): a pathogenic parasite of the red alga *Palmaria mollis* (Rhodophyceae). Can. J. Bot., *63*, 404–408.

Vishniac, H.S. (1955). The morphology and nutrition of a new species of *Sirolpidium*. Mycologia, *47*, 633–645.

Webber, F.C. (1967). Observations on the structure, life history and biology of *Mycosphaerella ascophylli*. Trans. Br. mycol. Soc., *50*, 583–601.

Whittick, A. & South, G.R. (1972). *Olpidiopsis antithamnionis* n. sp. (Oomycetes, Olpidiopsidaceae), a parasite of *Antithamnion floccosum* (O.F. Müll.) Kleen from Newfoundland. Arch. Mikrobiol., *82*, 353–360.

14 ASPECTS OF THE PROGRESS OF MYCOTIC INFECTIONS IN MARINE ANIMALS

J.L. Polglase

D.J. Alderman

R.H. Richards

The wide variety of fungi responsible for diseases of animals in aquatic environments has been the subject of several recent reviews, arranged either by the host taxon or tissues affected (Lauckner 1980; Neish & Hughes 1980; Lightner 1981; Johnson 1983) or by the taxonomy of the fungal pathogen (Alderman 1982). With this volume of literature already extant, a further review may seem redundant. However, none of these reviewers considered the way the fundamental physiological, immunological and behavioural differences in the animal groups affected interact with different fungal groups. These differences produce the final patterns of the distribution and development of disease and form the subject of this chapter.

Detailed study of aquatic environments by air breathing investigators presents problems, particularly when the extent of the marine environment is considered. Thus, our knowledge of marine mycoses is often very limited. This is exacerbated by the specialized requirements of some pathogenic fungi which can make their isolation and culture difficult, and by the limited number of competent investigators, which has frequently resulted in inadequate characterization of fungal parasites. Space limitations oblige the selection of only the best characterized host/fungus interactions for discussion here. The reviews cited, together with Reichenbach-Klinke & Elkan (1965 a,b) and Migaki & Jones (1983) may be consulted for additional information.

This review considers fungi infecting marine animals. However, marine animals may be divided into at least three major groups based upon physiological variations. These are: (i) the invertebrates, where fine osmotic regulation is generally at a cellular, rather than a whole body, level (Robertson 1964; Mantel & Farmer 1983), with the result that the osmotic and ionic balance of the tissue fluids is essentially that of sea water; (ii) the fish, where the whole body ionic and osmotic regulation produces an internal milieu more closely approximating to fresh than salt water (Hoar & Randall 1969); (iii) marine mammals, where homeothermy and further sophistications of the homeostatic mechanisms, render the animal's internal milieu yet more diverse from sea water. Marine reptiles could also be considered as a separate group under these criteria, but there is, as yet, virtually no information on mycoses

within this taxon (Reichenbach-Klinke & Elkan 1965 b; Glazebrook 1983).
On this basis it could be considered that only those fungi growing
systemically in marine invertebrates or forming superficial infections
in the other groups are truly growing in a marine environment. Two
major fungal groups fall within these limits, the Thraustochytriales and
the marine Oomycetes.

Although a number of authors (Alderman & Gras 1969; Franc & Arvy 1970;
Quick 1972) have implicated thraustochytrid or thraustochytrid-like
organisms as pathogens/saprophytes in marine molluscs, no definite
evidence has been available until recently. In contrast, members of the
group were well known as plant pathogens (Johnson & Sparrow 1961;
Gaertner 1979). Polglase (1980) reported a fatal ulcerative dermal
necrosis in the octopus, *Eledone cirrhosa* (Lamarck), with which both
thraustochytrids and the related labyrinthulids were associated.
Subsequent work (Polglase 1981) indicated that the pathogen is *Ulkenia
amoeboidea* Gaertner (Figure 14.1a). Simultaneously, Jones (1981) and
Jones & O'Dor (1983) found thraustochytrids in gill lesions of the squid
Illex illecebrosus LeSueur. Subsequently McLean & Porter (1982)
reported that the disease of the nudibranch, *Tritonia diomedea* Bergh,
known as Yellow Spot, was due to thraustochytrids growing within the
lesions. No identification of the fungus was made in either case. A
sharp contrast emerges, however, between the response of the nudibranch
and the cephalopods to the fungal invasion. *T. diomedea* encapsulates
the fungal cells, causing necrosis and McLean & Porter (1982)
speculated that few viable spores can result. The infection would
therefore seem to be controlled and may well be eliminated. The
thraustochytrids in squid produced only a minimal response; while in
octopus, only partial encapsulation was achieved at best. Octopus of
all ages and stages of maturity are affected both in the wild and
captivity. Thus, a true difference in response level exists, with the
infection in the nudibranch representing a more "balanced" relationship
between host and fungus.

Polglase (1981) has also reported cases of ulcerative dermal lesions in
caged rainbow trout (*Salmo gairdneri* Richardson), in which
thraustochytrids were implicated. The trout were in an artificially
stressed situation, as part of an experimental regime (Brown 1983) and
the thraustochytrids were eliminated from the lesions as they healed.
Thraustochytrids are only known from marine and brackish environments,
so, it seems possible that the osmotically dilute internal milieu of a
fish could be beyond their range, leading to their easy elimination, in
contrast to invertebrate infections. Further work on this group is
clearly needed, if their significance as parasites, of the Mollusca in
particular, but also of other animals and marine plants, is to be
understood.

The Oomycetes, ubiquitous in fresh and sea water, produce some of the
most serious fungal infections of aquatic invertebrates. In the marine
environment, this is particularly evident in the eggs and developing
larvae of crustaceans. Unestam (1973), Lightner (1981) and Alderman
(1982) provide detailed reviews. In the wild, Oomycetes such as

Lagenidium callinectes Couch are commonly found infecting the periphery of the egg masses of the blue crab, *Callinectes sapidus* Rathbun, of the north-west Atlantic. Sindermann (1970) estimated that up to 50% of the eggs in an egg mass were killed. Penetration of the eggs by *L. callinectes* is an active process to which there is no response and which swiftly results in total destruction (Couch 1942; Rogers-Talbot 1948). In aquaculture systems, *Lagenidium* will kill not only eggs but larvae of a variety of decapods, producing mortalities of up to 100% (Lightner 1981). Larvae appear to be able to make only a minimal response to invasion and once a hypha has penetrated the exoskeleton, the fungal mycelium gradually invades and replaces all tissues (Figure 14.1b), particularly striated muscle (Lightner & Fontaine 1973). Small adult crustaceans, such as copepods, may also be affected in a similar fashion after population blooms (Höhnk & Vallin 1953). In juvenile and adult crustaceans, in contrast to the situation in eggs, an inflammatory response is normally mounted against the oomycete infection and the enzyme cascade leading to melanin production is triggered (Soderhall & Smith 1983). However, this is generally insufficient to stop the infection becoming systemic. Death may result either from tissue destruction or from inability to moult, because mycelial growth, inflammation and melanization have anchored the new shell to the old (Delves-Broughton & Poupard 1976; Fisher *et al.* 1978).

Oomycetes found in fresh water are of different genera to those which occur in sea water, since salt tolerances generally limit the distribution of members of this group to one aquatic medium or the other (Harrison & Jones 1975). The Oomycetes infecting fish, of which *Saprolegnia* (Pickering & Willoughby 1982) is the prime example, belong to the "fresh water" genera, but provide an interesting comparison. In adult fish, damage to outer layers of the integument is necessary before an infection can become established (Neish & Hughes 1980). The infection generally remains superficial whatever its severity, and even in adult fish, there is virtually no inflammatory response unless secondary bacterial pathogens are present. Death is generally due to osmotic imbalance owing to leakage of salts through the lesions (Richards & Pickering 1978). Systemic infections are reported in young fish. Although some fresh water Oomycetes can survive in brackish water, their ability to reproduce is severely impaired (Harrison & Jones 1975) and such infections of fish as have been reported from the sea are generally on anadromous or catadromous species. An atypical case (Noga & Dykstra 1986) recently reported is of estuarine menhaden, *Brevoortia tyrannus* [Latrobe]. These had concurrent infections with *Saprolegnia* and *Aphanomyces*, of which *Aphanomyces* is believed to be the pathogen (Dykstra pers. comm.), which penetrated systemically from superficial lesions. The authors believed that the fungal infections were secondary to another trauma. It is of interest to speculate that the unfavourable environment of the estuary, combined with the weakened resistance of the traumatized fish, converted a superficial secondary infection into a lethal systemic invasion.

Oomycetes are true water moulds, in contrast to the Ascomycotina. However, one member of this group, the pyrenomycete, *Trichomaris*

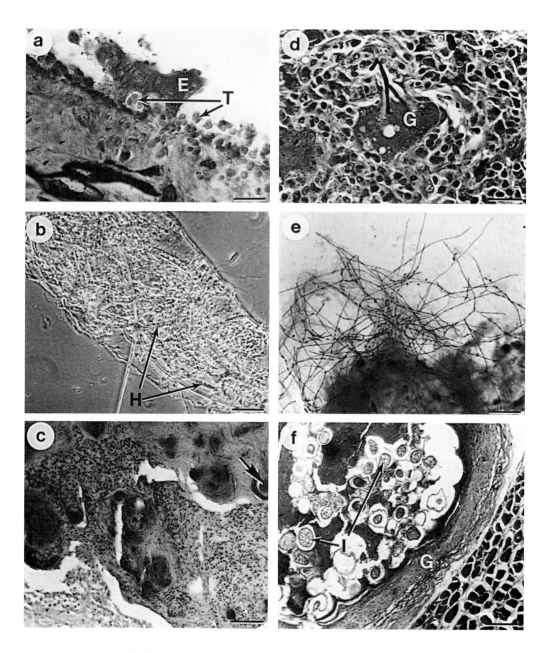

Figure 14.1a *Eledone cirrhosa* epidermis (E) attacked and invaded by
Ulkenia amoeboidea cells (T). No host reaction. Bar = 75 μm.
Figure 14.1b Juvenile lobster *Homarus gammarus* gill invaded by
Lagenidium sp. hyphae (H). No host reaction. Bar = 250 μm.
Figure 14.1c Adult lobster wound infection by *Fusarium solani*. Marked
host encapsulation response (arrowed). Bar = 350 μm.

invadens Hibbits, Hughes et Sparks, has been shown to be responsible for the disease known as black mat syndrome in the Tanner Crab, *Chionoecetes bairdi* Rathbun, and related species (Sparks & Hibbits 1979; Hibbits *et al.* 1981). Black mat syndrome takes its name from the highly pigmented encrustations on the crab carapace, composed of hyphae and fruiting bodies. Until the investigation by Sparks and co-workers, it was thought that this was the entire extent of the infection, which was not therefore considered to be of major significance. However, the use of specific fungal strains enabled Sparks & Hibbits (1979) to show that hyaline hyphae penetrated the exoskeleton and ramified within the epidermis. In more advanced infections penetration and destruction of other tissues was common. The preferred site of growth was connective tissue. This is a common factor in many mycelial fungal infections and appears to reflect preferential growth of hyphae along lines of physical weakness (Sparks & Hibbits 1979; Richards *et al.* 1978). Mix & Sparks (1980) noted raised "eosinophilic granulocyte" levels in the circulating haemolymph but record no other response to the fungus. It is unlikely that crabs infected with *T. invadens* would be able to moult, but the degree of penetration and proliferation of the fungus in many cases indicate severe disease. *T. invadens* has not been grown in culture, which indicates that it is likely to be one of the very few obligate fungal parasites of aquatic animals.

Species of Deuteromycotina (Fungi Imperfecti) are the higher fungi which are most frequently encountered as marine pathogens and an interesting contrast may be made between the two form genera of most significance, *Fusarium* and *Exophiala*. *Fusarium* has long been known as a plant pathogen, but its significance as a crustacean parasite is increasingly being recognized. In both fresh water and marine crustaceans, *Fusarium* infections take the form of a chronic, localized erosion of the exoskeleton, often accompanied by penetration of encapsulated and melanized mycelium deep into underlying tissues (Figure 14.1c). Details of the host response to *Fusarium* infections have been described by Hose *et al.* (1984). Death may be due to a number of factors such as failure of moult, secondary infections or possibly the production of toxins (Alderman 1982). Evidence that a wound is needed for infection to occur is provided by both Alderman (1982) and Hose *et al.* (1984) who found, in different species, that only the introduction of mycelium or conidia, in relatively small numbers to artificial wounds, produced infections which resembled those occurring naturally. Injection of spores either failed to produce an infection or killed the animal if too many were used. Thus, *Fusarium* infections of Crustacea are primarily superficial, only penetrating deeply in advanced cases, as with oomycete infections of

Figure 14.1d Rainbow trout kidney invaded by *Exophiala* sp. with host giant cell response (G). Bar = 125 µm.
Figure 14.1e *Cladosporium* invading wound in dermis of *Eledone cirrhosa*. Bar = 250 µm.
Figure 14.1f *Ichthyophonus* infection of herring. Marked host granulomatous (G) response surrounding many fungal cells (I). Bar = 250 µm.

fish, but unlike the Oomycetes, they engender a major response from the host. Like *Fusarium*, *Exophiala* infections, so far only described from fish, are found in both fresh and salt water. Again *Exophiala* triggers the immune system of the host, producing classic granulomata formation (Richards *et al.* 1978), but, as with *Fusarium*, this also appears only to limit and not to eradicate the fungus (Figure 14.1d). Unlike *Fusarium*, however, infections are generally deep seated from the start, with the posterior kidney as the most common site of infection. The route by which the fungus enters the host is unknown, but Richards *et al.* (1978) have speculated on either retrograde entry through the ureter or via the swim bladder or gut. Spread of the fungus is again primarily along anatomical lines of least resistance and death is generally due to the involvement, via the blood system, of other organs.

Both *Fusarium* and *Exophiala* also exist in damp soils and grow commonly on many other substrates. Thus, these serious marine pathogens are not necessarily confined to aquatic environments. The dematiaceous hyphomycete, *Cladosporium sphaerospermum* Penzig (Figure 14.1e) which has recently been isolated from an octopus (Polglase *et al.* 1984) provides a more extreme example of the ability of some fungi to use aquatic animals as merely one of many substrates for growth. *Cladosporium* is primarily terrestrial as are many of the fungi reported as infecting marine mammals (Migaki & Jones 1983). However, the majority of recorded mycoses in marine mammals are from captive individuals, which cannot be considered to be living normally. Under these circumstances, the significance of a given fungal infection is difficult to assess. The exception to this is lobomycosis, which has now been recorded repeatedly from wild dolphin species. The disease is caused by an incompletely characterized dimorphic fungus, *Loboa loboi* (Fonseca et Leão) Aj. *et al.*, which has so far only been isolated from dolphins and from man in Central America (Migaki & Jones 1983). In both man and dolphins, lobomycosis is a chronic indolent disease of the dermis, which occasionally involves the lymph nodes (Chandler *et al.* 1980). In the cetaceans, those parts of the body most frequently lifted out of the water are most commonly affected and infection is believed to follow skin abrasion (Migaki & Jones 1983). The distribution of lesions in dolphins may, therefore, be due to the fact that those parts of the body most frequently exposed to air are most frequently damaged or may be an indication of air-borne/air-water interface distribution for this fungus. In a similar manner spores of the terrestrial fungi *Phoma herbarum* Westend. and *Phialophora* spp., floating at the air water interface, may cause infection in young fish, when they are ingested as the fry inflate their swim bladder for the first time (Alderman 1982). Despite a major granulomatous response, neither man nor dolphins appears to be able to eradicate the fungus. *Loboa loboi* therefore appears to be specialized for a mammalian rather than a marine environment. Further epidemiological studies on this condition would be of considerable interest.

Finally, no discussion of marine fungal disease would be complete without considering *Ichthyophonus*, because of its unparalleled impact on fish populations (McVicar 1982), even though the taxonomic status of this organism is still uncertain (Alderman 1982). Infection appears to occur when fish feed on diseased tissue but copepods have also been found infected with *Ichthyophonus* (Lightner 1981). These may therefore be secondary hosts. *Ichthyophonus* encysts in fish muscle and other tissues producing a major granulomatous inflammatory response involving giant cells (McVicar 1982) (Figure 14.1f). Death is commonly due to predation of a weakened individual.

Some aspects of the complexity of the interactions between fungi and their animal hosts in the marine environment have been considered here. In general, with the exception of *Ichthyophonus*, it can be seen that fungi are of subsidiary importance to bacteria as pathogens of fish (cf. Unestam 1973) and aquatic mammals (Migaki & Jones 1983). In contrast, in invertebrates, the fungi are of at least equal, if not greater importance, than the bacteria. This undoubtedly reflects differences in the immune systems of the two groups, particularly in terms of recognition, but also in response. Support for this hypothesis is found in the many fungal infections reported from immunocompromized captive marine mammals (Migaki & Jones 1983). Fungi such as *Fusarium*, which are known to produce toxins or exoenzymes recognizable by the host (National Research Council 1983), appear to produce a much higher level of response in invertebrates than other taxa, where toxin production is unknown or uncertain. However, much of the data necessary to assess fully the implications of this are lacking, as are details of the host's response (cf. Hose *et al.* 1984). With the development of commercial invertebrate aquaculture, where disease problems are much more severe than in the wild, economic necessity is now providing the impetus for research into these areas.

REFERENCES

Alderman, D.J. (1982). Fungal Diseases of Aquatic Animals. In Microbial Diseases of Fish, ed. R.J. Roberts. Society for General Microbiology, Special Publication, 9, 189-242. London: Academic Press.

Alderman, D.J. & Gras, P. (1969). Gill diseases of Portuguese oysters. Nature, Lond., 224, 616-617.

Brown, L.A. (1983). Husbandry, Physiological and Disease Factors Associated with Surgical Gonadectomy of Salmonids. Ph.D. Thesis, University of Stirling.

Chandler, F.W., Caplan, W. & Ajello, L. (1980). A colour atlas and textbook of the histopathology of mycotic diseases. London: Wolfe Medical Publications.

Couch, J.N. (1942). A new fungus on crab eggs. J. Elisha Mitchell scient. Soc., 58, 158-162.

Delves-Broughton, J. & Poupard, C.W. (1976). Disease problems of prawns in recirculation systems in the U.K. Aquacult., 7, 201-217.

Fisher, W.S., Nilson, E.H., Steenbergen, J.F. & Lightner, D.V. (1978). Microbial diseases of cultured lobsters - a review. Aquacult., 14, 115-140.

Franc, A. & Arvy, L. (1970). Données sur l'evolution de la "maladie des brancies" chez les Ouitres et sur son agent causal

Thanatostrea polymorpha Franc et Arvy 1969. Bull. biol. Fr. Belg., <u>104</u>, 3-19.

Gaertner, A. (1979). Some fungal parasites found in the diatom populations of the Rosfjord Area (south Norway) during March 1979. Veröff. Inst. Meeresforsch. Bremerh., <u>18</u>, 29-33.

Glazebrook, J.S. (1983). Diseases of Captive and Wild Sea Turtles in Northern Australia. Ph.D. Thesis, James Cook University.

Harrison, J.L. & Jones, E.B.G. (1975). The effect of salinity on sexual and asexual sporulation in members of the Saprolegniaceae. Trans. Brit. mycol. Soc., <u>65</u>, 389-394.

Hibbits, J., Hughes, G.C. & Sparks, A.K. (1981). *Trichomaris invadens* gen. et sp. nov. An Ascomycete parasite of the Tanner Crab (*Chionoecetes bairdi*) Rathbun (Crustacea, Brachyura). Can. J. Bot., <u>59</u>, 2121-2128.

Hoar, W.S. & Randall, D.J. (1969). Fish Physiology Volume I. Excretion, Ionic Regulation and Metabolism. New York: Academic Press.

Höhnk, W. & Vallin, S. (1953). Epidemisches Absterben von *Eurytemora* in Bothnischen Meerbusen, Verursacht durch *Leptolegnia baltica* nov. spec. Veröff. Inst. Meeresforsch. Bremerh., <u>2</u>, 215-223.

Hose, J.E., Lightner, D.V., Redman, R.M. & Danald, D.A. (1984). Observations on the pathogenesis of the imperfect fungus, *Fusarium solani* in the California Brown Shrimp, *Penaeus californiensis*. J. Invert. Path., <u>44</u>, 292-303.

Johnson, P.T. (1983). Diseases caused by Viruses, Rickettsiae, Bacteria and Fungi. In The Biology of the Crustacea, Vol. 6, Pathobiology, ed. A.J. Provenzano, pp. 1-78. New York: Academic Press.

Johnson, T.W. & Sparrow, F.K. (1961). "Fungi in Oceans and Estuaries". Weinheim: J. Cramer.

Jones, G. (1981). Thraustochytrid pathogens. Bull. Brit. mycol. Soc., <u>16</u> (Suppl. 1), 5-6.

Jones, G. & O'Dor, R.K. (1983). Ultrastructural observations on a Thraustochytrid fungus parasitic in the gills of squid (*Illex illecebrosus* LeSueur). J. Parasit., <u>69</u>, 903-911.

Lauckner, G. (1980). In Diseases of Marine Animals. Volume I, General Aspects, Protozoa to Gastropoda, ed. O. Kinne, pp. 75-400. Chichester: John Wiley.

Lightner, D.V. (1981). Fungal Diseases of Marine Crustacea. In Pathogenesis of Invertebrate Microbial Disease, ed. E.W. Davidson, pp. 451-484. Totowa, New Jersey: Allanheld, Osmun.

Lightner, D.V. & Fontaine, C.T. (1973). A new fungus disease of the white shrimp *Penaeus setiferus*. J. Invert. Path., <u>22</u>, 94-99.

Mantel, L.H. & Farmer, L.L. (1983). Osmotic and Ionic Regulation. In The Biology of the Crustacea, ed. L.H. Mantel & D.E. Bliss, <u>5</u>, pp. 53-161. New York: Academic Press.

McClean, N. & Porter, D. (1982). The yellow spot disease of *Tritonea diomedea* Bergh (Mollusca: Gastropoda: Nudibranchia) : Encapsulation of the Thraustochytriaceous parasite by host amoebocytes. J. Parasit., <u>68</u>, 243-252.

McVicar, A.H. (1982). *Ichthyophonus* infections of fish. In Microbial Diseases of Fish, ed. R.J. Roberts, pp. 243-269. Society for

General Microbiology, Special Publication 9. London:
Academic Press.

Mix, M.C. & Sparks, A.K. (1980). Tanner Crab, *Chionoecetes bairdi*
Rathbun, haemocyte classification and an evaluation of using
differential counts to measure infection with a fungal
disease. J. Fish Dis., 3, 285-293.

Migaki, G. & Jones, S.R. (1983). Mycotic Diseases in Marine Mammals. In
Pathobiology of Marine Mammal Diseases, Volume 2, ed.
E.B. Howard, pp. 1-27. Boca Raton. Florida: CRC Press Inc.

National Research Council (1983). Protection Against Trichothecene
Mycotoxins. Washington D.C.: National Academy Press.

Neish, G.A. & Hughes, G.C. (1980). Fungal Diseases of Fishes. Neptune,
New Jersey: TFH Publications.

Noga, E.J. & Dykstra, M.J. (1986). Oomycete fungi associated with
ulcerative mycosis in menhaden, *Brevoortia tyrannus*
(Latrobe). J. Fish Dis., 9, 47-53.

Pickering, A.D. & Willoughby, L.G. (1982). *Saprolegnia* Infections of
Salmonid Fish. In Microbial Diseases of Fish, ed.
R.J. Roberts. Society for General Microbiology, Special
Publication 9, pp. 271-297. London: Academic Press.

Polglase, J.L. (1980). A preliminary report on the Thraustochytrid(s)
and Labyrinthulid(s) associated with a pathological
condition in the Lesser Octopus, *Eledone cirrhosa*. Botanica
mar., 23, 699-706.

Polglase, J.L. (1981). Thraustochytrids as potential pathogens of marine
animals. Bull. Br. mycol. Soc., 16 (Suppl. 1), 5.

Polglase, J.L., Dix, N.J. & Bullock, A.M. (1984). Infection of skin
wounds in the Lesser Octopus, *Eledone cirrhosa* by
Cladosporium sphaerospermum. Trans. Brit. Mycol. Soc., 82,
577-580.

Quick, J.A. (1972). A new Thraustochytridiaceous fungus endoparasitic on
the American oyster *Crassostrea virginica* Gmelin in Florida.
Soc. Invert. Path., Newsletter, 4, 13.

Reichenbach Klinke, H.H. & Elkan, E. (1965 a). Principal Diseases of
Lower Vertebrates, Volume II, Diseases of Amphibia. Hong
Kong: TFH Publications Inc. Ltd.

Reichenbach Klinke, H.H. & Elkan, E. (1965 b). Principal Diseases of
Lower Vertebrates, Volume III, Diseases of Reptiles. Hong
Kong: TFH Publications Inc. Ltd.

Richards, R.H., Holliman, A. & Helgason, S. (1978). Naturally occurring
Exophiala salmonis infection in Atlantic Salmon (*Salmo
salar* L.). J. Fish Dis., 1, 357-359.

Richards, R.H. & Pickering, A.D. (1978). Frequency and distribution
patterns of *Saprolegnia* infections in wild & hatchery
reared Brown Trout, *Salmo trutta* (L.) & Char, *Salvelinus
alpinus* (L.). J. Fish Dis., 1, 69-82.

Robertson, J.D. (1964). Osmotic and ionic regulation. In Physiology of
the Mollusca Vol. 1, ed. K.M. Wilbur & C.M. Yonge,
pp. 283-308. London: Academic Press.

Rogers-Talbot, R. (1948). The fungus *Lagenidium callinectes* Couch (1942)
on eggs of the blue crab in Chesapeake Bay. Biol. Bull.
mar. biol. Lab., Woods Hole, 95, 214-228.

Sindermann, C.J. (1970). Principal Diseases of Marine Fish and
        Shellfish. New York: Academic Press.

Soderhall, K. & Smith, V.J. (1983). The Prophenoloxidase Activating
        System - a Complement like Pathway in Arthropods? In
        Infection Processes of Fungi, eds. J. Aish & D.W. Roberts.
        Rockefeller Foundation Study Bellagio, Italy, March 21-25.

Sparks, A.K. & Hibbits, J. (1979). Black mat syndrome: an invasive
        mycotic disease of the Tanner Crab, *Chionoecetes bairdi*. J.
        Invert. Path., 34, 184-191.

Unestam, T. (1973). Fungal Diseases of Crustacea. Rev. med. vet. Mycol.,
        8, 1-20.

15 *PETERSENIA POLLAGASTER* (OOMYCETES): AN INVASIVE FUNGAL PATHOGEN OF *CHONDRUS CRISPUS* (RHODOPHYCEAE)

F.I. Molina

INTRODUCTION

This chapter presents the first report of *Petersenia pollagaster* (Petersen) Sparrow as an invasive pathogen of cultivated *Chondrus crispus* Stackh. and describes several stages of its endobiotic development. Except for the studies of Arasaki and co-workers (1947, 1962, 1968) on a serious fungal disease of maricultured *Porphyra* attributed to *Pythium porphyrae* Takahashi, Ichinotani et Sasaki, fungal infections of commercially important seaweeds have so far not been reported to reach epidemic proportions and their study has been largely neglected. The only exception is the study of Kazama & Fuller (1970) who examined the ultrastructure of *Porphyra perforata* J. Ag. artificially infected with zoospores of *Pythium marinum* Sparrow and noted the cytological effects on penetrated host cells including disruption of organelles, dissolution of floridean starch grains and rapid collapse of the cells. Kohlmeyer (1979) has identified several key areas for the study of parasitic algicolous fungi including life history, pathogenesis, epidemiology, and fungal growth within the host.

The lagenidioid oomycete, *Petersenia pollagaster* was first described as *Pleotrachelus* Zopf by Petersen in 1905 who found it on fronds of *Ceramium rubrum* (Hudson) J. Ag. from several Danish localities (Petersen 1905). Subsequently, the fungus has been reported only twice: Sparrow (1934) recollected the species in Denmark in 1934 on the same red algal host and established the genus *Petersenia* for it on the basis of the biflagellate nature of the zoospores; Johnson & Howard (1968) found it on *C. rubrum* from Iceland. Karling (1981) considered this species to be in the family Sirolpidiaceae owing to the similarity of its thallus with those of species of *Sirolpidium* Petersen. Virtually nothing is known of the biology, life history, or ontogeny of *P. pollagaster* and even less is known of its cytological effects on the host algae.

An epiphytotic, presumably of fungal origin, was reported in the *Chondrus crispus* culture facilities of Marine Colloids, Ltd. in lower East Pubnico, Nova Scotia in the autumn of 1980. The cultured strain, designated as 'T4', is a high k-carrageenan-producing male gametophyte. Outbreaks of the fungal disease occurred in the autumns of 1981 through to 1984. Preliminary examination of infected thalli revealed that the fungus associated with the disease was endobiotic, holocarpic, and produced biflagellate zoospores (Hughes unpublished).

MATERIALS AND METHODS
Collection of materials
        Diseased specimens were collected from the mariculture
facilities of Marine Colloids, Ltd. located at Pubnico Harbour, Nova
Scotia (latitude 43°42'N, longitude 65°48'W) and packed in ice for
shipment to Vancouver. Uninfected fronds were likewise collected for
comparative purposes.

Laboratory cultures of the host and the parasite
        Algal cultures were maintained at 15° C in "Instant Ocean"
(30%oo) supplemented with 0.05 mmol N, 0.005 mmol K, and 0.5 cm^3
organic micronutrients 1^{-1} (Chen & Taylor 1978). Irradiance level was
kept at 38.0 µE cm^{-2} sec^{-1} for 8 h and alternated with a dark period of
16 h. For infection of healthy thalli, tips measuring 1.0-2.0 cm were
excised and seeded into glass dishes with diseased fronds in 100 cm^3 of
culture medium. Although cultures were not axenic, care was taken to
minimize contamination by first washing both the inoculum and the
infected fronds with a strong stream of sterile sea water in a laminar
flow hood. Half of the medium was decanted and replenished with an
equal amount of fresh "Instant Ocean" and uninfected tips were added
every 2-3 days to maintain both the host and the parasite in culture.

Light microscopy
        Infected tips were incubated overnight in 4',6-diamidino-2-
phenylindole (DAPI) in pH 8.0 sea water at a final concentration of
1.0 µg cm$^{3-1}$. Freehand sections were made the following day, rinsed,
and mounted in filtered sea water. Formaldehyde-killed specimens were
likewise incubated in the fluorochrome to serve as controls. Specimens
were examined with a Leitz Dialux epifluorescence microscope. Zoospores
were photographed with a Leitz Ortholux microscope equipped with
phase-contrast optics and a Nikon Microflex AFM. For light microscopy,
specimens were fixed for 3 h at room temperature according to the method
of Karnovsky (1966). After three rinses in buffer, they were dehydrated
in a graded series of methanol, embedded in JB4 methacrylate and
sections (1.0-1.5 µm thick) cut with glass knives on a Sorvall
microtome. For the detection of cellulose, sections were treated with
the following stains: 1) 0.1% aqueous calcofluor white for 1 min;
2) 0.1% congo red in 50% ethanol for 1 min followed by differentiation
in 0.2% KOH in 80% ethanol; 3) zinc chloriodide according to the
procedure of Stevens (1974). Sections were also stained with acridine
orange for the detection of nuclear fluorescence following the procedure
given by McCully *et al.* (1978).

Electron microscopy
        Specimens were fixed as for routine light microscopy and
postfixed in 1% osmium tetroxide in the same buffer for 24 h. Specimens
were dehydrated in a graded methanol series to propylene oxide and
infiltrated with Epon or Polybed 812. Sections were cut with a diamond
knife, stained in saturated methanolic uranyl acetate followed by

Reynold's lead citrate and viewed with a Zeiss EM9A transmission electron microscope.

RESULTS AND OBSERVATIONS
Laboratory cultures of the host and the parasite

The cycle which consists of infection of healthy thalli, endobiotic development, and emergence of zoospores was completed in 48-72 h at 20° C. Infection did not occur in cultures kept at 10° C, although zoospores remained viable for a few days at this temperature and could be induced to infect inoculated thalli by raising the temperature. Lesions measuring 0.25 mm diam developed on the tips within 28 h of inoculation and increased in size to 0.85 mm diam after 72 h. At this stage, exit tubes which had penetrated the cortical layer were visible. The formation of lesions was limited to the last two bifurcations of the host alga and the number of lesions on growing tips increased only when healthy material was not added within 48 h from the time of zoospore release. Progressive changes in infected thalli are

Figures 15.1a-d *Chondrus crispus* 'T4' thalli. Bars = 1 mm.
Figure 15.1a Uninfected thallus.
Figures 15.1b-d Progressive stages of infection with *Petersenia pollagaster* which are limited to the growing tips of the host.

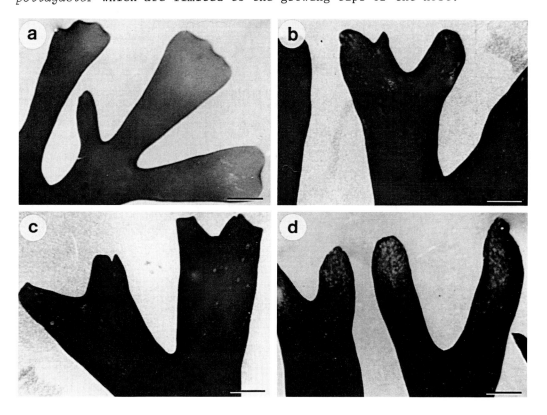

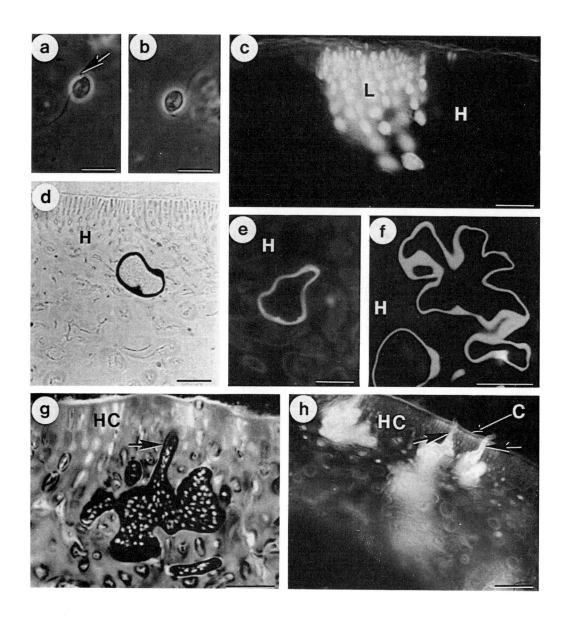

<u>Figures 15.2a-h</u> *Petersenia pollagaster*. Light microscopy.
<u>Figures 15.2a,b</u> Zoospores, the second flagellum (arrowed) is behind the planont in both micrographs. Phase contrast. Bars = 10 μm.
<u>Figure 15.2c</u> Infected thallus incubated in DAPI. Damage to host tissue (H) is shown by fluorescence of necrotic lesions (L). Freehand section. Bar = 20 μm.
<u>Figures 15.2d-f</u> Positive staining reactions with zinc chloriodide (Fig. 15.2d), calcofluor white (Fig. 15.2e), and congo red (Fig. 15.2f) indicate the presence of cellulose in sporangial walls of the parasite. H = host tissue, methacrylate embedment. Bars = 30 μm.

illustrated in Figures 15.1b-d.   Figure 15.1a which shows an uninfected specimen is included for comparison.

Attempts to grow the fungus on artificial media developed by Vishniac (1955) and Fuller *et al*. (1964) yielded negative results.   Similarly, heat-killed uninfected thalli were not colonized by the fungus.

Light microscopy
Zoospores were reniform to elongate, 3.0-3.5 μm wide by 4.0-4.5 μm long (Figures 15.2a,b), and heterokont with flagella attached within a groove that ran along the length of the spore.   Prolonged illumination of the planont caused the flagella to be absorbed.

Treatment of sections embedded in methacrylate with congo red and calcofluor white resulted in fluorescence of fungal walls (Figures 15.2e,f).   Zinc chloriodide also gave a positive staining reaction (Figure 15.2d).   These cytochemical staining procedures indicated the presence of cellulose in sporangial walls.   Sporangia appeared highly lobed, multinucleate, and each produced a single exit tube that penetrated the cortical tissue of the host (Figures 15.2g,h).

Four stains were initially screened for fluorescence differentiation between live and dead cells in order to determine the effect of the infection on host tissues: acridine orange, trypan blue, fluorescein diacetate, and DAPI.   Reliable differentiation between live and dead cells was obtained only with DAPI (Figure 15.2c); the fluorescence of host cells and formaldehyde-killed tissue were identical.   Acridine orange, fluorescein, and trypan blue could not be detected in the cells even after 48 h of incubation.

Electron microscopy
The parasite invaded mainly the cortical and subcortical tissues of the host and remained intracellular throughout its endobiotic development.   Early stages were devoid of a wall, were sparingly lobed, and were characterized by an abundance of mitochondria at the periphery of the cell (Figure 15.3a).   Intimate contact was maintained between the fungus and the invaded algal cell as shown by the close apposition of the parasite and the host membranes (Figures 15.3a,c).   This close proximity was exhibited by all stages observed prior to sporangium formation.   Later in development the parasite formed cytoplasmic processes or extensions and became more deeply lobed to form an increased surface area for contact with the host protoplasm

Figure 15.2g   The highly lobed, multinucleate sporangium produces a single exit tube (arrowed) penetrating the cortical tissue of the host (HC).   Methacrylate embedment, stained in acridine orange.   Bar = 30 μm.
Figure 15.2h   Penetration of the host cortex (HC) and cuticle (c) by the exit tubes of sporangia (arrowed).   Freehand section, stained in calcofluor white.   Bar = 50 μm.

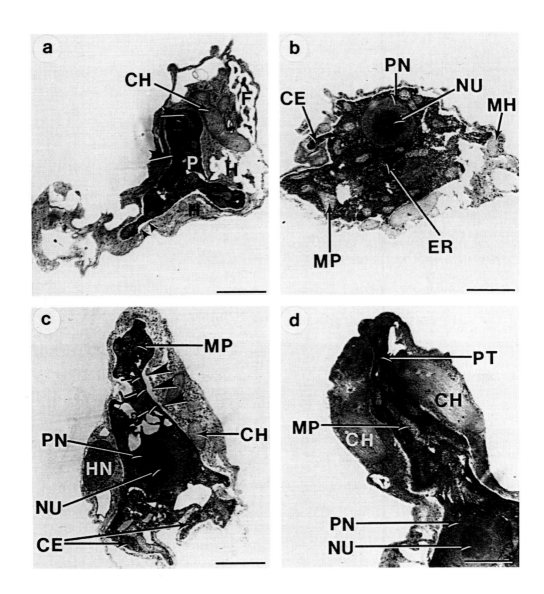

Figures 15.3a-d *Petersenia pollagaster* thalli in *Chondrus crispus*.
Transmission electron microscopy. Bars = 2.0 μm.
Figure 15.3a A naked parasite protoplast (P) with numerous peripheral
mitochondria (large arrowheads) surrounded by a host membrane (small
arrowheads). H = host cell, CH = chloroplast, F = floridean starch.
Figures 15.3b,c Formation of cytoplasmic extensions (CE) by the
parasite to produce a deeply lobed protoplast. The parasite nucleus
(PN) has a very prominent nucleolus (NU). Arrows and arrowheads in
Fig. 15.3c trace the parasite and host membranes, respectively.
HN = host nucleus, CH = chloroplast, MP = parasite mitochondrion, MH = host
mitochondrion, ER = endoplasmic reticulum.

(Figure 15.3b). Fungal mitochondria aggregated and migrated into these cytoplasmic processes prior to invasion of neighbouring host cells (Figures 15.3c,d, 4a). Previously uninfected cells were invaded via a symplastic route and this invasion seemed to be caused by the dissolution of pit plugs (Figure 15.4a). Repeated divisions of the parasite nucleus resulted in a multinucleate protoplast. At this stage some of the host organelles, including mitochondria and chloroplasts, were still intact and subsequent penetration of adjacent host cells by the fungus resulted in a confluent host cytoplasm (Figure 15.4b). An amorphous wall, whose electron opacity varied with the developmental stage was eventually synthesized by the fungus. The holocarpic thallus reached a maximum length of 100 μm. Nuclei measured 1.4-1.7 μm diam, were dispersed throughout the cytoplasm and became associated with several mitochondria with tubular cristae. Two types of membrane-bounded inclusions were present in developing sporangia: 1) regularly shaped electron-opaque bodies which ranged from 0.3 to 1.0 μm diam; 2) less electron-opaque bodies of variable shape which measured 0.8-2.0 μm (Figures 15.4d,f). The host cell wall, remaining chloroplasts, and connections with adjacent cells were still present at this stage (Figure 15.4d). Beaked nuclei were occasionally found in sporangia and became associated with a pair of centrioles oriented at right angles to each other (Figure 15.4c). Flagella were formed within the sporangium (Figures 15.4d,e).

DISCUSSION

The results of the cytochemical staining procedures with calcofluor white, congo red, and zinc chloriodide, the biflagellate nature of the zoospores, and the holocarpic, endobiotic thallus all warrant the assignment of the pathogen to *P. pollagaster*, Lagenidiales. This study describes the only known occurrence of a lagenidioid oomycete on a gigartinalean host and also constitutes the first report on a serious fungal epiphytotic in maricultured *C. crispus*.

In order to determine the effects of the infection on host tissue, four fluorochromes were tested for vital staining. Of these stains DAPI yielded the most consistent and reproducible results for the differentiation of live from dead cells. Similar results have been obtained in animal systems by van der Linden & Deelder (1984) who employed the dye as a probe for *in vitro* death of schistosomula. DAPI has been found to enter rapidly into dead cells and bind with DNA (Tanke *et al.* 1982). This penetration is said to be dependent on damage of the cell membrane. The concentration of DAPI used in the present study (1.0 μg cm$^{3-1}$) has been shown to have no toxic effect either on individual lymphocytes (Tanke *et al.* 1982) or on schistosomula (van der Linden & Deelder 1984) even after 48 h incubation. This fluorochroming technique should therefore be applicable in the study of

Figure 15.3d Aggregation and migration of parasite mitochondria (MP) into penetration tips (PT) precedes invasion of adjacent host cells. CH = chloroplast, NU = nucleolus, PN = parasite nucleus.

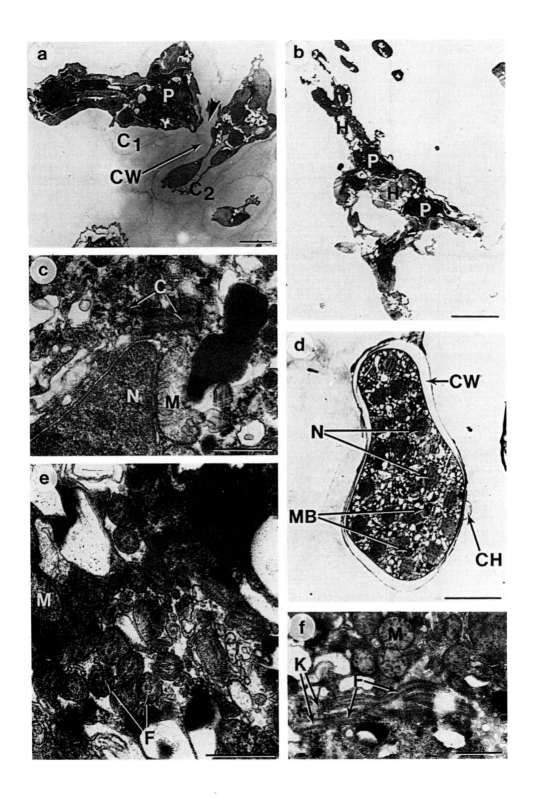

host-parasite interactions, especially for the demonstration of cell
death in the host.

There is presently a paucity of information on the ultrastructural
relationships of oomycetous fungal parasites with their host algae,
particularly in maricultured species. Studies of host-parasite
interactions have until now dealt mainly with the relationship of the
haustorium with the host cell and formation of papillae during the
penetration process (Bracker & Littlefield 1973; Aist 1976). The host-
parasite interface of *P. pollagaster* and *C. crispus* differs markedly
from that of haustorial parasites in the absence of haustoria in the
fungus and papillae in the host. Instead, the formation of cytoplasmic
processes by the parasite results in an increased surface area for very
close contact with the host and probably plays a role in nutrient
procurement. Whether the host membrane which surrounds the parasite is
plasmalemmal or vacuolar has not been determined. The intimate contact
of the parasite with the host cytoplasm and the symplastic spread of
the fungal thallus is reminiscent of *Pythium marinum* infections of
Porphyra perforata (Kazama & Fuller 1970). The development of
P. pollagaster differs from *Pythium marinum* in the resultant confluence
of invaded host cells and the apparent delay in the disruption of host
organelles. Aggregation and migration of mitochondria inside
penetration tips are indicative of the penetration process' energy
requirement. The persistence of chloroplasts is also remarkable and
indicates that the parasite may not be able to utilize all of the host's
cellular components.

ACKNOWLEDGEMENTS
This study was supported by a National Research Council

Figures 15.4a-f *Petersenia pollagaster*. Transmission electron
micrographs.
Figure 15.4a Penetration of adjacent host cells (C_1 & C_2) appears to
occur through the dissolution of pit plugs (arrowhead). P = parasite
cell, CW = cell wall. Bar = 5 µm.
Figure 15.4b Invasion of adjacent host cells (H) by the parasite (P) is
via a symplastic route and leads to the confluence of host cells.
Bar = 10 µm.
Figure 15.4c Centrioles (C) oriented at 90° with each other associate
with nuclei during sporogenesis. N = nucleus, M = mitochondrion.
Bar = 0.5 µm.
Figure 15.4d Holocarpic fungal thallus surrounded by an electron
transparent wall (CW). The cell is multinucleate (N) and contains many
membrane bounded electron-opaque bodies (MB). CH = chloroplast.
Bar = 5 µm.
Figure 15.4e Transverse section through flagella in a sporangium.
F = flagellum, M = mitochondrion. Bar = 0.5 µm.
Figure 15.4f Longitudinal section through flagella in a sporangium
showing the association of flagellar shafts (F) with the kinetosomes
(K). Bar = 1.0 µm.

Contract with the University of British Columbia (G.C. Hughes, principal investigator) and a Grant-in-Aid from the Sigma Xi Scientific Research Society to the author. I would like to thank Dr. J.S. Craigie, Atlantic Regional Laboratory, for supplying the *C. crispus* specimens and Dr. T. Bisalputra for laboratory facilities.

REFERENCES

Aist, J.R. (1976). Cytology of penetration and infection - fungi. In Encyclopedia of Plant Physiology, eds. R. Heitefuss & P.H. Williams, pp. 197-221. Berlin, Heidelberg, N.Y.: Springer.

Arasaki, S. (1947). Studies on the rot of *Porphyra tenera* by a *Pythium*. Bull. Jap. Soc. sci. Fish., 13, 74-90.

Arasaki, S. (1962). Studies on the artificial cultures of *Porphyra tenera* Kjellm. III. On the red wasting disease of *Porphyra*, especially on the physiology of the causal fungus *Pythium* sp. nov. J. agr. Lab., Abiko, Japan, 3, 87-93.

Arasaki, S., Akino, K. & Tomiyama, T. (1968). A comparison of some physiological aspects in a marine *Pythium* on the host and on the artificial medium. Bull. Misaki Mar. Biol. Inst. Kyoto Univ., 12, 203-206.

Bracker, C.E. & Littlefield, L.J. (1973). Structural concepts of host-pathogen interfaces. In Fungal Pathogenicity and the Plant's Response, eds. R.J.W. Byrde & C.V. Cutting, pp. 159-313. N.Y.: Academic Press.

Chen, L.C.M. & Taylor, A.R.A. (1978). Medullary tissue culture of the red alga *Chondrus crispus*. Can. J. Bot., 56, 883-886.

Fuller, M.S., Fowles, B.E. & McLaughlin, D.J. (1964). Isolation and pure culture study of marine phycomycetes. Mycologia, 56, 745-756.

Johnson, T.W. Jr. & Howard, K.L. (1968). Aquatic fungi of Iceland: species associated with algae. J. Elisha Mitchell scient. Soc., 84, 305-311.

Karling, J.S. (1981). Predominantly Holocarpic and Eucarpic Simple Biflagellate Phycomycetes. Vaduz: J. Cramer.

Karnovsky, M.S. (1966). A formaldehyde-glutaraldehyde fixative of high osmolality for use in electron microscopy. J. Cell Biol., 27, 137A-138A.

Kazama, F. & Fuller, M.S. (1970). Ultrastructure of *Porphyra perforata* infected with *Pythium marinum*, a marine fungus. Can. J. Bot., 48, 2103-2107.

Kohlmeyer, J. (1979). Marine pathogens among Ascomycetes and Deuteromycetes. Experientia, 35, 437-440.

Linden, P.W.G. van der & Deelder, A.M. (1984). *Schistosoma mansoni*: a diamidinophenylindole probe for *in vitro* death of schistosomula. Exp. Parasit., 57, 125-131.

McCully, M.E., Goff, L.J. & Adshead, P.C. (1978). Preparation of algae for light microscopy. In Handbook of Phycological Methods, ed. E. Gantt, pp. 263-283. Cambridge: Cambridge University Press.

Petersen, H.E. (1905). Contributions à la connaissance des Phycomycètes marins (Chytridineae Fischer). Oversigt. K. danske vidensk. Selsk. Forhandl., 1905, 439-488.

Sparrow, F.K. (1934). Observations on marine phycomycetes collected in Denmark. Dansk. bot. Ark., 8, 1-24.

Stevens, R.B. (1974). Mycology Guidebook. Seattle: University of Washington Press.

Tanke, H.J., Linden, P.W.G. van der & Langerak, J. (1982). Alternative fluorochromes to ethidium bromide for automated read out of cytotoxicity tests. J. Immun. Meth., 52, 91-96.

Vishniac, H.S. (1955). Marine mycology. Trans. N.Y. Acad. Sci. Ser. II, 17, 352-360.

D.L. Kingham

L.V. Evans

INTRODUCTION
        In symbioses involving the ascomycete fungus *Mycosphaerella ascophylli* Cotton there is invariable and permanent systemic infection of two intertidal brown algae, viz. *Pelvetia canaliculata* (L.) Dcne et Thur. and *Ascophyllum nodosum* (L.) Le Jol. Previous studies of these symbioses have revealed little of the true nature of the inter-relationship, and physiological investigations are few in number (Drew 1969; Kremer 1973). Light and electron microscopy, culturing and ecological studies, and physiological and analytical investigations have therefore been carried out (Kingham 1976), in an attempt to gain a better understanding of the *M. ascophylli*-brown algal association. Particular attention has been paid to the *P. canaliculata* symbiosis, as previous work has mainly centred upon the fungus in *A. nodosum* (Webber 1959, 1967; Kohlmeyer & Kohlmeyer 1972).

RESULTS AND DISCUSSION
Light and electron microscope studies
        The fine, septate hyphae of the *Mycosphaerella ascophylli* mycelium ramified throughout the seaweeds *Pelvetia canaliculata* and *Ascophyllum nodosum*, including the receptacles. They formed a network within the mucilage-filled intercellular regions (Figures 16.1a,b), and extended from a few cells behind the apex to the attaching holdfast. The hyphae were 0.8-1.6 µm in diameter and were branched and anastomosed, often forming a network around individual algal cells. Electron microscopy revealed a thick cell wall surrounding a highly convoluted plasma membrane, together with normal cellular organelles, with the exception of Golgi bodies. Although haustoria were not observed within algal cells, fungal hyphae (intra-membranous haustoria) occurred within algal cell walls (Figures 16.1c,d), where they generally occupied a median position. Intra-hyphal hyphae also occurred (Figure 16.1e). These hyphae-within-hyphae may have resulted from invasion of dead hyphae by finer hyphae growing out from adjacent mycelium (Farley *et al.* 1975), or, they could have been a mycoparasite, especially in the pseudothecial stroma, where they occurred within living hyphae. Hyphae of *M. ascophylli* also extended from *A. nodosum* into the holdfasts of epiphytic *Fucus* plants (Figure 16.1f), although a control mechanism was probably in operation, since penetration of these plants did not occur to any great distance.

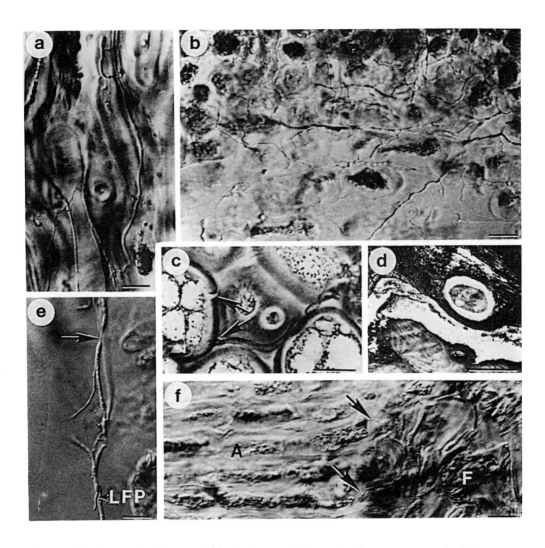

Figure 16.1a   L.S. *P. canaliculata* medulla showing *M. ascophylli* hyphae
and anastomosing branches.   Na$_2$CO$_3$-treated and squashed.   Interference
contrast.   L.M.   Bar = 20 µm.

Figure 16.1b   L.S. *A. nodosum* sporeling showing many hyphae in cortex.
As Fig. 16.1a but stained in cotton blue in lactophenol.   L.M.
Bar = 20 µm.

Figure 16.1c   Section of cortex of *P. canaliculata* receptacle showing
hyphae following algal cell contours and penetrating walls (arrowed).
Glycol methacrylate/toluidine blue.   L.M.   Bar = 20 µm.

Figure 16.1d   Intramembranous haustorium within wall of *P. canaliculata*
cell.   T.E.M.   Bar = 2 µm.

Figure 16.1e   Hyphae in medulla of *P. canaliculata* showing a fine
intrahyphal hypha linking two older hyphae (arrowed), and an outgrowing
intrahyphal hypha forming an external lateral fusion peg (LFP).   As
Fig. 16.1a, L.M.   Bar = 20 µm.

Fungal pseudothecia occurred only on the receptacles of *A. nodosum* but in *P. canaliculata* they were also found on receptacle stalks, on non-reproductive apices, and occasionally even on older thalli. They were seen as black specks, representing the darkened surrounds of the ostiole. Hyphae of the pseudothecial stroma and associated mycelium were 1.25-2.5 μm in diameter, and normal ascomycete septa with a median pore were present in these, as in vegetative hyphae. Ascospore formation was essentially as described for other Ascomycotina. Like the pseudothecium, the pycnidium was confined to the outer layers of the algal cortex, and had contact with surrounding vegetative hyphae. Histochemical studies showed that the pseudothecial mucilage was composed mainly of a strongly acidic polysaccharide, mainly sulphated, with large amounts of PAS-positive carbohydrates and some protein.

Culturing and ecological studies

Algal oospores/zygotes were rarely infected at the time of release. Less than 5% of *P. canaliculata* zygotes under 0.75 mm long were infected, whereas infection had occurred in all zygotes over 5 mm (*ca* 15 months) in length. *A. nodosum* zygotes under 3 mm were uninfected. All naturally-occurring mature plants of *P. canaliculata* and *A. nodosum* (but not other fucoids) were infected, but two species closely resembling these in morphology and ecology which grow on the west coast of America, viz. *Pelvetia fastigiata* Dcne et Thur. and *Pelvetiopsis limitata* (Setchell) Gardner, were both found to be uninfected. This may be the result of the fungus never reaching this coast, since *A. nodosum* plants from the east coast are infected (Kohlmeyer & Kohlmeyer 1972).

Physiological studies

Uptake of ^{14}C-glucose and ^{14}C-mannitol was found to be much greater in *P. canaliculata* than in *A. nodosum*, and there was very little uptake by *Fucus spiralis* L. (Table 16.1). The amount of radioactivity which could be washed out from *P. canaliculata* and *A. nodosum* was also very small compared with *F. spiralis*. (In *P. canaliculata* and *A. nodosum* the majority of label was found in the neutral fraction as mannitol, although a moderately high proportion of radioactivity was also incorporated into amino acids). *P. fastigiata* and *P. limitata* on the other hand were unable to utilize exogenous carbohydrates (Table 16.1) and gave results resembling those for *F. spiralis* rather than the closely related *P. canaliculata* and *A. nodosum*. Drew (1969) also found that 9 of the 11 species of brown algae he investigated were unable to utilize exogenous glucose and mannitol and Bidwell & Ghosh (1962, 1963) found the same for *Fucus vesiculosus* L. These results, together with those found in the present study, strongly indicate that it is not *P. canaliculata* and *A. nodosum* but the fungus *M. ascophylli* which is able to utilize exogenous carbohydrates. The failure of the algae to do this may be due to the absence of the necessary uptake

Figure 16.1f Hyphae (arrowed) growing between *A. nodosum* (A) cortex cells and out into *Fucus* holdfast tissue (F). As Fig. 16.1a. L.M. Bar = 20 μm.

mechanisms in the plasma membrane, thus preventing glucose from coming into contact with the appropriate enzymes for conversion to mannitol. Failure to utilize glucose by the green alga *Dunaliella tertiolecta* Butch. was attributed to membrane impermeability (Kwon & Grant 1971).

Further evidence for the involvement of *M. ascophylli* in glucose metabolism was provided by an experiment where uptake of ^{14}C-glucose and the conversion of this to mannitol were compared in developing receptacles and non-reproductive apices of *P. canaliculata*. In the former, with a lower hyphal density than the latter, although the small amount of glucose which did not wash out was converted to mannitol, this was 13 times less than that produced by the vegetative (control) tissue. This ratio corresponded approximately with the relative hyphal density in the two tissues.

The fungicide Nystatin and the glucose analogue 2-deoxy-D-glucose (2-DG) were used to determine whether differentially inhibiting or eliminating the fungus affected glucose and mannitol uptake and utilization. Incubation of *P. canaliculata* with Nystatin for 4.5 days at concentrations up to 250 units cm $^{-3}$ did not affect the alga but produced an almost parallel reduction in both glucose and mannitol

Table 16.1 Uptake of exogenous carbohydrates by various brown algae

	Incubation time (h)	Exogenous carbohydrate	Total uptake (dps mg $^{-1}$ DW)	% of Total washed out (after 3 h)
P. canaliculata	3	G	403	2
	3 kc	G	54	91
	4	M	284	6
A. nodosum	3	G	242	2
	3 kc	G	85	91
	4	M	149	4
F. spiralis	3	G	64	59
	3 kc	G	85	93
	4	M	113	85
P. fastigiata	5	G	26	57
	5	M*	15	51
P. limitata	5	G	72	57
	5	M*	36	55

G = ^{14}C-glucose; M = ^{14}C-mannitol; kc = killed control.
All incubations in light except *.

uptake with increasing concentration (Table 16.2). In addition,
following incubation at the highest concentration, the amount
of ^{14}C-mannitol retained after washing and glucose to mannitol
conversion were greatly reduced compared with the untreated control
(Table 16.2), presumably the result of specific anti-fungal activity by
the Nystatin. Nystatin has a wide range of fungal activity, and there
is evidence of some activity against algae (Cartwright 1975). Its mode
of action (Harman & Masterson 1957; Cartwright 1975) essentially
involves irreversible binding to a sterol in the cell membrane, with
subsequent disfunction, so that uptake of exogenous carbohydrates is
eliminated. It is likely that after 4.5 days incubation in the higher
concentration of Nystatin the fungus had been killed, or that its
membranes were no longer functional. Since photosynthesis rate was not
reduced, the algal membranes on the other hand were presumably
unaffected. Thus the results from the Nystatin pretreatment experiments
provide further evidence that uptake and utilization of exogenous
carbohydrates can be attributed to the fungus in this association.

Preincubation in 2% 2-DG for 17 h resulted in no change in the
photosynthesis rate or mannitol uptake (Table 16.2), but glucose uptake
was reduced by 87% and glucose to mannitol conversion by 88%. The
percentage washed out increased to 24%. 2-DG is an analogue of
D-glucose and is competitively taken up by the same membrane carrier
system involving phosphorylation (Van Steveninck 1968). The results
with glucose and mannitol seen here suggest that 2-DG treatment inhibits
glucose uptake either by competition for ATP or at the phosphohexo/
isomerase level, but mannitol which does not require subsequent
phosphorylation or any conversion is unaffected. A different site, or
mode of uptake is therefore envisaged for the two substances, and the
results with Nystatin indicate that both are functions of the fungus,
after diffusion through the algal intercellular mucilage has occurred.
None of the results implicates the algal partner in uptake and the
necessary uptake systems appear to be lacking in the fucoid algae. The
glucose uptake system involved in the fungus is believed to be a
carrier-linked active phosphorylative transport system, which appears to
be sensitive to factors such as exogenous glucose concentration and
temperature. In mannitol uptake by the fungus an active mediated
carrier transport system is believed to occur.

In a further investigation into the possible role of the alga in uptake
and metabolism, it was found that in naturally-occurring *P. canaliculata*
zygotes the amount of ^{14}C-mannitol produced or retained after standard
incubation methods was in the ratio photosynthesis : glucose uptake :
mannitol uptake, 100 : 10 : 18 for 'infected' zygotes and 100 : 4.5 : 8 for
'uninfected' zygotes. After incubation in ^{14}C-glucose, total neutral
extracts from 'uninfected' zygotes contained 25% of the radioactivity in
mannitol, 8% in glucose and the remainder stayed at the origin.
Following incubation in ^{14}C-mannitol, 25% occurred in mannitol, a little
as glucose and a disaccharide and the remainder stayed at the origin.
When 'infected' zygotes were incubated in glucose or mannitol, 99% of
the radioactivity occurred as mannitol. The incorporation results from
different sets of incubations were however very varied and whilst

Table 16.2 Uptake of ^{14}C-glucose and ^{14}C-mannitol by *P. canaliculata* following pre-incubation in Nystatin and 2-DG

Treatment	Photo-synthesis (dps mg $^{-1}$ DW)	^{14}C-glucose uptake			^{14}C-mannitol uptake		
		% washed out	soluble (dps mg $^{-1}$ DW)	% redn*	% washed out	soluble (dps mg $^{-1}$ DW)	% redn
Nystatin							
Control	9093	1	886	0	1	335	0
20 units cm $^{-3}$	9246	1	851	0	1	312	0
100 units cm $^{-3}$	10720	3	530	38	3	230	26
250 units cm $^{-3}$	10812	66	52	94	39	17	95
2-DG							
Control	9317	1	792	0	3	439	0
Treated	9267	24	102	88	2	436	0

* = % reduction in amount of mannitol produced from ^{14}C-glucose.
photosynthesis: 3h pulse, 15 min chase.
glucose and mannitol: 5h incubation, 3h wash.

uninfected zygotes did appear to show conversion of glucose to mannitol, the possibility that this was due to the presence of bacterial and other contaminants cannot be precluded. However, if the conversion is carried out by the zygotes it may indicate a pre-adaptation of *P. canaliculata* and *A. nodosum* to a symbiotic existence, since the uptake mechanisms are inherent. Sterile cultures of zygotes are necessary to clarify this.

None of the experiments gave conclusive proof of any carbohydrate movement between the symbionts, and application of the Inhibition Technique and autoradiographic methods to demonstrate this were not successful.

Figures 16.2a-d GLC traces of neutral extracts.
a, *P. canaliculata* with inositol (RI = 1.0) internal standard (dotted). (Attenuation = 20K). RI = 0.78 (mannitol); 1.01; 1.11 (α – volemitol); 1.21 (β – volemitol). b, *A. nodosum* (20K). RI = 0.78 (mannitol); 1.17. c, *Pelvetia fastigiata* (20K). RI = 0.78 (mannitol); 0.86; 1.92; 2.11. d, *Pelvetiopsis limitata* (5K). RI = 0.78 (mannitol); 1.16; 1.40; 1.76. (M = mannitol; V = volemitol; I = inositol).

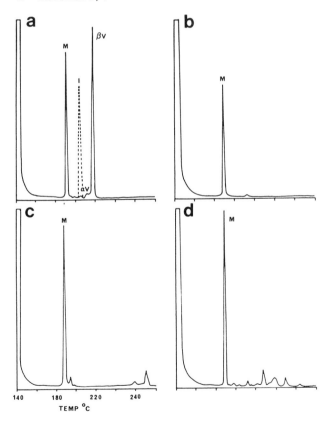

Analytical studies

Studies using gas-liquid chromatography (GLC) of neutral
fractions of the algae showed a dominant mannitol peak (RI = 0.78) in all
extracts (Figures 16.2a-d).   In addition, a larger β-volemitol peak
(RI = 1.21) was present in the *P. canaliculata* extract (Figure 16.2a),
but was absent from *A. nodosum*, *P. fastigiata* and *P. limitata*
(Figures 16.2b-d) (see also Kremer 1976).   Immediately after silylation
of the extract from *P. canaliculata* volemitol was present as a dominant
α-peak (RI = 1.11) and a smaller β-peak (RI = 1.21).   However, with time
the α-peak decreased with a corresponding increase in the β-peak, as
found by Holligan & Drew (1971).

GLC traces of neutral extracts from apical vegetative tissue of mature
plants of *P. canaliculata* and sporelings over 5 mm long showed mannitol
to volemitol peaks in the ratio 0.75-0.9 : 1 (Figure 16.3a), in agreement
with Holligan & Drew (1971).   However, zygotes less than 5 mm long
(Figure 16.3b) and pieces of thallus treated with the fungicide Nystatin
(Figure 16.3c) showed an increase in this ratio to 1.7 : 1 and 1.5 : 1
respectively.   Kremer (1973) proposed that the occurrence of volemitol
in *P. canaliculata* is connected with the presence of *M. ascophylli*.
Since Kremer demonstrated that volemitol is labelled after 30 sec
photosynthesis in sodium [^{14}C] bicarbonate, it must be an algal rather
than a fungal product.   Since small uninfected zygotes and mature tissue
treated with Nystatin (in quantities which destroy the fungus) produced
volemitol in smaller amounts (compared with mannitol) than adult
P. canaliculata plants, the presence of the fungus appeared to stimulate
the photosynthetic production of volemitol by *P. canaliculata*, and
volemitol may therefore be an important carbohydrate in this symbiosis.
The present results, showing that volemitol is present in mature apical
vegetative *P. canaliculata* tissue in greater quantities than mannitol
(Figure 16.3a), are in agreement with Holligan & Drew (1971), but are at
variance with the results of Lindberg & Paju (1954) and Quillet (1957).

In his definition of biotrophy Starr (1975) required that both organisms
remain living and associated and that at least one of the symbionts
should derive nutrients from the other.   Although the *M. ascophylli*-
brown algal symbiosis fulfils the former criterion the nutritional
aspect is not readily demonstrated directly.   However, several features,
such as intramembranous haustoria, suggest that the fungus is
nutritionally dependent upon the alga at least to some extent.
M. ascophylli has never been found in a free living state, although
various clones of the fungus have now been grown successfully in
culture and asexual sporulation achieved in many isolates (Higgins
1984).   Such specific association with two brown algal species suggests
a nutritional requirement only fulfilled by these species.

Although *M. ascophylli* appears to produce mannitol, the most common
fungal polyol as well as the main photosynthetic product of brown algae,
the present investigation indicates that a mannitol uptake mechanism
does not occur in the cell membranes of fucoid seaweeds, including
P. canaliculata and *A. nodosum*.   It is unlikely therefore that
carbohydrate transfer from alga to fungus, in the manner demonstrated

for lichens, occurs in the form of mannitol, and the results support
this conclusion. However, it is possible that the re-wetting of a
desiccated thallus by the incoming tide may lead to a temporary increase
in the permeability of algal membranes (Simon 1974; Farrar & Smith
1976) resulting in the release of mannitol which would then become
available for uptake by the fungus. Again, nutrients may be released
from the algal cells in some other form and become available to the
fungus, or, the fungus may utilize brown algal intercellular, wall or
storage polysaccharides such as sulphated polysaccharide, alginic acid
or laminaran. Mannitol, as well as being a major metabolite of the
alga is however assumed to serve as the soluble fungal product, since
exogenously supplied glucose is converted to mannitol and unconverted
exogenous mannitol is accumulated by the fungus.

The only indication of any nutritional relationship between
M. ascophylli and its brown algal partner was the apparent stimulation
by the fungus of photosynthetic volemitol production by *P. canaliculata*.
There was also no evidence that exogenous carbohydrates taken up by the
fungus subsequently became available to the alga, and no advantage
appeared to be conferred by the fungus on the alga, even at a juvenile
stage.

In the absence of information on the nutritional aspects of
M. ascophylli associations it is difficult to fit them into the general
scheme of symbiosis. Kohlmeyer & Kohlmeyer (1972) referred to them as
mycophycobioses, and they defined mycophycobiosis as "a permanent
symbiotic association between a systemic marine fungus and a marine

Figures 16.3a-c   GLC traces of neutral extracts of
P. canaliculata.   a, untreated mature thallus, with
M. ascophylli. (Attenuation = 20K).   b, sporeling less than
5 mm long (10K).   c, mature thallus preincubated with
Nystatin (20K). RI = 0.78 (mannitol); 1.01; 1.11 (α -
volemitol); 1.21 (β - volemitol).

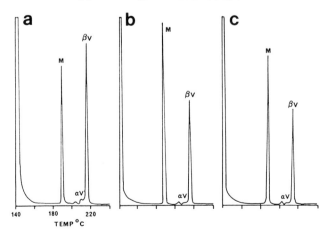

alga, in which the habit of the alga dominates". From the present
study, fungal dependence on the alga is implied, but there is no
evidence for any benefits to the alga from the presence of the fungus.
If, as Smith (1975) suggested in connection with lichens "mutualism were
defined in terms of mutual exploitation rather than benefit", then the
M. ascophylli associations also might be accepted as examples of
mutualistic symbiosis.

REFERENCES

Bidwell, R.G.S. & Ghosh, N.R. (1962). Photosynthesis and metabolism of
      marine algae. IV. The fate of ^{14}C-mannitol in *Fucus
      vesiculosus*. Can. J. Bot., 40, 803-807.
Bidwell, R.G.S. & Ghosh, N.R. (1963). Photosynthesis and metabolism of
      marine algae. V. Respiration and metabolism of ^{14}C-labelled
      glucose and organic acids supplied to *Fucus vesiculosus*.
      Can. J. Bot., 41, 155-163.
Cartwright, R.Y. (1975). Antifungal drugs. J. Antimicrobial
      Chemotherapy, 1, 141-162.
Drew, E.A. (1969). Uptake and metabolism of exogenously supplied sugars
      by brown algae. New Phytol., 68, 35-43.
Farley, J.F., Jersild, R.A. & Niederpreum, D.J. (1975). Origin and
      ultrastructure of intra-hyphal hyphae in *Trichophyton
      terrestre* and *T. rubrum*. Arch. Mikrobiol., 106, 195-201.
Farrar, J.F. & Smith, D.C. (1976). Ecological physiology of the lichen
      Hypogymnia physodes. III. The importance of the rewetting
      phase. New Phytol., 77, 115-125.
Harman, J.W. & Masterson, J.G. (1957). The mechanism of Nystatin action.
      Irish J. Med. Sci., 378, 249-253.
Higgins, J. (1984). Distribution, growth and differentiation of the
      endosymbiotic ascomycete fungus *Mycosphaerella ascophylli*
      Cotton isolated from the fucoid alga *Ascophyllum nodosum*
      (L.) LeJolis growing in different intertidal environments.
      M.Sc. thesis, University of Maine, Orono, U.S.A.
Holligan, P.M. & Drew, E.A. (1971). Routine analysis by gas-liquid
      chromatography of soluble carbohydrates in extracts of plant
      tissues. II. Quantitative analysis of standard carbohydrates
      and the separation and estimation of soluble sugars and
      polyols from a variety of plant tissues. New Phytol., 70,
      271-297.
Kingham, D.L. (1976). Studies relating to the fucacean endomycobiont
      Mycosphaerella ascophylli. Ph.D. thesis, University of
      Leeds, U.K.
Kohlmeyer, J. & Kohlmeyer, E. (1972). Is *Ascophyllum nodosum* lichenized?
      Botanica mar., 15, 109-112.
Kremer, B.P. (1973). Untersuchungen zur Physiologie von Volemit in der
      marinen Braunalge *Pelvetia canaliculata*. Mar. Biol., 22,
      31-35.
Kremer, B.P. (1976). Distribution and biochemistry of alditols in the
      genus *Pelvetia*. Br. phycol. J., 11, 239-342.
Kwon, Y.M. & Grant, B.R. (1971). Assimilation and metabolism by
      Dunaliella tertiolecta. I. Uptake by whole cells and

　　　metabolism by cell free systems. Pl. Cell Physiol., 12, 29-39.

Lindberg, B. & Paju, J. (1954). Low-molecular carbohydrates in algae. IV. Investigation of *Pelvetia canaliculata*. Acta Chem. Scand., 8, 817-820.

Quillet, M. (1957). Volemitol et mannitol chez les Pheophycees. Bull. due lab. Maritime de Dinard, 43, 119-124.

Simon, E.W. (1974). Phospholipids and plant membrane permeability. New Phytol., 73, 377-420.

Smith, D.C. (1975). Symbiosis and the biology of lichenised fungi. Symp. Soc. exp. Biol., 29, 373-407.

Starr, M.P. (1975). A general scheme for classifying organismic associations. Symp. Soc. exp. Biol., 29, 1-20.

Van Steveninck, J. (1968). Transport and transport associated phosphorylation of 2-deoxy-D-glucose in yeast. Biochim. Biophys. Acta, 163, 386-394.

Webber, F.C. (1959). Marine fungi. Ph.D. thesis, University College of Wales, Aberystwyth, U.K.

Webber, F.C. (1967). Observations on the structure, life history and biology of *Mycosphaerella ascophylli*. Trans. Br. mycol. Soc., 50, 583-601.

17 ARE FUNGAL DISEASES SIGNIFICANT IN THE MARINE ENVIRONMENT?

D.J. Alderman

J.L. Polglase

The answer to the title of this chapter is not immediately obvious. The reasons for this are of interest and are discussed below. Major reviews on marine diseases include Johnson & Sparrow (1961), Alderman (1976, 1982), Sindermann (1977), Lightner (1981, 1983) and Johnson (1983). The publications covered by these reviews are widely dispersed, consisting largely of reports of unusual cases rather than intensive studies of individual diseases. The difficulties of interpretation of the significance of the role of disease organisms in natural populations of marine animals whose own ecology is poorly understood are obvious. Thus, Johnson & Sparrow (1961), commenting on *Ichthyophonus* infections of herring, said "until herring *per se* is intensively studied synecologically, there probably will be little or no confirmed data on the role of hydrographic factors in (*Ichthyophonus*) disease". Only where the suspect populations are sedentary (e.g. bivalve molluscs), will the effect of diseases of wild populations be readily evident. As a result, only those diseases which affect species of major economic importance, particularly species in intensive aquaculture, have been investigated sufficiently to enable their significance to be understood, and significance inevitably means economic significance.

Attitudes on fungal diseases have changed noticeably over the past 25 years. Thus, in the first detailed review covering this area (Johnson & Sparrow 1961), *Lagenidium* infestations of crustaceans were listed under "apparently minor diseases of marine animals". The more recent reviews of Sindermann (1977), Lightner (1981) and Johnson (1983) show *Lagenidium* to be a most serious pathogen of marine crustaceans. The best known is *L. callinectes* Couch, which has been found to be pathogenic to the eggs and larvae of a wide range of marine crustaceans, particularly the blue crab, *Callinectes sapidus* Rathbun, the Dungeness crab, *Cancer magister* Dana, most *Penaeus* species and the lobsters, *Homarus americanus* Milne Edwards and *H. gammarus* L. In commercial hatcheries, *L. callinectes* forms a major obstacle to successful operation and rapidly attacks various larval stages in hatchery tanks for crabs, prawns and lobsters. Mortalities can reach 100% in 48 h and develop most rapidly when hatchery conditions are poor with dead eggs and larvae forming reservoirs of fungal infection. Its effects are not limited to intensive aquaculture facilities, since Sindermann (1977) reports prevalence up to 95% with *L. callinectes* infections of crab egg masses in natural populations of blue crab on the east coast of the U.S.A. Infections are mainly confined to the outer part of the egg masses and are responsible for egg mortalities up to 50%. Infection levels such as this must clearly have a major effect on the blue crab populations of such regions. Two other species of *Lagenidium* have been described as pathogens of marine crustaceans. Bian *et al.* (1977) described *L. scyllae* Bian, Hatai et Egusa causing a lethal disease of the larvae and eggs of the mangrove crab, *Scylla serrata* Forsskål, in experimental hatchery facilities whilst *L. chthamalophilum* Johnson was reported from barnacle eggs (Johnson 1958).

Other fungal diseases of Crustacea are less well understood. Amongst the Oomycetes, *Haliphthoros* seems to have almost as great a pathogenic potential to crustaceans as does *Lagenidium*. *H. milfordensis* Vishniac, largely under experimental hatchery conditions, has been found to infect a range of crustacean eggs and larvae, plus the eggs of the oyster drill, *Urosalpinx*. In lobsters it has been shown to be capable of producing >95% mortalities in larvae in a hatchery (Fisher *et al.* 1975). The very similar *H. philippinensis* Hatai, Bian, Baticados et Egusa (Hatai *et al.* 1980), described from the larvae of *Penaeus monodon* Fabricius, has equal pathogenic potential, but neither of these fungi is well known in natural populations.

Two species of *Atkinsiella* are known from crustacean eggs. *A. dubia* (Atkins) Vishniac was discovered parasitizing the eggs of *Pinnotheres pisum* L. and *Goneplax rhomboides* Pennant (Atkins 1954 a) and was experimentally demonstrated to infect the eggs of several other marine crustaceans. *A. hamanaensis* Bian et Egusa was, like *L. scyllae*, parasitic on the larvae and ova of the mangrove crab (Bian & Egusa 1980). These authors believed that *A. hamanaensis* was a more effective pathogen than *L. scyllae*, but no field data are available on the effects of *Atkinsiella* spp. in natural populations.

Clam and oyster larvae in laboratory conditions on the east coast of the United States were the subject of sporadic mycoses owing to *Sirolpidium zoophthorum* Vishniac (Davis *et al.* 1954). Members of the genus *Pythium* have occasionally been recorded from marine crustaceans. Atkins (1955) found *P. thalassium* Atkins on pea crab egg masses and showed it to be capable of infecting viable eggs of a range of marine crustaceans. Under experimental rearing conditions, the marine prawn, *Palaemon serratus* (Pennant), was occasionally found to be subject to an epidemic systemic fungal disease caused by what the authors felt to be *P. afertile* Kanouse et Humphrey (Delves-Broughton & Poupard 1976). Infection could be transmitted by feeding fungus-contaminated normal diet, producing infections of up to 50% within 17 days.

Two members of the Leptolegniales have also been associated with crustacean mortalities. The first, *Leptolegniella marina* (Atkins) Dick, has been identified from the eggs and gills of *P. pisum*, in *Barnea candida* L. and *Cardium echinatum* L. (Atkins 1954 b), and on the eggs of the blue crab *Callinectes sapidus* (Johnson & Pinschmidt 1963). Again, these were largely from experimental conditions, so that significance in the natural environment is unknown. In contrast, *Leptolegnia baltica* Höhnk et Vallin is known, not from experimental conditions, but from its association with massive mortalities of the copepod, *Eurytemora hirundoides* (Nordquist) (Vallin 1951; Höhnk & Vallin 1953) in the Baltic. The species is oligohaline and was recorded as producing a systemic infection in the copepod, resulting in mortalities so great that the water surface was turbid with dead floating copepods, which formed a slimy coating on plankton nets.

Johnson & Sparrow (1961) commented on the "nearly barren status of knowledge of marine plankton epiphytotics and epizootics". Although Höhnk & Vallin (1953) did not produce evidence that *L. baltica* was the cause of *E. hirundoides* mortalities, there are other reports that indicate the occurrence of widespread fungal infestations of zoo- and phyto-plankton. Sparrow (1936) recorded up to 88% infection of two common diatoms by *Ectrogella perforans* H.E. Petersen. Aleem (1953) reported a chytridiaceous fungus to be the cause of a "disease" of *Navicula* spp. in deep-water sediments, and Jiroveck (1956) found 5% prevalence of *Olpidium* spp. in a marine rotifer. Artemchuk & Zelzinskaia (1969) reported observations of fungal infection of the cladoceran *Penilia avirostris* (Dana) by an organism identified as *Hyphochytrium peniliae* Artemchuk et Zelzinskaia. Mortalities occurred in the Black Sea over a five-year period in late summer, reducing the susceptive populations by nearly 98% in three weeks. More recently, Gaertner (1979), in the Rosfjord, southern Norway, found the diatom *Thalassiosira nordenskioeldii* Cleve parasitized by a marine chytrid and *Thalassionema nitzchioides* Grunow parasitized by a thraustochytrid probably *Schizochytrium*. The number of parasitized cells reached 35% over a very short period.

The significance of the Thraustochytriales and Labyrinthulales as aquatic pathogens is best discussed here. The Thraustochytriales are ubiquitous lower marine fungi which can readily be cultured from marine animal and marine algal surfaces. In the past five years, three cases of true pathogenicity by thraustochytrid fungi have been reported. Polglase (1980) described a progressive, erosive epidermal infection of the lesser octopus, *Eledone cirrhosa* Lamarck. This was first recognized in aquarium conditions and was subsequently found on newly-caught animals. Jones & O'Dor (1983) found similar infections of squid in Canadian waters where the infection was more systemic than in the superficial octopus lesions. Finally, McLean & Porter (1982) described a yellow pustule disease from a nudibranch mollusc in which the thraustochytrid involved was responsible for a purely deep-seated infection. Polglase & Brown (unpublished) have also detected thraustochytrid involvement in skin lesions of marine farmed rainbow trout. Between 1930 and 1938, several papers were published describing a wasting disease in eel grass (*Zostera* spp.) in the United States. This was occasioned by the observation of widespread dieback of *Zostera* stands on various parts of the American coast. Some authors (Duncan & Cotton 1933) suggested pollution as a cause, whilst Tutin (1938) postulated a largely environmental cause; but Renn (1935, 1936) and

Young (1937) attributed the problem to invasion by *Labyrinthula* spp.
Certainly, *Labyrinthula* is always present and is readily isolable from
moribund stands of *Zostera* and *Spartina*, but, since the period of
interest in the 1930s, there has been no published evidence which would
prove *Labyrinthula* to be a primary pathogen of these marine grasses.

The extensive culture for food of the red alga *Porphyra* in Japan (where
it is called nori), is subject to a fungal disease caused by *Pythium*
spp. which causes problems, especially at deeper warm water sites where
significant proportions of the crop may be destroyed (Korringa 1976).

Hyphomycete infections of marine animals comprise one reasonably
well-known disease caused by *Fusarium* in crustaceans in intensive
aquaculture and a number of case reports of infection, largely in marine
fish, involving *Cladosporium* and *Exophiala* spp. A characteristic of
hyphomycete infections is that the fungi concerned are not obligately
marine and are capable of producing similar pathologies in the
freshwater environment. Thus, *Fusarium* infections are also known from
freshwater crustaceans (Alderman & Polglase 1985), as are *Exophiala*
infections in freshwater fish (Alderman & Feist 1985). The majority of
marine crustacean infections are interpreted as wound infections and are
attributed to *F. solani* (Mart.) Sacc., which has been isolated from a
wide range of crustaceans, principally crabs, lobsters and penaeid
shrimps from many parts of the world. *F. solani* infections in penaeids
mainly involve the gills and the host melanization reactions produce
gross signs in the form of blackening of the affected gills which gives
the name "black gill disease". Hose *et al.* (1984) discussed the disease
and its pathogenesis in *Penaeus californiensis* Holmes. Differences in
susceptibility to *F. solani* infections were reported; *P. japonicus* Bate
and *P. californiensis* being most susceptible, *P. stylirostris* Stimpson
moderately susceptible, and *P. vannamei* Boone and *P. monodon* (Fabricius)
least susceptible. One hundred per cent of a group of artificially-
wounded *P. californiensis* experimentally exposed to *F. solani* conidia
became infected whilst 50% of a wounded, but not experimentally-exposed
group, also developed infection from conidia which were naturally
present in the experimental water supply. Gross signs of infection were
apparent in all experimentally-infected animals by 14 days at 24° C and
by 24 days 50% mortality had occurred. From these results it is evident
that *F. solani*, although an opportunistic wound parasite, is a very
effective and highly pathogenic opportunist. Under intensive artificial
culture conditions, minor wounds are frequent and since *Fusarium* conidia
are common in the aquatic environment it is not surprising that *Fusarium*
infections are frequently reported from marine crustacean aquaculture
sites. Infections in wild stock have not been reported, but have been
observed in fresh water crayfish (Alderman & Polglase unpublished) and
undoubtedly occur in wild marine crustaceans. *Cladosporium* spp.
infections occur more rarely and largely in the form of case reports.
Thus, Polglase *et al.* (1984) reported *Cladosporium* infections of wounds
on the lesser octopus, *Eledone cirrhosa*, whilst Reichenbach-Klinke
(1956) implicated *Cladosporium* spp. from lesions in cod (*Gadus
morhua* L.). *Cladosporium* was also isolated by Miller & Flemming (1983)
infecting the gills of lobsters (*H. americanus*) which were infested with

a nemertean parasite.

As with *Fusarium*, *Exophiala* infections have been reported both from
marine and fresh water animals. Blazer & Wolke (1979) reported an
Exophiala from five genera of captive fish in a marine aquarium. These
were single cases for each species, two animals showing external dermal
masses attributed to the fungus and two with internal gross lesions.
Transmission was achieved by intra-peritoneal injection of spores and
hyphae, but not by exposure of fish to spores and/or hyphae in water.
Richards *et al.* (1978) described an outbreak of *E. salmonis* Carmichel
infection in Atlantic salmon, *Salmo salar* L., held in a sea cage in
Scottish waters. During a four-month period, 10% of 1,250 fish died in
a single sea cage, all deaths being associated with infection with
E. salmonis. Identical fish in adjacent cages were not affected and the
authors postulated that contaminated food may have been the source of
infection. These *Exophiala* lesions are essentially chronic granulomas.
Many fish pathologists have observed similar lesions which, on
histopathological observation, have been found to contain fungal hyphae.
These observations are largely made on fixed material and case reports
are rarely published. A recent report of this type of lesion was made
by Anders (1984), who described mouth granulomas in smelt (*Osmerus
eperlanus* L.) in which septate fungal hyphae were demonstrated.
Adequate descriptions of such lesions are rare, perhaps owing to the
problems of obtaining fresh material suitable for isolation and
identification from commercial catches of wild fish stocks.

Ostracoblabe implexa Bornet et Flahault, a mycelial fungus of uncertain
affinity, is responsible for fungal infection of the shells of the
European oyster *Ostrea edulis* L. (Alderman & Jones 1971). The damage
caused by the shell invasion results in a reduced sale value for the
oyster, which is normally sold fresh on the half-shell and there is a
reduction in the condition index of the animal. The incidence of
infection is high on shallow, warmer water oyster beds, particularly
when poor husbandry results in large amounts of old, infected shells
remaining on the ground to infect young animals. Under normal
conditions of husbandry, the economic significance of the disease is
small.

If scientific interest were to be regarded as an indication of
significance, then *Ichthyophonus* (provided that *Ichthyophonus* is
accepted as a fungus) infections of marine fish might be regarded as of
low significance in recent years, since few papers have been published
on it in the 20 years following Johnson & Sparrow's review in 1961.
Prior to that, Sindermann (1956, 1958) and Sindermann & Scattergood
(1954) had described extensive mortalities of herring in the Gulf of
St Lawrence which were attributed to *Ichthyophonus*. Mortalities were on
a scale sufficient to produce reports describing floating shoals of dead
fish on the water surface, together with fouling of fishing nets with
dead fish and large numbers of dead fish being washed up on the nearby
shores of the Gulf. From earlier reports, Sindermann (1963, 1970)
believed that at least six earlier epizootics of *Ichthyophonus* could be
identified since 1900 and regarded it as the major limiting factor to

western North Atlantic herring populations. Despite these very strong indications that *Ichthyophonus* is of major significance, little further interest in it was expressed until it was recognized that some natural populations of plaice in northern Scottish waters were and had been for many years subject to a >55% annual mortality owing to *Ichthyophonus* (McVicar 1981, 1982). Despite the fact that plaice is an important commercial species, the existence of a disease capable of producing such mortalities had not been recognized scientifically until the mid-1970's (McVicar 1982). Although *Ichthyophonus* infection produces its most important population effect in flat fish, infections in species such as herring have a significant economic effect in rendering the infected fish unsaleable. Infected herring are unsuitable for smoking owing to post-mortem germination of *Ichthyophonus*, which results in muscle necrosis. In Scotland, fish merchants have used the term "greasers" to describe and reject infected herring since at least 1913 (McVicar 1982). Although Sindermann (1970) identified at least seven epizootics of *Ichthyophonus* in the western North Atlantic up to the beginning of the 1960's, no further such epizootics have been reported. Have further epizootics, other than the Scottish one, occurred and not been noted?

The sole ascomycete parasite of marine animals is *Trichomaris invadens* Hibbits, Hughes et Sparks, on the Alaskan tanner crab *Chionoecetes bairdi* Rathban (Hibbits *et al.* 1981). The fishery for this crab was worth more than $50,000,000 in 1977. *T. invadens* infection is known as black mat disease and incidences up to 75% have been recorded in natural populations (Van Hyning & Scarborough 1973). No proof of the pathogenicity of *T. invadens* has been published so far, but Sparks & Hibbits (1979) commented that the way in which the fungus penetrates and proliferates in many internal tissues encourages the assumption that it is a serious and perhaps virulent disease.

In addition to those mycoses in which the fungal pathogen is known and has at least been named, there are apparently significant diseases which are reported to be of fungal aetiology, for which no description of the supposed pathogens has been published. The first of these is the sponge wasting disease which reached epizootic proportions in 1938 in the Bahamas, the West Indies and Florida, only to subside in 1940. Galtsoff (1942) reported the disease was of fungal aetiology, but gave no description of the supposed pathogen. Mortalities were 75-90% in many areas, causing a brief but severe economic impact on the commercial sponge industry. A second, poorly understood disease is black line disease in star corals which has been reported in several Atlantic reefs, including Bermuda, Barbados, Florida and Venezuela. Ramos-Flores (1983) describes the presence of fungal hyphae growing within both the polyp and the hard part of the coral. The fungus has not yet been identified, nor has there been any demonstration of its pathogenicity.

From consideration of the information presented above, a number of points are immediately apparent. Firstly, fungal diseases are of major significance to crustaceans, particularly under conditions of intensive aquaculture where they form the major obstacles to successful hatchery operation. In the natural environment, fungi are responsible for major epizootics which, as in the case of *Leptolegnia baltica* with copepods and *Ichthyophonus hoferi* Plehn et Muhlsow in marine fish, can cover large areas of the surface of the sea with the bodies of their victims. Although the effects of these diseases in the natural environment are evident, the causes of such major outbreaks are not understood, because they have not been adequately investigated. Equally apparent from the above review is the fact that there are considerable numbers of infections such as "black mat disease" of tanner crabs in which significant fungal pathogenicity is suggested by investigators, but who for various reasons were unable to provide confirmatory evidence. With these cases and those in which it is impossible to identify the fungal pathogens, it is simply not possible to determine whether the fungus observed is simply a secondary saprophyte, an opportunist parasite or a primary pathogen. Certainly, the information available on hyphomycete infections tends to indicate that many if not all are opportunists, as with the crustacean wound infections produced by *Fusarium* spp. The ubiquitous presence of *Fusarium* spores, however, means that many of the opportunities provided will be taken up in the form of pathogenic effects on susceptive animals.

Apart from animals in aquaculture and in aquaria, pathological investigations of fungal diseases of marine animals are hampered by the impossibility of identifying hyphal fragments in fixed or frozen tissues of miscellaneous specimens. However, from the available information on this type of material, together with the limited number of documented drastic mortalities, it is reasonable to conclude that fungal diseases are important in the marine environment and in many cases can act as major limitations on natural and cultured populations of marine animals. Their significance to phyto- and zoo-plankton remains to be shown, but from already recognized mortalities, the potential is clearly great.

REFERENCES
Alderman, D.J. (1976). Fungal diseases in marine animals. In Recent
        Advances in Aquatic Mycology, ed. E.B.Gareth Jones,
        pp. 223-260. London: Paul Elek.
Alderman, D.J. (1982). Fungal diseases of aquatic animals. In Microbial
        Diseases of Fish, ed. R.J. Roberts. Society for General
        Microbiology, Special Publication, 9, 189-242. London:
        Academic Press.
Alderman, D.J. & Feist, S.W. (1985). *Exophiala* infection of kidney of
        rainbow trout recovering from proliferative kidney disease.
        Trans. Br. mycol. Soc., 84, 157-159.
Alderman, D.J. & Jones, E.B.G. (1971). Shell disease of oysters.
        Fishery Investigations, London, Ser. II, 26, no. 8, 1-19.
Alderman, D.J. & Polglase, J.L. (1985). *Fusarium tabacinum* (Beyma) Gams.
        as a gill parasite in the crayfish *Austropotamobius pallipes*
        Lereboullet. J. Fish Dis., 8, 249-252.
Aleem, A.A. (1953). Marine fungi from the west coast of Sweden. Arkiv
        für Bot., Ser. 2, 3, 1-33.

Anders, K. (1984). Preliminary results of histopathological and epidemiological studies on two "new" diseases of smelt (*Osmerus eperlanus* L.). Bull. Eur. Assoc. Fish Pathol., 4, 62-63.

Artemchuk, N.J. & Zelzinskaia, L.M. (1969). The marine fungus *Hyphochytrium peniliae* n. sp. infecting the zooplanktonic crustacean *Penilia avirostris* (Dana). Mik. Fitopat., 3, 356-359.

Atkins, D. (1954 a). A marine fungus *Plectospira dubia* n. sp. (Saprolegniaceae) infecting crustacean eggs and small Crustacea. J. mar. biol. Ass. U.K., 33, 721-732.

Atkins, D. (1954 b). Further notes on a marine member of the Saprolegniaceae, *Leptolegnia marina* n. sp. infecting certain invertebrates. J. mar. biol. Ass. U.K., 33, 613-625.

Atkins, D. (1955). *Pythium thalassium* n. sp. infecting the egg mass of the pea crab *Pinnotheres pisum*. Trans. Br. mycol. Soc., 38, 31-46.

Bian, B.Z. & Egusa, S. (1980). *Atkinsiella hamanaensis* sp. nov. isolated from cultivated ova of the mangrove crab, *Scylla serrata* (Forsskål). J. Fish Dis., 3, 373-386.

Bian, B.Z., Hatai, K.P.G.L. & Egusa, S. (1977). Studies on the fungal diseases in crustaceans. I. *Lagenidium scyllae* sp. nov. isolated from cultivated ova and larvae of the mangrove crab (*Scylla serrata*). Trans. mycol. Soc. Jap., 20, 115-124.

Blazer, V.S. & Wolke, R.E. (1979). An *Exophiala*-like fungus as the cause of a systemic mycosis of marine fish. J. Fish Dis., 2, 145-152.

Davis, H.C., Loosanoff, V.L., Weston, W.H. & Martin, C. (1954). A fungus disease in clam and oyster larvae. Science, N.Y., 120, 36-38.

Delves-Broughton, J. & Poupard, C.W. (1976). Disease problems of prawns in recirculation systems in the U.K. Aquacult., 7, 201-217.

Duncan, F.M. & Cotton, A.D. (1933). Disappearance of *Zostera marina*. Nature, Lond., 132, 483.

Fisher, W.S., Nilson, E.H. & Schlesser, R.A. (1975). Effect of the fungus *Haliphthoros milfordensis* on the juvenile stages of the American lobster *Homarus americanus*. J. Invert. Path., 26, 41-45.

Gaertner, A. (1979). Some fungal parasites found in the diatom populations of the Rosfjord area (south Norway) during March 1979. Veröff. Inst. Meeresforsch. Bremerh., 18, 29-33.

Galtsoff, P.S. (1942). Wasting disease causing mortality of sponges in the West Indies and Gulf of Mexico. Proceedings of the 8th American Science Congress, 1940, 3, 411-421.

Hatai, K., Bian, B.Z., Baticados, C.A. & Egusa, S. (1980). Studies on the fungal diseases in crustaceans. II. *Haliphthoros philippinensis* sp. nov. isolated from cultivated larvae of the jumbo tiger prawn. Trans. mycol. Soc. Jap., 21, 47-55.

Hibbits, J., Hughes, G.C. & Sparks, A.K. (1981). *Trichomaris invadens* gen. sp. nov. An ascomycete parasite of the tanner crab *Chionoecetes bairdi* Rathbun (Crustacea, Brachyura). Can. J. Bot., 59, 2121-2128.

Höhnk, W. & Vallin, S. (1953). Epidemisches Absterben von *Eurytemora*
in Bothnischen Meerbusen, verursacht durch *Leptolegnia*
baltica nov. spec. Veröff. Inst. Meeresforsch. Bremerh., 2,
215-223.

Hose, J.E., Lightner, D.V., Redman, R.M. & Danald, D.A. (1984).
Observations on the pathogenesis of the imperfect fungus,
Fusarium solani, in the Californian brown shrimp, *Penaeus*
californiensis. J. Invert. Path., 44, 292-303.

Jiroveck, O. (1956). Parasiten der Rotatorien in plankton javanscher
Seen. Arch. Hydrobiol. Suppl., 23, 105-108.

Johnson, P.T. (1983). Diseases caused by Viruses, Rickettsiae, Bacteria
and Fungi. In The Biology of Crustacea, 6, Pathobiology,
ed. A.J. Provenzano, pp. 1-78. New York: Academic Press.

Johnson, T.W. Jr. (1958). A fungus parasite in the ova of the barnacle
Chthalamus fragilis denticulata. Biol. Bull. mar. biol.
Lab., Woods hole, 114, 205-214.

Johnson, T.W. Jr. & Pinschmidt, W.C. Jr. (1963). *Leptolegnia marina*
Atkins in blue crab ova. Nova Hedwigia, 5, 413-418.

Johnson, T.W. Jr. & Sparrow, F.K. (1961). Fungi in Oceans and Estuaries.
Weinheim: J. Kramer.

Jones, G.M. & O'Dor, R.K. (1983). Ultrastructural observations on a
Thraustochytrid fungus parasitic in the gills of squid (*Ilex*
illecebrosus LeSueur). J. Parasit., 69, 903-911.

Korringa, P. (1976). Farming marine organisms low in the food chain.
Developments in Aquaculture and Fisheries Research I.
Amsterdam: Elsevier.

Lightner, D.V. (1981). Fungal diseases of marine Crustacea. In
Pathogenesis of Invertebrate Microbial Diseases, ed.
E.W. Davidson, pp. 451-484. Totowa, New Jersey: Allanheld,
Osmun.

Lightner, D.V. (1983). Diseases of cultured penaeid shrimp. In
Mariculture, ed. J.P. McVey, pp. 289-320. Boca Raton,
Florida: CRC Press.

McLean, N. & Porter, D. (1982). The yellow spot disease of *Tritonia*
diomedea Bergh (Mollusca, Gastropoda, Nudibranchia):
encapsulation of the Thraustochytriaceous parasite by host
amoebocytes. J. Parasit., 68, 243-252.

McVicar, A.H. (1981). An assessment of *Ichthyophonus* disease as a
component of natural mortality in plaice populations in
Scottish waters. International Council for the Exploration
of the Sea. Demersal Fish Committee, Doc. G: 49, 1-7.

McVicar, A.H. (1982). *Ichthyophonus* infections of fish. In Microbial
Diseases of Fish, ed. R.J. Roberts, pp. 243-269. Society for
General Microbiology, Special Publication 9. London:
Academic Press.

Miller, J.D. & Flemming, L.C. (1983). Fungi associated with an
infestation of *Pseudocarcinonemertes homari* on *Homarus*
americanus. Trans. Brit. mycol. Soc., 80, 9-12.

Polglase, J.L. (1980). A preliminary report on the Thraustochytrid(s)
and Labyrinthulid(s) associated with a pathological
condition in the lesser octopus *Eledone cirrhosa*. Botanica
mar., 23, 699-706.

Polglase, J.L., Dix, N.J. & Bullock, A.M. (1984). Infection of skin wounds in the lesser octopus *Eledone cirrhosa* by *Cladosporium sphaerospermum*. Trans. Brit. mycol. Soc., 82, 577-580.

Ramos-Flores, T. (1983). Lower marine fungus associated with black line disease in star corals (*Monastrea annularis* E. & S.). Biol. Bull. mar. biol. Lab., Woods Hole, 165, 429-435.

Reichenbach-Klinke, H.H. (1956). Über einige bisher unbekannte Hyphomyceten bei Verscheidenen Süswasser und Meeresfischen. Mycopath. Mycol. appl., 7, 333-347.

Renn, C.E. (1935). A mycetozoan parasite of *Zostera marina*. Nature, Lond., 134, 416.

Renn, C.E. (1936). The persistence of the eel-grass disease and parasite on the American Atlantic coast. Nature, Lond., 138, 507-508.

Richardson, R.H., Holliman, A. & Helgason, S. (1978). *Exophiala salmonis* infection in Atlantic salmon, *Salmo salar* L. J. Fish Dis., 1, 357-368.

Sindermann, C.J. (1956). Diseases of fish in the western North Atlantic. IV. Fungus disease and resultant mortalities in the Gulf of St. Lawrence in 1955. Maine Department of Sea and Shore Fisheries, Res. Bull., 25, 1-23.

Sindermann, C.J. (1958). An epizootic in Gulf of St. Lawrence fishes. Transactions of the 23rd North American Wildlife Conference, 349-360. Washington: Wildlife Management Institute.

Sindermann, C.J. (1963). Disease in marine populations. Transactions of the 28th North American Wildlife Conference, 336-356. Washington: Wildlife Management Institute.

Sindermann, C.J. (1970). Principle Diseases of Marine Fish and Shell-fish. New York: Academic Press.

Sindermann, C.J., ed. (1977). Disease Diagnosis and Control in North American Marine Aquaculture. Amsterdam: Elsevier.

Sindermann, C.J. & Scattergood, L.W. (1954). Diseases of fishes of the western North Atlantic. II. *Ichthyosporidium* disease of the sea herring (*Clupea haerengus*). Maine Department of Sea and Shore Fisheries, Research Bulletin, 19, 1-40.

Sparks, A.K. & Hibbits, J. (1979). Black mat syndrome, an invasive disease of the tanner crab, *Chionoecetes bairdi*. J. Invert. Path., 34, 184-191.

Sparrow, F.K. Jr. (1936). Biological observations on the marine fungi of Woods Hole waters. Biol. Bull. mar. biol. Lab., Woods Hole, 70, 236-263.

Tutin, T.G. (1938). The antecology of *Zostera marina* in relation to its wasting disease. New Phytol., 37, 50-71.

Vallin, S. (1951). Plankton mortality in the northern Baltic caused by a parasitic water mould. Reports, Institute of Fresh Water Research, Drottningholm, 32, 139-148.

Van Hyning, J.M. & Scarborough, A.M. (1973). Identification of fungal encrustation on the shell of the snow crab (*Chionoecetes bairdi*). J. Fish. Res. Bd. Can., 30, 1738-1739.

Young, E.L. (1937). Notes on the labyrinthulan parasite of the eel-grass *Zostera marina*. Bull. Mt. Desert Island Biol. Lab., 1937, 33-35.

J. Kohlmeyer

INTRODUCTION

Filamentous higher marine fungi are a heterogeneous assemblage that may be divided into primary and secondary marine species (Kohlmeyer 1974; Kohlmeyer & Kohlmeyer 1979). This hypothesis proposes that primary marine fungi evolved directly from marine ancestors and did not leave the marine environment, whereas secondary marine species derived from terrestrial ancestors and returned secondarily into marine habitats. The first part of this chapter presents a new classification for marine representatives of the division Ascomycota, based on recent treatises by Barr (1979 a,b, 1983) and Eriksson (1981, 1984). In the second part, characters considered ancient or evolved are discussed by applying phylogenetic principles, and examples of primary and secondary marine Ascomycotina are presented.

CLASSIFICATION OF FILAMENTOUS MARINE ASCOMYCOTINA

Since publication of *Marine Mycology* (Kohlmeyer & Kohlmeyer 1979) 12 new genera, 32 new species, 3 new varieties and 14 new combinations of marine Ascomycotina have been proposed. The main changes are the inclusions of classes and subclasses and additional families, following Barr (1983). Consequently, a rearrangement of the genera, most of those formerly included in Sphaeriales and Dothideales, was necessary (Table 18.1). I recognize 80 genera of Ascomycotina with marine members, with a total of 183 species and 3 varieties. Thirty-one of the genera are monotypic. Among the four classes and subclasses of Ascomycotina (= Ascomycota, Barr 1983) Ascosphaeromycetes are absent in marine environments, and Laboulbeniomycetes has only one marine species. The bulk of marine representatives is in the Ascomycetes (53 genera with 120 species and 3 varieties), followed by the Loculoascomycetes (27 genera with 63 species). The Halosphaeriales *ad int*. (Eriksson 1984) include about half of all marine genera in the Ascomycetes with 75 species and 2 varieties, i.e. 63% of all species in this class. Eight genera with 16 species of the Ascomycetes cannot be assigned to higher categories. In the Loculoascomycetes, the Pleosporales includes the largest number of marine taxa (9 genera with 29 species), followed by the Melanommatales (7 genera with 12 species), and the Dothideales (6 genera with 15 species).

THE POSSIBLE ORIGIN OF MARINE ASCOMYCOTINA

In the absence of a fossil record of higher marine fungi

extant marine species must be examined for characters that can be
considered archaic. Obligate parasitism is a fundamental attribute of
primitive fungi (Raper 1968; Savile 1968) and the antiquity of the host
reflects that of the parasite, because both evolved together. The asci,
as main diagnostic elements, and the ascospores are the most important
morphological characters to judge the relative age of genera within the
marine Ascomycotina. Thus, four features, viz., mode of nutrition,
phylogenetic antiquity of host, and the asci and ascospores will mainly
be used to separate primary from secondary marine fungi and the
phylogenetic scheme of origin (Figure 18.1) is based on these
characters.

Algae are phylogenetically the oldest group of host plants of marine
Ascomycotina and primary marine fungi can be expected to occur in
particular among algicolous species. Of course, not all fungi on algae
are primitive types. Symbiotic species of Loculoascomycetes occurring
in the intertidal zone, i.e. *Leiophloea pelvetiae* (Suth.) Kohlm. et
Kohlm., *Mycosphaerella apophylaeae* Kohlm. and *M. ascophylli* Cotton
(Figure 18.2), are considered secondary marine species. Primary marine
Ascomycotina live as parasites or possibly as symbionts on permanently
submerged algae. Examples are *Chadefaudia*, *Haloguignardia*,

Figure 18.1 Possible origin of marine filamentous higher fungi;
downward arrows indicate secondary marine fungi: hypothetical ancestors
are marked with asterisks.(revised from Kohlmeyer & Kohlmeyer 1979)

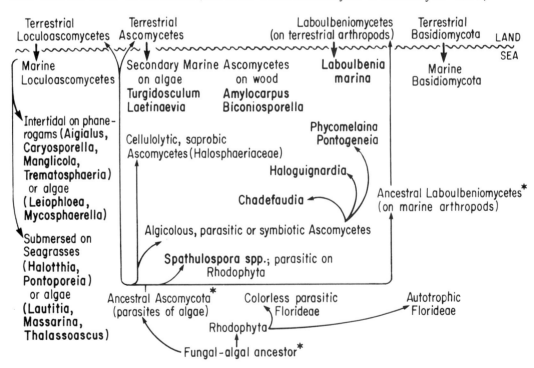

Figure 18.2 *Mycosphaerella ascophylli*, a bitunicate ascomycete with fissitunicate ascus dehiscence; endotunicas filling ostiole after ascospore release. Phase contrast. Bar = 20 μm.

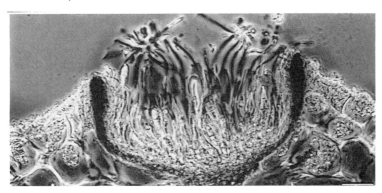

Figures 18.3a–c Thin-walled asci. Light microscopy, bars = 10 μm.
Figure 18.3a *Haloguignardia decidua*.
Figure 18.3b *Lignincola tropica*.
Figure 18.3c *Halosarpheia fibrosa*, base of deliquescing ascus.

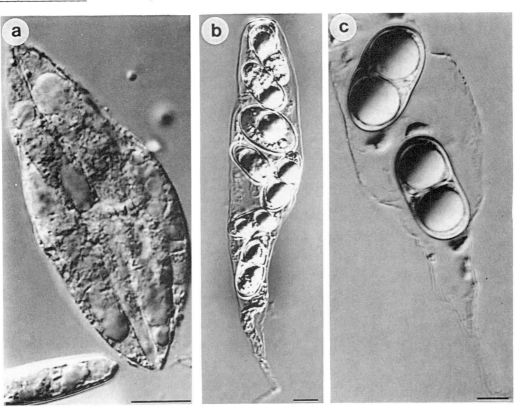

Phycomelaina, *Pontogeneia* and *Spathulospora* spp. (Figure 18.1).

Members of the Halosphaeriales are also primary marine, although this group may have developed later than the obligate parasites on algae, because most of the representatives of this order are saprobes on lignocellulose, originally a nonmarine substrate. They may have acquired the ability to produce cellulolytic enzymes in the course of evolution, after growing initially on algal products. Other representatives of the class Ascomycetes have close affinities with terrestrial relatives and are most likely secondary inhabitants of marine habitats, e.g. lignicolous species in the Eurotiales (*Amylocarpus*, *Eiona*), the Sordariales (*Biconiosporella*, *Zopfiella*) and the Hypocreales (*Halonectria*, *Heleococcum*, *Hydronectria*, *Nectriella*). The Laboulbeniomycetes, with 1730 species inhabiting mostly insects in terrestrial habitats, has only one marine representative, definitely a secondary marine species like its host, an intertidal beetle.

All marine species of the Loculoascomycetes can be considered secondary marine (Figure 18.1). Most of them live intertidally on phanerogams, e.g. mangrove roots, or on submerged seagrasses, both secondary invaders of marine habitats. More difficult to explain is the existence of totally submersed Loculoascomycetes as parasites of Phaeophyta, e.g. *Massarina cystophorae* (Cribb et Herbert) Kohlm. et Kohlm., *Thalassoascus lessoniae* Kohlm. and *T. tregoubovii* Ollivier, and Rhodophyta, e.g. *Lautitia danica* (Berlese) Schatz. They may turn out not to be characteristic members of Loculoascomycetes.

Among the morphological characters of the Ascomycotina, the ascus is one of the most important diagnostic features used in modern systematics (Reynolds 1981; Parguey-Leduc & Janex-Favre 1982, 1984). Denison & Carroll (1966) postulated that primitive marine Ascomycotina had deliquescing asci, and that the active discharge of ascospores by fissitunicate asci (Figure 18.2) developed as a mechanism for aerial dissemination, a view accepted by Kohlmeyer (1974), Kohlmeyer & Kohlmeyer (1979), and Eriksson (1981). Another adaptation to terrestrial life is the development of dark ascospore pigments that protect the propagules against radiation and desiccation (Savile 1968). Thus, parasitism on a marine algal host, combined with the possession of unitunicate, deliquescing asci and hyaline ascospores marks an ascomycete as most likely primary marine. In contrast, the combination of a marine phanerogam host or substrate with the possession of fissitunicate (or functionally bitunicate) asci and dark-coloured

Figures 18.4a-f Deliquescing (Figs b, c) and fissitunicate (Figs a, d-f) asci. Light microscopy, bars = 20 µm.
Figure 18.4a *Trematosphaeria lignatilis.*
Figure 18.4b *Bathyascus avicenniae.*
Figure 18.4c *Arenariomyces triseptatus.*
Figure 18.4d *Trematosphaeria lignatilis.*
Figure 18.4e *Mycosphaerella salicorniae.*
Figure 18.4f *Caryosporella rhizophorae.*

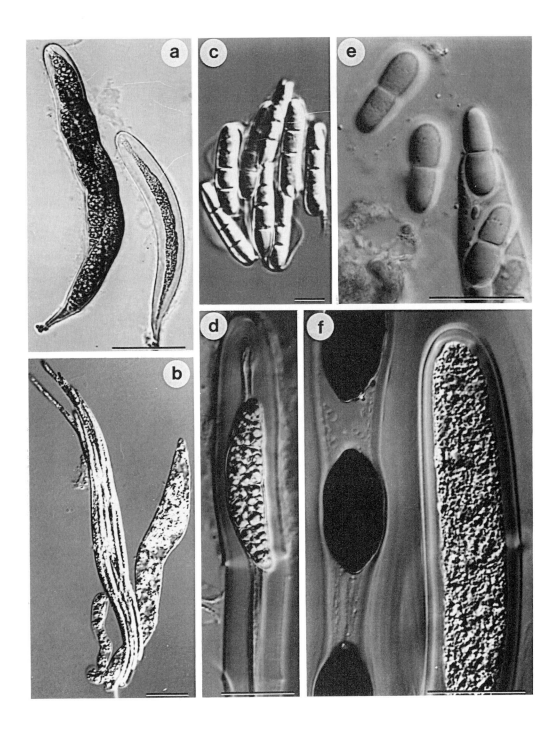

ascospores characterizes *a priori* a secondary marine fungus. Examples of primary marine fungi, algal parasites with thin-walled asci and hyaline ascospores, are *Haloguignardia decidua* Cribb et Cribb (Figure 18.3a), *Pontogeneia codiicola* (Dawson) Kohlm. et Kohlm. (Figure 18.5d) and *Spathulospora lanata* Kohlm. (Figure 18.5a). Other primary marine species, but probably phylogenetically younger, are the Halosphaeriaceae *Lignincola tropica* Kohlm. (Figure 18.3b), *Halosarpheia fibrosa* Kohlm. et Kohlm. (Figures 18.3c, 5b), *Arenariomyces triseptatus* Kohlm. (Figures 18.4c, 5c) and *Bathyascus avicenniae* Kohlm. (Figure 18.4b). Secondary marine species are ascomycetes with fissitunicate asci on intertidal or submerged phanerogams, e.g. on mangroves (*Trematosphaeria lignatilis* Kohlm., Figures 18.4a,d; *Caryosporella rhizophorae* Kohlm., Figure 18.4f; *Aigialus grandis* Kohlm. et Schatz, Figure 18.5e), marsh plants (*Mycosphaerella salicorniae* (Auersw.) Petrak, Figure 18.4e), or seagrasses (*Pontoporeia biturbinata* (Dur. et Mont.) Kohlm., Figure 18.5f).

CONCLUSIONS

More than half of the orders of Ascomycotina listed by Barr (1983; as Ascomycota) have no marine representatives and, because of the specialized nutritional requirements of some of their species, it is unlikely that they will be found in submerged marine habitats. For instance, Ascosphaerales are associated with pollen in honeycombs of bees, Coryneliales are mostly parasites on leaves of Podocarpaceae, Elaphomycetales are ectomycorrhizal, hypogeal fungi, the Cyttariales are parasites on species of *Nothofagus*, and the Clavicipitales are mainly entomogenous or parasitic on higher plants. *Claviceps*, a genus of the last order, occurs on species of *Spartina* in salt marshes, but not in submerged parts. Other orders, for which at present no marine members are known, could occur as saprobes on wood in the marine environment, viz. Hypodermatales, Xylariales, Phacidiales, Ostropales, and Caliciales. The Onygenales could be expected on feathers and bones in the beach wrack, but probably not as true marine fungi. A representative of the Microascales, *Microascus senegalensis* v. Arx has been isolated from mangrove mud (Senegal, v. Arx 1975; Mexico, Kohlmeyer unpublished), but its status as a true marine fungus has to be verified. The absence of Pezizales among marine fungi could be explained by the usually larger size of the cup fungi with their exposed hymenia. However, members of other orders with exposed hymenia have conquered the marine environment, namely, *Dactylospora haliotrepha* (Kohlm. et Kohlm.) Hafellner (*Lecanorales*) and *Laetinaevia marina* (Boyd) Spooner (Helotiales), if only at or above the upper intertidal

Figures 18.5a-f. Ascospores. Light micrographs, bars = 10 μm.

Figure 18.5a	*Spathulospora lanata.*
Figure 18.5b	*Halosarpheia fibrosa.*
Figure 18.5c	*Arenariomyces triseptatus.*
Figure 18.5d	*Pontogeneia codiicola.*
Figure 18.5e	*Aigialus grandis.*
Figure 18.5f	*Pontoporeia biturbinata.*

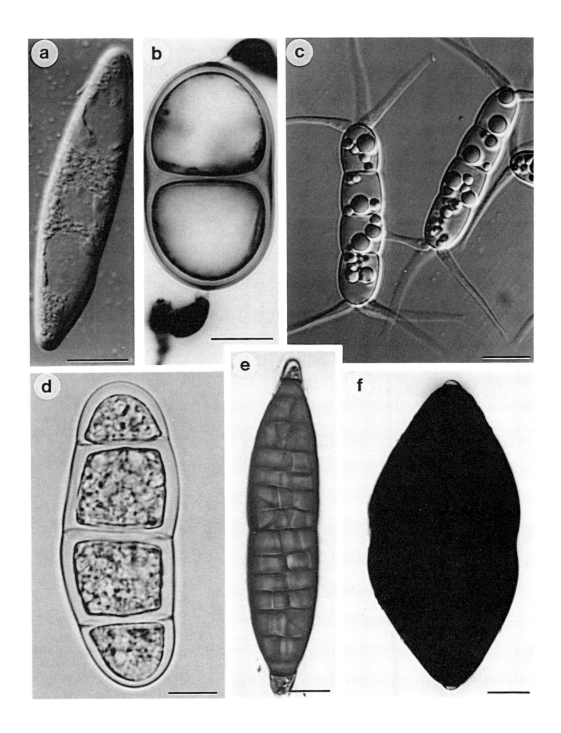

Table 18.1 Classification of the filamentous marine Ascomycotina
(Classification based on Barr (1979 a,b, 1983), Eriksson (1981, 1984),
and partly Hawksworth *et al.* (1983); a question mark indicates uncertain
placement of a genus or its marine species under preceding higher taxon)

Division Ascomycota (*sensu* Barr 1983)
   Class Laboulbeniomycetes
        Laboulbeniales
           Laboulbeniaceae
              Laboulbenia Mont. et Robin (1 marine sp.)
   Class Ascomycetes
     Subclass Parenchymatomycetidae
        Eurotiales
           Trichocomaceae
              Amylocarpus Currey (1 sp.)
              Eiona Kohlm. (1 sp.)
        Spathulosporales
           Spathulosporaceae
              Spathulospora Cavaliere et Johnson (5 spp.)
        Diaporthales
           Gnomoniaceae
              Gnomonia Ces. et De Not.? (1 marine sp.)
        Sordariales
           Lasiosphaeriaceae
              Biconiosporella Schaumann (1 sp.)
              Zopfiella Winter (2 marine spp.)
           Nitschkiaceae
              Groenhiella Koch, Jones et Moss (1 sp.)
     Subclass Edaphomycetidae
        Helotiales
           Dermateaceae
              Laetinaevia Nannf. (1 marine sp.)
        Phyllachorales
           Hyponectriaceae
              Buergenerula Sydow (1 marine sp.)
           Phyllachoraceae
              Haloguignardia Cribb et Cribb (5 spp.)
              Phycomelaina Kohlm. (1 sp.)
              Polystigma DC. (1 marine sp.)
        Phyllachorales *incertae sedis*
              Adomia Schatz (1 sp.)
        Hypocreales
           Hypocreaceae
              Halonectria Jones (1 sp.)
              Heleococcum Jørgensen (1 marine sp.)
              Hydronectria Kirschstein (1 marine sp., 1 marine var.)
              Nectriella Nitschke (1 marine sp.)
        Trichosphaeriales
           Trichosphaeriaceae
              Chaetosphaeria Tul. et C. Tul. (1 marine sp.)

Table 18.1 (continued)

Lecanorales
 Physciaceae
  Dactylospora Körber (1 marine sp.)
Halosphaeriales *ad int.* (Eriksson 1984)
 Halosphaeriaceae
  Aniptodera Shearer et Miller (1 sp.)
  Arenariomyces Höhnk (2 spp.)
  Bathyascus Kohlm. (3 spp.)
  Carbosphaerella Schmidt (2 spp.)
  Ceriosporopsis Linder (6 spp.)
  Chadefaudia Feldmann (6 spp.)
  Corollospora Werderm. (5 spp.)
  Haligena Kohlm. (3 spp.)
  Halosarpheia Kohlm. et Kohlm. (10 marine spp.)
  Halosphaeria Linder (5 spp.)
  Halosphaeriopsis Johnson (1 sp.)
  Kohlmeyeriella Jones, Johnson et Moss (1 sp.)
  Lignincola Höhnk (3 spp.)
  Lindra Wilson (4 spp., 1 var.)
  Lulworthia Sutherland (at least 5 spp., 1 var.)
  Marinospora Cavaliere (2 spp.)
  Nais Kohlm. (1 sp.)
  Nereiospora Jones, Johnson et Moss (2 spp.)
  Nautosphaeria Jones (1 sp.)
  Nimbospora Koch (3 spp.)
  Ocostaspora Jones, Johnson et Moss (1 sp.)
  Ondiniella Jones, Johnson et Moss (1 sp.)
  Remispora Linder (5 spp.)
  Trailia Sutherland (1 sp.)
  Trichomaris Hibbits, Hughes et Sparks (1 sp.)
Ascomycetes *incertae sedis*
  Abyssomyces Kohlm. (1 sp.)
  Banhegyia Zeller et Toth (1 sp.)
  Crinigera Schmidt (1 sp.)
  Oceanitis Kohlm. (1 sp.)
  Orcadia Sutherland (1 sp.)
  Pontogeneia Kohlm. (7 spp.)
  Savoryella Jones et Eaton (2 spp.)
  Torpedospora Meyers (2 spp.)
Class Loculoascomycetes
 Subclass Loculoparenchymatomycetidae
  Dothideales
   Dothideaceae
    Mycosphaerella Johansen (8 marine spp.)
    Sphaerulina Sacc. (2 marine spp.)
   Dothideales *incertae sedis*
    Helicascus Kohlm. (1 sp.)
    Manglicola Kohlm. (1 sp.)
    Paraliomyces Kohlm. (1 sp.)

Table 18.1 (continued)

                    Thalassoascus Ollivier (2 spp.)

     Subclass Loculoedaphomycetidae
       Pleosporales
         Arthopyreniaceae
           Leiophloea S.F. Gray (1 marine sp.)
         Massarinaceae
           Massarina Sacc. (1 marine sp.)
         Phaeosphaeriaceae
           Didymella Sacc. (3 marine spp.)
           Lautitia Schatz (1 sp.)
           Leptosphaeria Ces. et De Not. (13 marine spp.)
           Microthelia Körber (1 marine sp.)
           Ophiobolus Riess? (1 marine sp.)
           Phaeosphaeria Miyake (1 marine sp.)
         Pleosporaceae
           Pleospora Rabenh. (7 marine spp.)
       Melanommatales
         Didymosphaeriaceae
           Didymosphaeria Fuckel (3 marine spp.)
           Keissleriella v. Höhnel (1 marine sp.)
         Massariaceae
           Caryosporella Kohlm. (1 sp.)
           Halotthia Kohlm. (1 sp.)
           Pontoporeia Kohlm. (1 sp.)
         Melanommataceae
           Trematosphaeria Fuckel (3 marine spp.)
         Melanommatales *incertae sedis*
           Aigialus Kohlm. et Schatz (2 spp.)
       Chaetothyriales
         Herpotrichiellaceae
           Herpotrichiella Petrak (1 marine sp.)
       Verrucariales
         Verrucariaceae
           Pharcidia Körber (3 marine spp.)
 Familia incertae sedis
         Mastodiaceae
           Kohlmeyera Schatz (1 sp.), a synonym of *Mastodia* Hook.
             et Harvey? (Eriksson 1981)
           Turgidosculum Kohlm. et Kohlm. (1 sp.)
Loculoascomycetes *incertae sedis*
           Amarenomyces O. Eriksson (1 sp.)

zone.

A relatively large number of genera with exclusively marine species
cannot be properly placed in the present system (Table 18.1;
Abyssomyces, Crinigera, Oceanitis, Orcadia, Pontogeneia, Torpedospora).
They include 13 species with unitunicate asci and hyaline ascospores.
Examples of these genera without relationships to terrestrial taxa
indicate that they belong to lines of development which originate from
marine ancestors which did not make the transition from an aquatic to a
terrestrial environment. In the absence of a fossil record of marine
Ascomycotina all phylogenetic theories are based on deductions.
Therefore, it is important to search for fossils that could have been
preserved together with their calcareous hosts, viz. coralline algae
(Kohlmeyer & Kohlmeyer 1979) or Foraminifera (Kohlmeyer 1984).

ACKNOWLEDGEMENT
This work was supported by N.S.F. Grant BSR-8407155.

REFERENCES
Barr, M.E. (1979 a). On the Massariaceae in North America. Mycotaxon, 9,
        17-37.
Barr, M.E. (1979 b). A classification of Loculoascomycetes. Mycologia,
        71, 935-957.
Barr, M.E. (1983). The ascomycete connection. Mycologia, 75, 1-13.
Denison, W.C. & Carroll, G.C. (1966). The primitive ascomycete: A new
        look at an old problem. Mycologia, 58, 249-269.
Eriksson, O. (1981). The families of bitunicate ascomycetes. Opera Bot.,
        60, 1-220.
Eriksson, O. (1984). Outline of the ascomycetes. Systema Ascomycetum, 3,
        1-72.
Hawksworth, D.L., Sutton, B.C. & Ainsworth, G.C. (1983). Ainsworth &
        Bisby's Dictionary of the Fungi. Kew, Surrey: Commonwealth
        Mycological Institute.
Kohlmeyer, J. (1974). On the definition and taxonomy of higher marine
        fungi. Veröff. Inst. Meeresforsch. Bremerh., Suppl. 5,
        263-286.
Kohlmeyer, J. (1984). Tropical marine fungi. Marine Ecol. (P.S.Z.N.I),
        5, 329-378.
Kohlmeyer, J. & Kohlmeyer, E. (1979). Marine Mycology. New York:
        Academic Press.
Parguey-Leduc, A. & Janex-Favre, M.C. (1982). La paroi des asques chez
        les Pyrénomycètes: étude ultrastructurale. I. Les asques
        bituniqués typiques. Can. J. Bot., 60, 1222-1230.
Parguey-Leduc, A. & Janex-Favre, M.C. (1984). La paroi des asques chez
        les Pyrénomycètes: étude ultrastructurale. II. Les asques
        unituniqués. Cryptogamie, Mycol., 5, 171-187.
Raper, J.R. (1968). On the evolution of fungi. In The Fungi: An
        Advanced Treatise, Vol. III, eds. G.C. Ainsworth &
        A.S. Sussman, pp. 677-693. New York: Academic Press.
Reynolds, D.R., ed. (1981). Ascomycete Systematics. The Luttrellian
        Concept. New York: Springer Verlag.

Savile, D.B.O. (1968). Possible interrelationships between fungal
          groups. In The Fungi: An Advanced Treatise, Vol. III, eds.
          G.C. Ainsworth & A.S. Sussman, pp. 649-675. New York:
          Academic Press.
von Arx, J.A. (1975). Revision of *Microascus* with the description of a
          new species. Persoonia, 8, 191-197.

TAXONOMIC STUDIES OF THE HALOSPHAERIACEAE – PHILOSOPHY AND
RATIONALE FOR THE SELECTION OF CHARACTERS IN THE
DELINEATION OF GENERA

E.B.G. Jones

R.G. Johnson

S.T. Moss

INTRODUCTION
        During the two decades 1960-1980 some 65 species belonging
to 15-28 genera, depending on the authority, have been assigned to the
Halosphaeriaceae. Many of these genera are assemblages of disparate
species and have been classified on the basis of ascocarp, ascus,
ascospore and appendage morphologies at the light microscope level.
However, these criteria have not been subjected to critical evaluation.

The first revision of the family was made by Kohlmeyer (1972) and
subsequently modified slightly by Kohlmeyer & Kohlmeyer (1979) who
reduced a number of genera to synonymy, e.g. *Remispora* to *Halosarpheia*,
Marinospora to *Ceriosporopsis*. Schmidt (1974) was the first to draw
attention to the confused state of the taxonomy of the genus
Corollospora, and suggested that it should be divided into two sections:
the "coronate" (*C. maritima* Werdermann (type species), *C. intermedia*
Schmidt, *C. lacera* (Linder) Kohlm. and *C. pulchella* Kohlm., Schmidt
et Nair) and the "cristatiae" (*C. comata* (Kohlm.) Kohlm. and
C. cristata (Kohlm.) Kohlm. The remaining species (*C. trifurcata*
(Höhnk) Kohlm. and *C. tubulata* Kohlm.) to be assigned to another
section or "possibly to *Carbosphaerella*".

Furtado & Jones (1980) accepted the revisions of Kohlmeyer (1972) for
practical reasons as did a number of other marine mycologists. Hughes'
(1975) critical review of the lignicolous and algicolous marine fungi
triggered the senior author to embark on a detailed study of other
characters for the delineation of genera within the Halosphaeriaceae.
These included scanning and transmission electron microscopy and light
histochemical observations. This gave rise to a series of exploratory
papers (Moss & Jones 1977; Jones & Moss 1978, 1980; R.G. Johnson 1980)
which highlighted the requirement for taxonomic revisions of the family
as many genera were considered to be polyphyletic. Shearer & Crane
(1980) concurred with these views in their treatment of species assigned
to *Haligena* and *Halosarpheia*.

Jones & Moss (1978, 1980) considered ascocarp structure of significance
at the familial level only and believed that genera could be separated
on the basis of ascospore structure and the ontogeny of the appendages.
This gave rise to the view that we did not consider other criteria,
e.g. ascocarp and ascus structures, in the delineation of genera. It
was Cavaliere (1966 a,b,c) and Cavaliere & T.W. Johnson (1966 a,b) who

indicated that ascocarp structure and morphology were not useful
characters for separating genera. It is unfortunate that many of the
genera they examined were assemblages of species bearing little affinity
with one another. In their studies they were unable to find similar
patterns of development and noted great variation within a species.
However, many of the patterns they observed conformed with the concept
of generic limits as proposed by Jones *et al.* (1983 a, cf. *Corollospora*
species).

Jones *et al.* (1983 a,b, 1984) and R.G. Johnson *et al.* (1985) in a series
of papers on the taxonomy of the Halosphaeriaceae examined 17 genera
(including 27 species) at the light microscope (including ascospore
histochemistry) and scanning and transmission electron microscope
levels. Five new genera have been described to accommodate species that
could not be assigned to any of the existing genera (*Kohlmeyeriella*,
Nereiospora, *Appendichordella*, *Ocostaspora* and *Ondiniella*) while five
genera (*Antennospora*, *Arenariomyces*, *Halosphaeriopsis*, *Marinospora* and
Remispora) are regarded as valid taxa. Although criteria for
delineation of genera have included ascospore development and the
ontogeny of the appendages, other characters have also been used,
especially ascocarp wall structure and the presence or absence of
catenophyses. However, separation of genera based on ultrastructural
characteristics is now a well accepted practice not only for the fungi,
but for other groups, e.g. algae: Eustigmatophyceae (Hibberd 1981);
Haptophyceae (Manton 1966). Pegler & Young (1981) proposed a more
natural arrangement of the Boletales on the basis of basidiospore
morphology and structure and concluded that the spores ".... offer good
evidence in support of phyletic relationships. The spore is the most
constant and the least susceptible to change, within narrow limits of
all basidiome structures".

Some workers have misunderstood our philosophy on the criteria employed
for separating genera (Nakagiri & Tubaki 1982), while others have urged
caution in the use of ascospore appendage development as the single
character for the differentiation of genera. Yet most keys to genera of
the Halosphaeriaceae use ascospore and appendage morphology to separate
genera (Cavaliere 1977; Kohlmeyer & Kohlmeyer 1979). Because of these
conflicting views we propose to examine in this chapter the many
characters available for separation of genera in the Halosphaeriaceae.

MATERIALS AND METHODS
Ascocarps of the fungi to be examined were collected: on
test panels which had been submerged in the sea in various parts of the
world; from driftwood from the sand dunes of Løkken, Grønhøj and
Blockhus, Denmark and San Juan Island, U.S.A.; *Spartina* spp. collected in
Langstone harbour, England and Norfolk, Virginia, U.S.A.; sporulating
cultures donated by colleagues. The wood and *Spartina* spp. culms were
maintained in plastic boxes at 20° C and 30° C and examined regularly for
the presence of ascocarps.

Except for ascocarp wall structure the techniques and procedures

employed for the preparation of material for scanning electron
microscopy were those described by Moss & Jones (1977). For
investigations of ascocarp structure specimens were frozen in Freon 13,
and then fractured with a super-cooled knife. Fractured ascocarps were
either maintained frozen and then freeze-dried, or allowed to attain
room temperature and post-fixed in 2% aqueous OsO4. Post-fixed material
was dehydrated to acetone and then critical point dried.

Material for the histochemical studies and transmission electron
microscopy were those described by R.G. Johnson (1980, 1982).

RESULTS
The following characters have been used to describe species
and will be considered in turn: ascocarp development; ascocarp
structure; periphyses; catenophyses; ascus structure; ascospore
appendage ontogeny.

Ascocarp development
Only a few studies of ascocarp initiation have been
undertaken in the Halosphaeriaceae: *Chadefaudia marina* G. Feldmann
(Feldmann 1957), *Halosphaeria mediosetigera* Cribb et Cribb and
Ceriosporopsis circumvestita (Kohlm.) Kohlm. (Kohlmeyer & Kohlmeyer
1966), *Ceriosporopsis halima* Linder (Wilson 1965) and *Lulworthia medusa*
(Ellis et Everh.) Cribb et Cribb (Lloyd & Wilson 1962). Kohlmeyer &
Kohlmeyer (1966) stated that a pseudoparenchymatous ball of cells grow
into a young ascocarp with a differentiated peridium which encloses a
sterile pseudoparenchyma. Ascogonial cells then originate in the
centre of the pseudoparenchyma, and these give rise to ascogenous
hyphae which in turn form the asci. They did not see croziers in either
H. mediosetigera or *C. circumvestita*. Kohlmeyer & Kohlmeyer (1979)
stated that ascocarp development is the same in species of
Ceriosporopsis, *Halosphaeria* and *Lulworthia* and was of the *Diaporthe*
type (Luttrell 1951). However R.G. Johnson (1982) showed ascocarp
development in *H. mediosetigera* to be significantly different from the
account given by Kohlmeyer & Kohlmeyer (1966).

In *Halosphaeriopsis mediosetigera* T.W. Johnson, perithecia are
initiated by coiled vegetative hyphae that differentiate into ascogonia
(Figure 19.1a). The ascogonium comprises a tip cell, a swollen
ascogonial cell and a stalk cell (Figure 19.1b). Although a tip cell
develops on the ascogonium no differentiated male organ was found.
This indicates a parthenogenetic development. Ascogenous hyphae develop
from the swollen ascogonial cell to form a domed hymenium, which
according to Cavaliere (1966 a,b,c) and Cavaliere & Johnson (1966 a,b)
indicates the presence of one ascogonium per perithecium.

Asci in *H. mediosetigera* are derived either as lateral branches of
ascogenous hyphae or through crozier formation (Figures 19.2a,b).
Conversely, Kohlmeyer & Kohlmeyer (1966) indicated the absence of
croziers in *H. mediosetigera* and *Ceriosporopsis circumvestita* whereas

crozier formation has been observed in *Lulworthia medusa* (Lloyd & Wilson 1962), *Ceriosporopsis halima* (Wilson 1965) and *Mycosphaerella ascophylli* Cotton (Webber 1967).

Ontogenetic investigations of ascocarp production in marine fungi have shown that a plectenchymatous ball of vegetative hyphae forms within which the fertile tissue develops (Lloyd & Wilson 1962; Wilson 1965; Kohlmeyer & Kohlmeyer 1966; Shearer & Miller 1977). In *H. mediosetigera*

Figures 19.1a-d *Halosphaeriopsis mediosetigera*, scanning electron micrographs of ascocarp development.
Figure 19.1a Coiled hyphae present in a culture prior to perithecium formation. Bar = 1 μm.
Figure 19.1b Perithecium initial with swollen ascogonium (AG) and tip cell (TC) produced from the coiled hyphae. Bar = 5 μm.
Figures 19.1c,d Development of investing hyphae around the ascogonium and vegetative hyphae (VH) to form prosenchyma and the perithecial wall respectively. Bars = 5 μm.

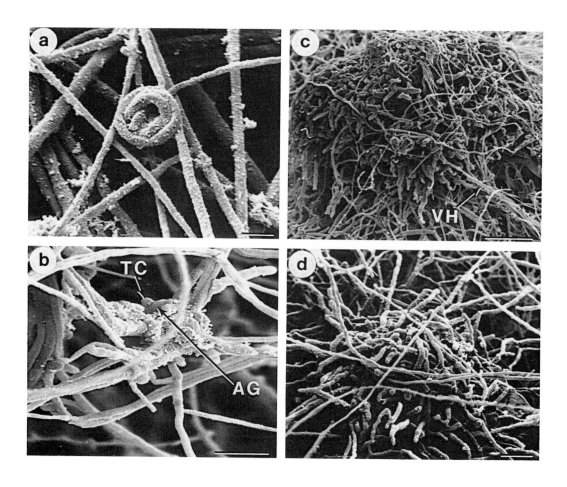

Figures 19.2a-c *Halosphaeriopsis mediosetigera*,
transmission electron micrographs of ascus development.
Figure 19.2a Longitudinal section of the terminal cells of
an ascogenous hypha. Each cell is binucleate (N) and
contains mitochondria (M), oil droplets (O) and volutin
bodies (V). Bar = 2 μm.
Figure 19.2b Crozier formation. Bar = 2 μm.
Figure 19.2c Ascus with delimited ascospores. Bar = 10 μm.

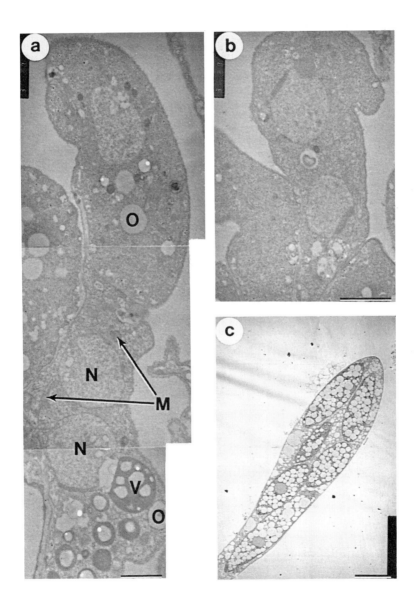

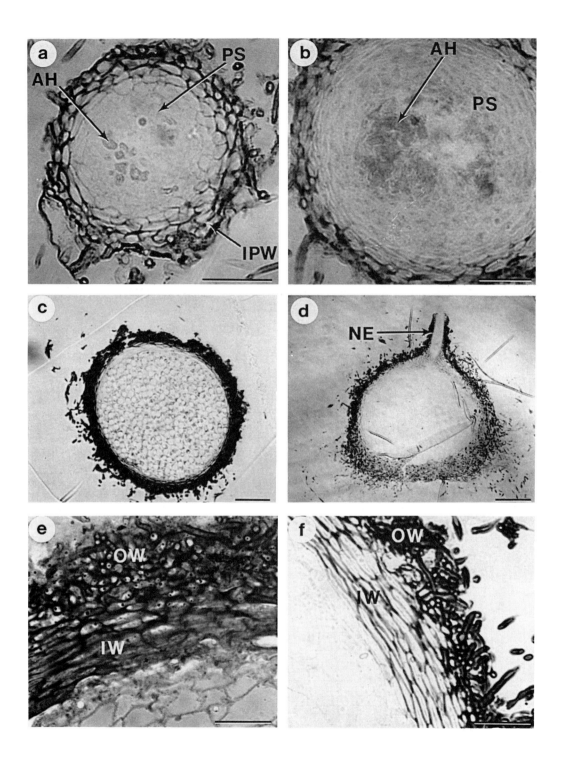

sterile ascocarp tissues which surround the fertile elements develop
from both the ascogonial stalk cells and the surrounding vegetative
hyphae (Figures 19.1c,d), a similar situation has been reported in
Neurospora crassa Shear et Dodge (Searle 1973).

Ascocarp development in *H. mediosetigera* (Figures 19.3a-d), once
surrounded by sterile elements, parallels the *Diaporthe*-type (Luttrell
1951) ontogenetic pattern elucidated for other marine fungi (Lloyd &
Wilson 1962; Wilson 1965; Kohlmeyer & Kohlmeyer 1966; Webber 1967;
Shearer & Miller 1977). Sterile tissues form the peridium and centrum
pseudoparenchyma. The thickness and texture of the peridium in
H. mediosetigera is dictated by the position of the perithecium with
respect to its substrate (Figures 19.3e,f). Concurrent with ascus
development the surrounding pseudoparenchyma is displaced and a
meristematic region of peridial cells develops at the top of the
ascocarp. Upward growth of the meristematic region forms a neck lined
by wide, lumened, pseudoparenchyma-like cells but lacking periphyses.
Similar neck development occurs in *L. medusa* (Lloyd & Wilson 1962),
C. circumvestita and *H. mediosetigera* (Kohlmeyer & Kohlmeyer 1966)
whereas in *C. halima* (Wilson 1965) and *Aniptodera chesapeakensis* Shearer
et Miller (Shearer & Miller 1977) periphyses are present.

Clearly there are sufficient differences in interpretation to require a
more detailed study of ascocarp development in genera of the
Halosphaeriaceae. A wider range of genera should be studied, in
particular *Carbosphaerella*, *Corollospora*, *Marinospora* and *Arenariomyces*
because these do have differences in ascocarp wall structure. With the
present lack of information on ascocarp development it is not possible
to use this character as a means of resolving differences between
genera.

         Ascocarp size, colour, texture and papillae
         Ascocarp size is considered significant only at the species
level, while colour of the perithecia is extremely variable even within

Figures 19.3a-f *Halosphaeriopsis mediosetigera*, light micrographs of
sectioned perithecia.
Figures 19.3a,b Early (Fig. 19.3a) and later (Fig. 19.3b) stages in
perithecium ontogeny showing ascogenous hyphae (AH) surrounded by the
pseudoparenchyma (PS) and perithecial wall initial (IPW). Bars = 10 μm.
Figure 19.3c Mature perithecium prior to spore release. The
pseudoparenchyma has disappeared and the centrum is full of ascospores.
Bar = 20 μm.
Figure 19.3d Perithecium following release of ascospores through the
neck (NE). Bar = 20 μm.
Figures 19.3e,f Walls of mature perithecia which have formed either on
the surface (Fig. 19.3e) or within (Fig. 19.3f) the substratum. In both
perithecia the wall comprises an outer (OW) and an inner (IW) layer,
although the outer layer is characteristically thicker in perithecia
produced on the surface of the substratum. Bars = 10 μm.

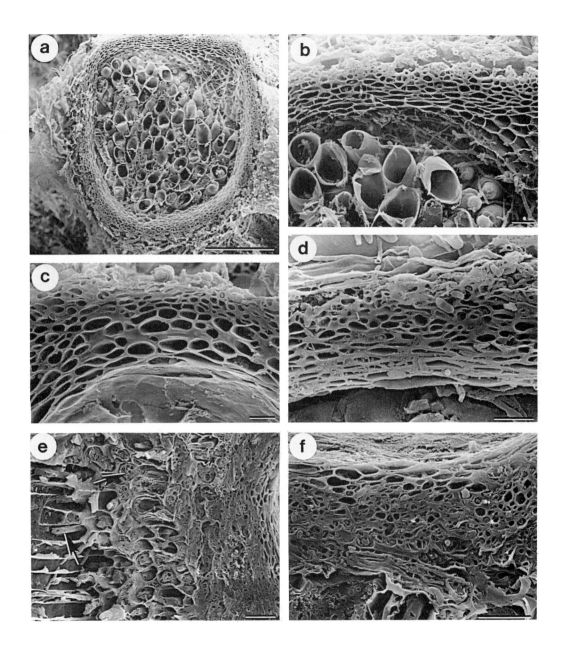

<u>Figures 19.4a-f</u> Scanning electron micrographs of fractured ascocarps.
<u>Figures 19.4a,b</u> *Halosarpheia retorquens*, perithecium containing mature
ascospores. The perithecial wall comprises an outer layer of thick-
walled cells and an inner layer of thin-walled cells. No discrete inner
layer of flattened cells is present although hyphae are numerous.
Fig. 19.4a, bar = 100 μm; Fig. 19.4b, bar = 10 μm.

a species. Colour frequently depends on whether the perithecia are
immersed (hyaline) or superficial (brown), a character well illustrated
in *Remispora stellata* Kohlm. The presence or absence of a neck is
variable as are the length and number per perithecium. None of these
characters can be used with any degree of confidence in separating
genera. Likewise the texture of perithecia is variable. Ascocarps
produced on sand grains can be carbonaceous, but when immersed in wood
may be membranous, as seen in *Lulworthia* species. Ascocarp morphology
and texture, but not structure are considered to be greatly affected by
environmental conditions and are not considered suitable criteria for
the differentiation of genera.

Ascocarp wall structure
        Cavaliere (1966 a,b,c) and Cavaliere & T.W. Johnson
(1966 a,b) considered ascocarp wall structure of limited value for the
separation of genera, largely because the genera they examined were
complexes of species. However, with the advent of scanning electron
microscopy, this character can be deemed to be of taxonomic significance
at the generic level. However, it will always be secondary to ascospore
appendages owing to the time involved in sectioning perithecia; spore
appendages give a more immediate visual answer. Ascocarp wall structure
has been used extensively in the separation of genera (Kohlmeyer &
Kohlmeyer 1979; R.G. Johnson *et al.* 1985; Nakagiri & Tubaki 1985) and
along with ascocarp development and spore appendage ontogeny must rank
as one of the major criteria for the characterization of the
Halosphaeriaceae.

From an analysis of the literature it would appear that genera can be
divided into 3 groups based on the structure of the ascocarp wall.
Group 1, those with an undifferentiated peridium usually composed of
3-5(-10) layers of cells that are elongate and thin-walled, e.g. *Nais*,
Nautosphaeria, *Lignincola*, *Ondiniella*, *Antennospora*, *Remispora*,
Bathyascus, *Aniptodera*. Group 2, those with a peridium that is
differentiated into 2 layers, e.g. *Corollospora*, *Haligena*, *Marinospora*,
Halosarpheia, *Lindra*, *Trailia*; these genera usually have an inner layer
of thin-walled, elongate, hyaline cells and each layer may be a number

Figure 19.4c *Ceriosporopsis circumvestita*, the perithecial wall
comprises an outer layer of small thick-walled cells and a middle layer
of larger thick-walled cells. The innermost layer of cells has been
compacted. Bar = 10 μm.
Figure 19.4d *Halosarpheia spartinae*, the wall of the perithecium
comprises a similar two-layered arrangement to that of *H. retorquens*.
Bar = 10 μm.
Figures 19.4e,f *Marinospora calyptrata*, perithecium embedded within a
wood substrate. The perithecial wall is not discrete and fungal hyphae
have invaded the wood cells (arrowed). Cells of the outer region of the
perithecial wall are thick-walled, irregular in shape and are not
organized into discrete cell layers. A middle layer of thin-walled and
an inner layer of flattened cells are present. Bars = 20 μm.

of cells thick. Group 3 comprises those genera with a 3-layered peridium, e.g. *Kohlmeyeriella*, *Halosphaeria*, *Nereiospora*.

The number of cell layers which comprises the ascocarp wall can be an unreliable character, e.g. *Halosphaeriopsis mediosetigera* has a peridium composed of 3 layers in nature, but 2 layers when grown under laboratory conditions. Also, environmental conditions can modify the wall structure; perithecia growing on sand grains may have much thicker, larger cells than those growing on wood, e.g. *Corollospora* (*sensu stricto*) species (Nakagiri & Tubaki 1985).

Our observations show that in *Halosarpheia retorquens* Shearer et Crane and *H. spartinae* (Jones) Shearer et Crane (Figures 19.4a,b,d)

Figures 19.5a-c *Halosphaeriopsis mediosetigera*, scanning electron micrograph of a fractured perithecial wall which is comprised of three layers; an outer layer of thick-walled elongated cells, a middle layer of less thick-walled flattened cells and an inner layer of thin-walled cells which becomes compacted at maturity. Bars = 10 μm.
Figure 19.5d *Ceriosporopsis circumvestita*, scanning electron micrograph of a transversely fractured neck showing periphyses (P). Bar = 10 μm.

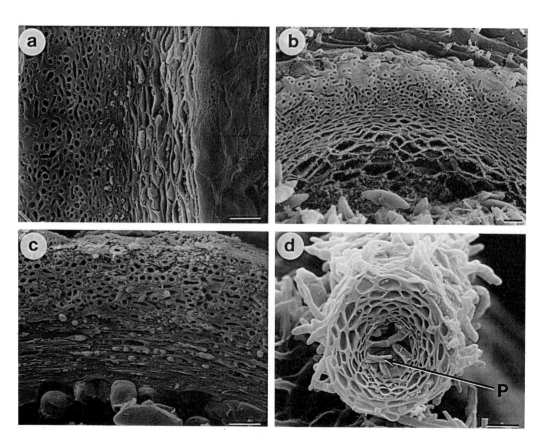

perithecial walls comprise two layers and this is in agreement with
published reports. In *Halosphaeriopsis mediosetigera*, *Ceriosporopsis
circumvestita* and *Marinospora calyptrata* (Kohlm.) Cavaliere
(Figures 19.4c,e,f; Figures 19.5a-c) three wall layers are discernible,
although the cell types making up the layers differ from genus to genus.
In each case at maturity, the inner layer is composed of compacted,
thin-walled, elongate cells. Further research is required to resolve
whether this layer is a true perithecial wall layer or flattened
prosenchymatous cells from the centrum. Wall layers varied in
morphology from discrete, e.g. *H. spartinae* (Figure 19.4d),
C. circumvestita (Figure 19.4c) and *H. mediosetigera* (Figures 19.5a-c)
to the outer wall merging with the wood tissue, e.g. *M. calyptrata*
(Figures 19.4e,f). Further, cells of the outer layer of the ascocarp
wall may be regular (*H. retorquens*, *C. circumvestita*) or irregular
(*H. mediosetigera*, *M. calyptrata*).

Despite the views expressed by Cavaliere (1966 a,b,c) and Cavaliere &
T.W. Johnson (1966 a,b) we are of the opinion that ascocarp wall
structure can be used to separate genera, not as a single character, but
in combination with others. We have highlighted this in the case of
Marinospora (R.G. Johnson *et al.* 1985), *Haligena*, *Appendichordella*
and *Ceriosporopsis* (R.G. Johnson *et al.* unpublished). Nakagiri & Tubaki
(1985) have also demonstrated that the ascocarp in species of
Corollospora has an inner layer of elongate cells and an outer layer of
columnar cells, whereas in species of *Arenariomyces*, a genus previously
assigned to the genus *Corollospora*, the ascocarp wall is undifferentiated.

Periphyses
This character can be used in conjunction with others to
delineate genera. Seven genera have species with well defined
periphyses (*Aniptodera*, *Halosarpheia*, *Appendichordella*, *Haligena*,
Chadefaudia, *Marinospora*) while the remaining genera lack periphyses or
have very reduced structures which may deliquesce at maturity. This
character requires further investigation as *Ceriosporopsis circumvestita*
has been shown to have periphyses (Figure 19.5d), contrary to the
observations of Kohlmeyer & Kohlmeyer (1966).

Catenophyses
True paraphyses are absent but the thin-walled
pseudoparenchymatous cells may break up to form catenulated structures
called catenophyses and these are found in species of eight genera
(*Aniptodera*, *Marinospora*, *Haligena*, *Halosarpheia*, *Remispora*, *Ondiniella*,
Lignincola and *Nais*), while in other genera (15) catenophyses are
absent. This character can be useful in separating genera, e.g.
Ceriosporopsis from *Marinospora*, *Remispora* from *Halosphaeria*,
Ondiniella from *Halosphaeria*.

Asci
Ascus structure in the Halosphaeriaceae is fairly constant

in that it is fusiform to clavate, without an apical apparatus, unitunicate, thin-walled, deliquesces early and develops in a layer of asci at the base of the centrum. Because of the uniformity in ascus structure this character is not considered suitable for the separation of genera (Figure 19.2c). However, differences do exist and these should be examined in the transmission electron microscope. In *Lignincola* spp. the ascus is often persistent and thick-walled near the apex. In species of *Aniptodera* the ascus is thickened at the apex with a simple pore (Farrant, Chapter 20). Likewise the asci of some *Halosarpheia* species may be thickened apically. Indeed, Kohlmeyer & Kohlmeyer (1979) have queried whether species of *Aniptodera*, *Nais* and *Lignincola* should be placed in separate genera. Studies in progress, at

Figures 19.6a–d Transmission electron micrographs of appendage ontogeny.
Figure 19.6a *Corollospora maritima*, developing ascospore (S) surrounded by the membrane complex (MC). Bar = 5 μm.
Figure 19.6b *Nereiospora comata*, appendage initials developing between the membrane complex (MC) and the ascospore wall which at this stage is comprised of only the episporium (EP). Bar = 0.5 μm.
Figures 19.6c,d *Nereiospora comata*, appendages of a mature ascospore contained within extensions of the membrane complex (MC). Bars = 1 μm.

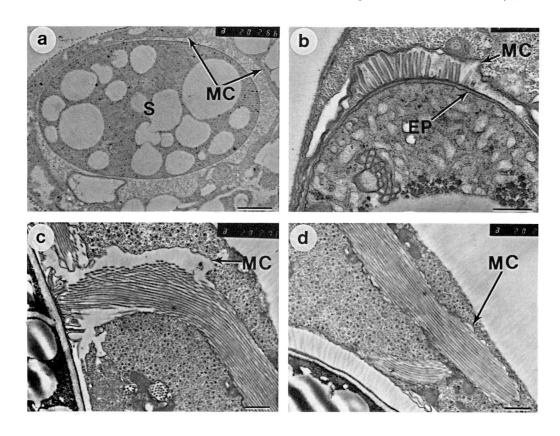

the transmission electron microscope level, should resolve this problem
and indicate whether these genera should be included in the
Halosphaeriaceae.

Ascospore colour, septation

These are not considered characters for the separation of
genera, but may be of importance at the species level. Ascospores of
Corollospora maritima are normally hyaline, but coloured forms have been
collected on sand grains from Kuwait (Zainal & Jones 1986).

Membrane complex

R.G. Johnson (1980) described ascospore initiation in
several species of the Halosphaeriaceae. The ascospore wall in the
Halosphaeriaceae is deposited between two delimiting membranes, the
inner membrane forms the spore plasma membrane, while the outer membrane
becomes thicker, more electron-opaque and loses its tripartite structure
following its coalescence with epiplasmic vesicles. This membrane is
the membrane complex (Figures 19.6a–d). It is present in a number of
genera and its function is speculative. We believe that it protects the
developing ascospores and prevents the appendages from dilating while
still within the perithecium. The membrane complex breaks down on
contact with sea water. Care should be taken in the interpretation of
scanning electron micrographs in order that this layer is not confused
with the exosporium. Only studies at the transmission electron
microscope level can distinguish between these two layers.

It is doubtful whether the membrane complex has any significance in
generic delineation, as it is present in all genera examined with the
exception of *Torpedospora radiata* Meyers, a species not regarded by
Kohlmeyer & Kohlmeyer (1979) as a member of the Halosphaeriaceae, and
Kohlmeyeriella, the only genus of Halosphaeriaceae without radiating
appendages.

Ascospore and appendage ontogeny

Jones & Moss (1978, 1980) and R.G. Johnson (1980) have
shown that studies of ascospores at the scanning and transmission
electron microscope levels can yield a number of characters useful in
the separation of genera (Table 19.1). They fall into two categories:
number of wall layers to the spore; appendage ontogeny. Three wall
layers are recognized: the mesosporium, an electron transparent
innermost layer; the episporium, a thin, electron-opaque layer; the
exosporium, an outer layer of variable electron opacity. In some genera
these layers may be differentiated into sublayers, e.g. 2 sublayers to
the mesosporium in species of *Carbosphaerella* and *Marinospora* whereas in
species of *Kohlmeyeriella* and *Arenariomyces* the episporium is
laciniated. Jones & Moss (1980) described four types of spore appendage
ontogeny, while R.G. Johnson (1980) considered there to be only three
types. There is no basic difference in these views, only a matter of
expressing or classifying the different forms. In this chapter we

recognize the following types; a) appendage exuded through a pore or
pores in the spore wall; b) direct growth of the spore wall;
c) fragmentation of an outer wall layer; d) a combination of b) and c);
e) outgrowth of spore wall and elaboration of an outer wall layer;
f) other types.

a) *Appendage exuded through a pore(s) in the spore wall.*   In
Halosarpheia species the hamate appendages are formed by release of
material from a pore in the spore wall (Jones & Moss 1978). This
material is not derived from the layers that form the spore wall, but
is formed internal to the mesosporium (Farrant unpublished).

b) *Direct growth of the appendage from the spore wall.*   The appendages
in this case may be formed from a well defined layer, e.g. in species
of *Ondiniella* and *Appendichordella* the episporium gives rise to the
appendages.   In other genera (*Antennospora*, *Nereiospora*, *Halosphaeria*)
the precise origin of the appendages is not known, but they are
probably outgrowths of the mesosporium (Figures 19.6b-d).

c) *Fragmentation of an outer wall layer.*   This type generally applies to
the fragmentation (*Halosphaeriopsis*, *Remispora*) of the exosporium.   It
is important to emphasize that appendages formed by fragmentation are
not to be grouped together, on the contrary it is a "generic" term with
an infinite variety of ways by which the appendages are formed.   Great
variation in the substructure of the exosporium has been demonstrated
(Jones *et al.* 1983 a,b, 1984; R.G. Johnson *et al.* 1985).

d) *Appendages formed by a combination of outgrowths of the spore wall
and fragmentation of the exosporium.*   In *Corollospora* species the polar
spines are formed as outgrowths of the meso- and epi-sporium.   These
are initially surrounded by the exosporium, which fragments or cleaves
at predefined points to form the equatorial appendages and the polar
caps to the spines (Jones *et al.* 1983 a).

e) *Appendages formed by a combination of outgrowths of the spore wall
and elaboration of an outer wall layer.*   In *Ceriosporopsis* the
exosporium is modified or thrown into folds to form appendages, while
in *Marinospora* the exosporium is corrugated.   In both genera the main
appendages arise as outgrowths of the spore wall, probably the
mesosporium.   In *Ceriosporopsis tubulifera* the end chamber is formed
from the exosporium and the appendage is formed within the chamber and
not as an outgrowth of the spore wall.

f) *Other types.*   In *Kohlmeyeriella* the polar appendages are formed as a
continuation of the episporium and the outermost layer of the
mesosporium.   Mucilage is formed within the appendages and subsequently

Table 19.1 Types of ascospore appendage ontogeny in the
species examined

Appendage ontogeny type	Species	Reference
Appendage exuded through a pore(s) in the spore wall	*Halosarpheia spartinae* (Jones) Shearer et Crane	Jones & Moss (1978)
Direct growth of the appendage from the spore wall	*Appendichordella amicta* R.G. Johnson, Jones et Moss	R.G. Johnson, Jones & Moss (1985)
	Antennospora quadricornuta (Cribb et Cribb) T.W. Johnson	Jones, R.G. Johnson & Moss (1984)
	Nereiospora comata Jones, R.G. Johnson et Moss	Jones, R.G. Johnson & Moss (1983)
	Halosphaeria appendiculata Linder	Jones, R.G. Johnson & Moss (1984)
	Arenariomyces trifurcatus Höhnk ex Jones, R.G. Johnson et Moss	Jones, R.C. Johnson & Moss (1983)
	Ondiniella torquata Jones, R.G. Johnson et Moss	Jones, R.G. Johnson & Moss (1984)
Fragmentation of an outer wall layer	*Haligena elaterophora* Kohlm.	R.G. Johnson, Jones & Moss (1985)
	Halosphaeriopsis mediosetigera T.W. Johnson	Jones, R.G. Johnson & Moss (1984)
	Remispora maritima Linder	R.G. Johnson, Jones & Moss (1984)
	Remispora pilleata Kohlm.	R.G. Johnson, Jones & Moss (1984)
	Remispora stellata Kohlm.	R.G. Johnson, Jones & Moss (1984)
	Remispora galerita Tubaki	R.G. Johnson, Jones & Moss (1984)
	Carbosphaerella leptosphaerioides Schmidt	R.G. Johnson, Jones & Moss (1984)
	Ocostaspora apilongissima Jones, R.G. Johnson et Moss	Jones, R.G. Johnson & Moss (1984)
Appendages formed by a combination of outgrowths of the spore wall and fragmentation of an exosporium	*Corollospora maritima* Werdermann	Jones, R.G. Johnson & Moss (1983)
	Corollospora intermedia Schmidt	Jones, R.G. Johnson & Moss (1983)
	Corollospora lacera (Linder) Kohlm.	Jones, R.G. Johnson & Moss (1983)
	Corollospora pulchella Kohlm., Schmidt et Nair	Jones, R.G. Johnson & Moss (1983)
Appendages formed by a combination of outgrowths of the spore wall and elaboration of an outer wall layer	*Ceriosporopsis halima* Linder	Jones, R.G. Johnson & Moss (1985)
	Ceriosporopsis tubulifera (Kohlm.) Kirk in Kohlm.	R.G. Johnson, Jones & Moss (1985)
	Marinospora calyptrata (Kohlm.) Cavaliere	R.G. Johnson, Jones & Moss (1984)
	Marinospora longissima (Kohlm.) Cavaliere	R.G. Johnson, Jones & Moss (1984)
Other types	*Kohlmeyeriella tubulata* Jones, R.G. Johnson et Moss	Jones, R.G. Johnson & Moss (1983)
	Lulworthia sp.	Unpublished observations

released as a drop which may aid in spore attachment. This type could be placed in either section b) or d), but at the present we wish to keep it separate. It is probable that *Lulworthia* species should also be placed in this type.

These types of spore appendage ontogeny have been used by Jones *et al.* (1983 a,b, 1984) and R.G. Johnson *et al.* (1985, 1986) to describe and to delineate genera, because we believe spore ontogeny to be a basic character and least modified by environmental conditions.

DISCUSSION

From these results a number of conclusions can be drawn. We believe that ascospore ontogeny is a primary character for the delineation of genera within the Halosphaeriaceae. A wide range of types has been described and the characters are constant for any one genus (e.g. *Corollospora*, *Remispora*). Characters regarded as of secondary importance include presence or absence of catenophyses and periphyses and ascocarp wall structure but these must be used in conjunction with other characters (e.g. spore appendage ontogeny). Persistent characters are considered more important than ephemeral characters. Characters considered of little value in the delineation of genera include: ascus structure and morphology (familial characters as they are constant for a wide range of genera); ascocarp and ascospore phenologies (colour, texture, size, etc.); the membrane complex (present in nearly all the genera examined).

Observations should be made at all developmental stages of a structure: ascocarp wall, e.g. the inner, thin-walled, elongate cells may be lost (deliquesce) or compressed during development; ascospore and appendage, e.g. in some *Halosarpheia* species the appendage material may be formed prior to ascospore wall formation; while in *Carbosphaerella* appendages may change during development. Ascocarp development in the Halosphaeriaceae requires further study as an insufficient number of genera has been examined. Characters, other than ascospore ontogeny, which are important at the generic and familial levels need to be identified.

Although in this chapter we recognize 6 types of appendage ontogeny it is those formed from the exosporium which exhibit greatest diversity. Our studies of the taxonomy of the Halosphaeriaceae have highlighted the wide degree in structure and elaboration of the exosporium. The structure of the exosporium may be amorphous with electron-opaque regions (*Ocostaspora*), laminate (*Halosphaeriopsis*) or amorphous with bundles of fibres (*Remispora*). Differentiation of this layer to form the appendages is as diverse as its structure. In *Corollospora* species the exosporium separates in predetermined regions to form sheet-like appendages, whereas in *Halosphaeriopsis mediosetigera* loss of part of the exosporium forms spirally arranged sheet-like appendages. The appendages in *Remispora* species form from both amorphous and fibrous components of the exosporium while in species of *Carbosphaerella* a fibrous appendage is released by loss of the amorphous component. All of these characters are useful in the separation of genera.

Finally, preliminary observations suggest that the structure of the ascus in *Aniptodera* differs from that of the majority of the genera currently assigned to the Halosphaeriaceae. Also greater variation exists in ascus structure in genera such as *Halosarpheia*, *Nais*, *Lignincola* than previously realized (Kohlmeyer 1984) and a study of these may yield information as to the evolution of the group.

With the development of new techniques for the study of marine Ascomycotina progress is being made to establish more natural groupings. Future studies should concentrate on the development and ontogeny of ascocarps and ascospores while establishment of genomic libraries may yield valuable confirmation as to the affinities and origins of genera assigned to the Halosphaeriaceae.

ACKNOWLEDGEMENTS
We are grateful to Drs. R. Cavaliere, P.W. Kirk, J. Kohlmeyer and C.A. Shearer for valuable discussions; to Dr. J. Kohlmeyer for reading parts of the manuscript; to various colleagues for the donation of material for study: Drs. J. Kohlmeyer, J. Koch, C.A. Shearer, K. Schaumann, P.W. Kirk, J. Schneider, M. Henningsson, R.A. Eaton; for laboratory facilities and hospitality: P.W. Kirk, J. Koch, C.A. Shearer, J. Kohlmeyer, A. Zainal, G.C. Hughes, the Director and staff of the Friday Harbor Laboratory; to Messrs. G. Bremer, C. Derrick for photographic assistance; to Dr. Gareth Rees for all his help with the field work, and to the Royal Society the Marshall and Orr Bequest for travelling funds for Professor E.B. Gareth Jones to travel to the U.S.A.

REFERENCES
Cavaliere, A.R. (1966 a). Marine ascomycetes: Ascocarp morphology and its application to taxonomy. I. *Amylocarpus* Currey, *Ceriosporella* gen. nov., *Lindra* Wilson. Nova Hedwigia, <u>10</u>, 387-398.

Cavaliere, A.R. (1966 b). *Marinospora*, a correction. Nova Hedwigia, <u>11</u>, 548.

Cavaliere, A.R. (1966 c). Marine ascomycetes: Ascocarp morphology and its application to taxonomy. IV. Stromatic species. Nova Hedwigia, <u>10</u>, 438-452.

Cavaliere, A.R. (1977). Marine flora and fauna of the northeastern United States. Higher fungi: Ascomycetes, deuteromycetes, and basidiomycetes. NOAA Tech. Rep., <u>398</u>, 1-49.

Cavaliere, A.R. & Johnson, T.W. (1966 a). Marine ascomycetes: Ascocarp morphology and its application to taxonomy. II. A revision of the genus *Lulworthia* Sutherland. Nova Hedwigia, <u>10</u>, 425-437.

Cavaliere, A.R. & Johnson, T.W. (1966 b). Marine ascomycetes: Ascocarp morphology and its application to taxonomy. V. Evaluation. Nova Hedwigia, <u>10</u>, 453-461.

Feldman, G. (1957). Un nouvel ascomycete parasite d'une algue marine: *Chadefaudia marina*. Rev. gén. Bot., 64, 140-152.

Furtado, S.E.J. & Jones, E.B.G. (1980). The colonisation of selected naturally durable timbers by marine fungi and borers. Bull. Liaison Com. Int. Perm. Mech. Pres. Mat. Mil. Mar., 8, 69-93.

Hibberd, D. (1981). Notes on the taxonomy of the algal classes Eustigmatophyceae and Tribophyceae (synonym Xanthophyceae). J. Linn. Soc. (Bot), 82, 93-119.

Hughes, G.C. (1975). Studies of fungi in oceans and estuaries since 1961. I. Lignicolous, caulicolous and foliicolous species. Oceanogr. Mar. Biol. Annu. Rev., 13, 69-180.

Johnson, R.G. (1980). Ultrastructure of ascospore appendages of marine ascomycetes. Botanica mar., 23, 501-527.

Johnson, R.G. (1982). Ultrastructure and histochemistry of the ontogeny of ascospores, and their appendages, in marine ascomycetes. Ph.D. thesis, C.N.A.A., Portsmouth Polytechnic.

Johnson, R.G., Jones, E.B.G. & Moss, S.T. (1985). Taxonomic studies of the Halosphaeriaceae: *Remispora* Linder, *Marinospora* Cavaliere and *Carbosphaerella* Schmidt. Botanica mar., 27, 557-566.

Jones, E.B.G. & Moss, S.T. (1978). Ascospore appendages of marine Ascomycetes: An evaluation of appendages as taxonomic criteria. Mar. Biol., 49, 11-26.

Jones, E.B.G. & Moss, S.T. (1980). Further observations on the taxonomy of the Halosphaeriaceae. Botanica mar., 23, 483-500.

Jones, E.B.G., Johnson, R.G. & Moss, S.T. (1983 a). Taxonomic studies of the Halosphaeriaceae: *Corollospora* Werdermann. J. Linn. Soc. (Bot), 87, 193-212.

Jones, E.B.G., Johnson, R.G. & Moss, S.T. (1983 b). *Ocostaspora apilongissima* gen. et sp. nov. A new marine pyrenomycete from wood. Botanica mar., 26, 353-360.

Jones, E.B.G., Johnson, R.G. & Moss, S.T. (1984). Taxonomic studies of the Halosphaeriaceae: *Halosphaeria* Linder. Botanica mar., 27, 129-143.

Kohlmeyer, J. (1972). A revision of Halosphaeriaceae. Can. J. Bot., 50, 1951-1963.

Kohlmeyer, J. (1984). Tropical marine fungi. Mar. Ecol., 5, 329-378.

Kohlmeyer, J. & Kohlmeyer, E. (1966). On the life history of marine ascomycetes: *Halosphaeria mediosetigera* and *H. circumvestita*. Nova Hedwigia, 12, 189-202.

Kohlmeyer, J. & Kohlmeyer, E. (1979). Marine Mycology: The Higher Fungi. New York: Academic Press.

Lloyd, L.S. & Wilson, I.M. (1962). Development of the perithecium in *Lulworthia medusa* (Ell. & Ev.) Cribb & Cribb, a saprophyte of *Spartina townsendii*. Trans. Br. mycol. Soc., 45, 359-372.

Luttrell, E.S. (1951). Taxonomy of the Pyrenomycetes. Univ. Mo. Stud., 24, 1-120.

Manton, I. (1966). Observations on scale production in *Prymnesium parvum*. J. Cell Sci., 1, 375-380.

Moss, S.T. & Jones, E.B.G. (1977). Ascospore appendages of marine ascomycetes: *Halosphaeria mediosetigera*. Trans. Br. mycol. Soc., 69, 313-315.

Nakagiri, A. & Tubaki, K. (1982). *Corollospora luteola*, a new marine ascomycete and its anamorph from Japan. Trans. mycol. Soc. Japan, 23, 101-110.

Nakagiri, A. & Tubaki, K. (1985). Teleomorph and anamorph relationships in marine Ascomycetes (Halosphaeriaceae). Botanica mar., 28, 485-500.

Pegler, D.H. & Young, T.W.K. (1981). A natural arrangement of the Boletales, with reference to spore morphology. Trans. Br. mycol. Soc., 76, 103-146.

Schmidt, I. (1974). Hohere meerespilze der Ostsee. Biol. Runds., 12, 96-112.

Searle, T. (1973). Life history of *Neurospora crassa* viewed by scanning electron microscopy. J. Bact., 113, 1015-1025.

Shearer, C.A. & Crane, J.L. (1980). Fungi of the Chesapeake Bay and its tributaries. VIII. Ascomycetes with unfurling appendages. Botanica mar., 23, 607-615.

Shearer, C.A. & Miller, M. (1977). Fungi of the Chesapeake Bay and its tributaries. V. *Aniptodera chesapeakensis* gen. et sp. nov. Mycologia, 69, 887-898.

Webber, F.C. (1967). Observations on the structure, life history and biology of *Mycosphaerella ascophylli*. Trans. Br. mycol. Soc., 50, 583-601.

Wilson, I.M. (1965). Development of the perithecium and ascospores of *Ceriosporopsis halima*. Trans. Br. mycol. Soc., 48, 19-33.

Zainal, A. & Jones, E.B.G. (1986). Occurrence and distribution of lignicolous marine fungi in Kuwait coastal waters. Proceedings 6th International Biodeterioration Symposium, in press.

20 AN ELECTRON MICROSCOPE STUDY OF ASCUS AND ASCOSPORE
    STRUCTURE IN *ANIPTODERA* AND *HALOSARPHEIA*, HALOSPHAERIACEAE

        C.A. Farrant

        INTRODUCTION
        Taxonomy and phylogeny of the marine Ascomycotina,
particularly the Halosphaeriaceae, Pyrenomycetes, have been subjects of
controversy since the monograph by Barghoorn & Linder (1944). Within
the Halosphaeriaceae particular taxonomic significance has been
attributed to ascospore characters including the form of the
appendages, septation, size and colour (Kohlmeyer & Kohlmeyer 1979).
Appendage ontogeny and structure have been extensively studied at the
ultrastructural level and are considered by some workers to be of
phyletic importance (Jones & Moss 1978; Johnson 1980; Shearer & Crane
1980; Jones *et al*. 1983, 1984; Jones, Moss & Cuomo 1983; Johnson *et al*.
1984). Conversely Kohlmeyer (1960, 1962, 1972) and Kohlmeyer &
Kohlmeyer (1964-1969) believed that ascocarp ontogeny was of greater
taxonomic significance at the generic level. Cavaliere & Johnson (1966)
considered that the perithecial or stromatic nature of the ascocarp and
the presence or absence of paraphyses were the only taxonomically
significant ascocarp characters although of less significance than
spore morphology.

Luttrell (1951) introduced a new concept into the taxonomy of the
Pyrenomycetes in which the unitunicate or bitunicate nature of the ascus
wall was considered to be a major taxonomic criterion and established
the Unitunicatae and Bitunicatae respectively. Chadefaud (1960) also
considered ascus structure to be a fundamental criterion but placed
emphasis on features of the apical apparatus. Beckett (1981) in
reference to the asci of terrestrial Ascomycotina stated that "the
presence or absence of certain apical structures, variation in their
morphology and structural details of the wall are now regarded as
important diagnostic features in modern schemes of classification". In
the marine Ascomycotina classification has followed Luttrell's scheme
with groupings based on the unitunicate or bitunicate nature of the
ascus wall. However, three genera of the Halosphaeriaceae contain
species characterized by some degree of differentiation of the ascus
apex; namely *Aniptodera chesapeakensis* Shearer et Miller, *Halosarpheia
fibrosa* J. et E. Kohlmeyer, *H. trullifera* (Kohlmeyer) Jones, Moss et
Cuomo and *Lignincola laevis* Höhnk.

This chapter presents ultrastructural information on the ascus and
ascospore structure of species from two of those genera of
Halosphaeriaceae described to possess differentiation of the apical

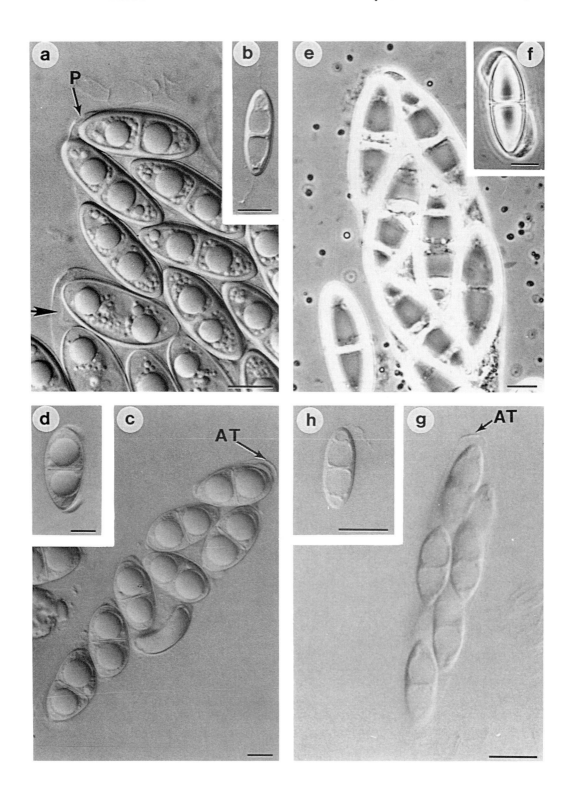

region of the ascus wall; namely *Aniptodera chesapeakensis*, *Halosarpheia trullifera*, *H. marina* (Cribb et Cribb) Kohlmeyer and *Halosarpheia* sp. and speculates on their taxonomic significance.

MATERIALS AND METHODS

Squash mounts of fresh material in sea water were used for light microscopy. Material for scanning electron microscopy (SEM) was fixed in osmium tetroxide, dehydrated and critical point dried. Specimens for transmission electron microscopy (TEM) were fixed in potassium permanganate, dehydrated and embedded in epoxy resin. Sections were examined without further staining in a JEOL 100S TEM.

RESULTS

The four species examined were characterized by two-celled, ellipsoidal ascospores produced within deliquescent (*Halosarpheia trullifera*, *Halosarpheia* sp.) or semi-persistent (*Aniptodera chesapeakensis*, *Halosarpheia marina*) asci. The ascospores of all species each possessed a single, thread-like appendage attached at each pole, although in *A. chesapeakensis* and *H. marina* many ascospores were found without appendages. Detailed ultrastructural studies of each species have shown discrete differences in appendage, ascospore and ascus wall structures. These results are presented separately for each species.

Aniptodera chesapeakensis

At the light microscope level an ascus apical apparatus and what appeared to be a subapically thickened wall were observed (Figure 20.1a). Scanning electron microscopy showed the apical apparatus to be a ring-shaped structure with a central depression (Figures 20.2a,d) but the subapical region of the ascus wall lacked any

Figures 20.1a-h Asci and ascospores. Figures 20.1a-d,g,h, Nomarski interference L.M.; Figures 20.1e,f, phase contrast L.M. Bars = 10 µm.
Figure 20.1a *Aniptodera chesapeakensis*, ascus with apical plug (P) and an apparent subapical thickening (arrowed).
Figure 20.1b *Aniptodera chesapeakensis*, released ascospore with polar thread-like appendages.
Figure 20.1c *Halosarpheia trullifera*, ascus with apical thickening (AT).
Figure 20.1d *Halosarpheia trullifera*, released ascospore with coiled polar thread-like appendages.
Figure 20.1e *Halosarpheia* sp., ascus with no resolvable apical apparatus.
Figure 20.1f *Halosarpheia* sp., released ascospore with coiled polar appendages.
Figure 20.1g *Halosarpheia marina*, ascus with apical thickening (AT).
Figure 20.1h *Halosarpheia marina*, released ascospore with polar thread-like appendages.

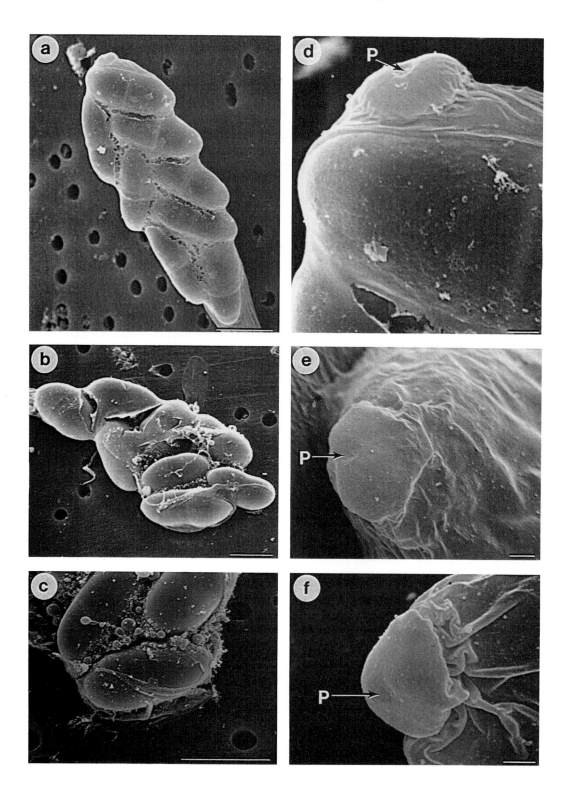

superficial differentiation. At the TEM level the mature ascus had an
apical apparatus (Figures 20.4a,b) comprised of an electron-opaque
granular region, the apical ring, the centre of which was occluded by
an electron-transparent plug. The ascus wall was comprised of a
50-70 nm thick, electron-opaque layer (Figure 20.5c) which was
continuous over the apical apparatus (Figures 20.4a,b). The plasma
membrane in the subapical region was retracted from the ascus wall
(Figure 20.4b) although elsewhere, including the apex, it was closely
adpressed to the wall. Both appendaged (Shearer & Crane 1980) and
unappendaged (Shearer & Miller 1977) released ascospores were observed
at the light microscope and SEM levels (Figures 20.1b,3a,b). The
appendaged ascospore possessed a single, 60-80 nm diam and > 100 μm long,
appendage attached at each pole. At the TEM level ascospores of
A. *chesapeakensis* possessed a wall comprised of an outer 50 nm thick,
electron-opaque episporium and a bipartite 700 nm thick mesosporium
(Figure 20.5a). No appendage was found associated with ascospores in
sectioned material.

Halosarpheia marina
     Light microscope observations of *H. marina* showed the ascus
to possess an apical apparatus (Figure 20.1g) and an apparent subapical
thickening of the ascus wall. Transmission (Figures 20.4g,h) and SEM
(Figures 20.2e,f) showed the apical structure of the ascus to be very
similar to that in *Aniptodera chesapeakensis*. It comprised an
electron-opaque ring-shaped thickening around a central electron-
transparent, biumbonate plug (Figure 20.4h). The ascus wall comprised a
70-80 nm thick electron-opaque layer which was continuous over the
apical apparatus. The ascus plasma membrane was closely adpressed to
the apical ring and lateral regions of the ascus wall but retracted from
the subapical region. The wall of the mature ascospore was comprised of
an outer 30-40 nm thick episporium and a 800-900 nm thick bipartite
mesosporium (Figure 20.5d). The appendages of ascospores within the
ascus and immediately after release from the ascus appeared as coiled
threads (Figures 20.3g,h).

Halosarpheia sp.
     At the light microscope level only the ascus of *Halosphaeria*
sp. (Figure 20.1e) from which ascospores had been released was found to
possess an apical thickening. However sectioned material examined at
the TEM level showed this thickening of the ascus wall to be present

Figures 20.2a-f Scanning electron micrographs of asci. Figures
20.2a-c, bars = 10 μm; Figures 20.2d-f, bars = 1 μm.
Figures 20.2a,d *Aniptodera chesapeakensis*, entire ascus (Fig. a)
showing ring-shaped apical thickening and plug (Fig. d. P).
Figures 20.2b,c *Halosarpheia trullifera*, entire (Fig. b) and apical
region (Fig. c) of partially deliquesced asci showing no apical
apparatus.
Figures 20.2e,f *Halosarpheia marina*, apical region of asci showing
ring-shaped thickening and central plug (P).

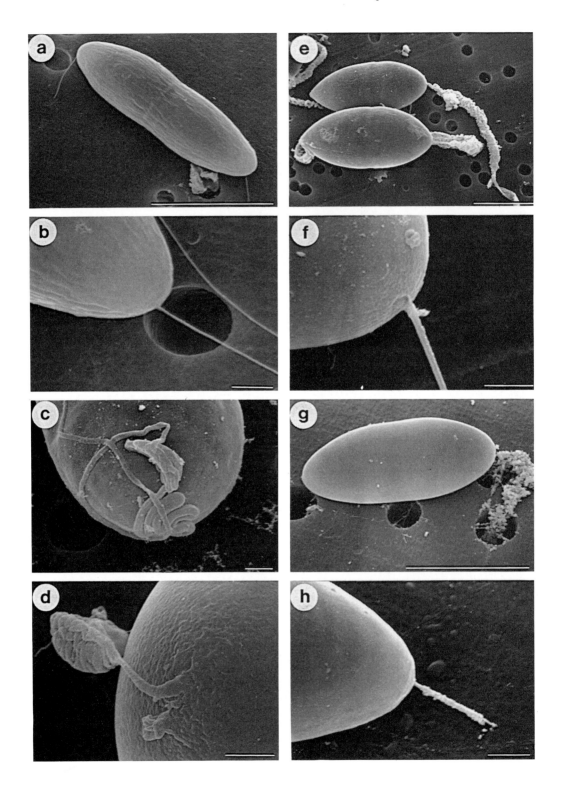

even prior to ascospore release (Figures 20.4e,f). The thickening
consisted of a granular apical ring with a central electron-transparent
plug which extended into a subapical fibrous region adjacent to the
ascus wall. This subapical thickening extended basally much further in
Halosarpheia sp. than in any other species investigated. The lack of
plasma membrane retraction from the subapical region of the ascus wall
distinguished this species from both *Aniptodera chesapeakensis* and
Halosarpheia marina. The ascus wall consisted of a 40 nm thick
electron-opaque layer which was continuous over the apical apparatus
(Figure 20.4f). The ascospores possessed a single appendage at each
pole (Figure 20.1f) which when released into sea water uncoiled into a
long thread, 80 nm diam and > 100 μm long (Figures 20.3e,f). The wall of
the mature ascospore comprised a 25 nm thick electron-opaque episporium
and an inner electron-transparent, 770-790 nm thick, mesosporium.

        Halosarpheia trullifera
        When examined under bright-field and phase-contrast light
microscopy and SEM (Figures 20.2b,c) the ascus of *H. trullifera* showed
no apical apparatus. However with Nomarski interference light
microscopy an apical apparatus was discernible (Figure 20.1c).
Transmission electron microscopy confirmed the Nomarski microscope
observations and showed a simple apical thickening. Owing to the very
limited material available, only a single ascus was serially sectioned
and examined at the TEM level. In this ascus no apical plug or ring was
observed but a thickening extended as a cap over the entire ascus apex
(Figure 20.4d) bounded externally by a 40 nm thick electron-opaque ascus
wall (Figure 20.4c). The ascus plasma membrane remained closely
adpressed to the apical thickening but less adpressed to the ascus wall.
There was no indication of a discrete and localized retraction of the
plasma membrane similar to that in *Aniptodera chesapeakensis* and
Halosarpheia marina. The ascospore was characterized by a single
cap-like appendage at each pole (Figure 20.1d) which at maturity
differentiated into a single, coiled thread-like appendage
(Figures 20.3c,d) attached within a collar at the ascospore pole.
Sectioned material showed the ascospore wall to be three-layered
(Figure 20.5b). The mesosporium was electron-transparent, 420-450 nm
thick, the episporium was a 300-320 nm thick electron-opaque layer but
unlike the other species investigated *H. trullifera* possessed a discrete
30 nm thick outer electron-opaque layer which appeared loosely attached
to the episporium. The collar around the base of the appendage appeared
as a discrete, electron-opaque lamellated structure.

Figures 20.3a-h Scanning electron micrographs of released ascospores
with polar, thread-like appendages. Figures 20.3a,e,g, bars = 10 μm;
Figures 20.3b-d,f,h, bars = 1 μm.
Figures 20.3a,b *Aniptodera chesapeakensis*.
Figures 20.3c,d *Halosarpheia trullifera*, appendages partly uncoiled.
Figures 20.3e,f *Halosarpheia* sp.
Figures 20.3g,h *Halosarpheia marina*.

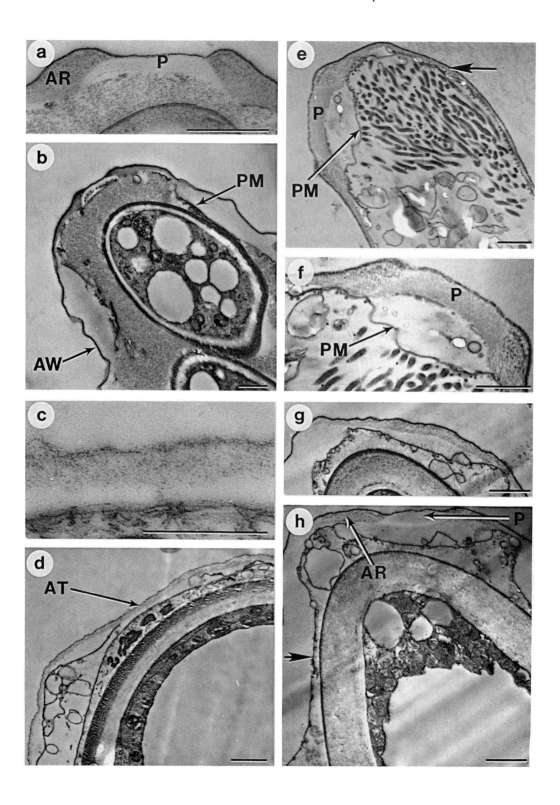

DISCUSSION
          Criteria used as a basis for a revised classification of
the Halosphaeriaceae by Jones *et al.* (1983, 1984) included both
ultrastructural and histochemical characters. Particular significance
was attributed to the structure of the ascospore wall and the ontogeny
and structure of ascospore appendages. Prior to the work reported in
this chapter the structure of the ascus apex had not been investigated
at the TEM level for any species of Halosphaeriaceae. However, an
indication of some differentiation of the ascus wall was reported at
the light microscope level by Kohlmeyer & Kohlmeyer 1977; Shearer &
Miller 1977; Patil & Borse 1982; Kohlmeyer 1984 but in spite of the
taxonomic importance of apically differentiated ascus structure in
terrestrial forms (Chadefaud 1960; Beckett 1981) little importance has
been assigned to this character in the Halosphaeriaceae.

Characters which presently delineate the Halosphaeriaceae are: the
absence of paraphyses; ascus without apical apparatus, rarely with a
simple pore, unitunicate, thin-walled or rarely thick-walled below the
apex, aphysoclastic, deliquescing at or before maturity, rarely
persistent; ascospores with ornamentations and/or gelatinous sheaths.

Based on these characters alone assignment of the species of *Aniptodera*
and *Halosarpheia* investigated to the Halosphaeriaceae is supported.
However species of *Aniptodera* and *Halosarpheia* together with
Lignincola (Kohlmeyer 1984) differ from other Halosphaeriaceae in that
they possess an ascus apical apparatus. The phyletic significance of an
apical apparatus in the marine Ascomycotina cannot be evaluated until
more species have been examined, and it is considered premature to
suggest either the removal of *Aniptodera*, *Halosarpheia* and *Lignincola*
from the Halosphaeriaceae or to emend the familial description. At the
generic level the absence of an apical plug in *Halosphaeria* is the only
only criterion which separates the genus from *Aniptodera*.

Figures 20.4a-h Transmission electron micrographs of ascus apices in
species of *Aniptodera* and *Halosarpheia*. Figures 20.4a,b,d-h, bars =
1 µm; Figure 20.4c, bar = 0.5 µm.
Figures 20.4a,b *Aniptodera chesapeakensis*, low (Fig. b) and higher
(Fig. a) magnifications showing the thickened ring (AR) with plug (P)
and the plasma membrane (PM) retracted from the subapical region of the
ascus wall (AW).
Figures 20.4c,d *Halosarpheia trullifera*, the ascus lacks a thickened
ring with plug but possesses a nondifferentiated apical wall thickening
(AT), shown at high magnification in Fig. c.
Figures 20.4e,f *Halosarpheia* sp. showing apical plug (P) and a
subapically extended apical ring (Fig. e, arrowed). The ascus
plasma membrane (PM) is retracted from the plug.
Figures 20.4g,h *Halosarpheia marina*, median (Fig. h) and nonmedian
(Fig. g) longitudinal sections of the apical apparatus showing the
plug (P), apical ring (AR) and the retracted ascus plasma membrane in
the subapical region (arrowed).

Electron microscopy of species of *Aniptodera* and *Halosarpheia* has shown both similarities and differences in ascospore wall layers.   In *Aniptodera chesapeakensis*, *Halosarpheia marina* and *Halosarpheia* sp. the ascospore walls were comprised of an episporium and a mesosporium, although in *A. chesapeakensis* and *H. marina* the mesosporium was bilaminate.   The ascospore wall of *H. trullifera* was distinct from the other species investigated in that it possessed an outer wall layer which is homologous to the exosporium in some other genera of the Halosphaeriaceae (e.g. *Corollospora*, *Halosphaeriopsis*).   However unlike the exosporium described for other members of the Halosphaeriaceae it does not form appendages, although it is readily separable from the episporium and is quite distinct from the membrane complex (*sensu*

Figures 20.5a–d   Transmission electron micrographs of ascus and ascospore walls.   Bars = 0.5 µm.
Figure 20.5a   *Aniptodera chesapeakensis*, ascospore wall comprised of a bilaminate mesosporium (M) and an outer episporium (E).
Figure 20.5b   *Halosarpheia trullifera*, ascospore wall with three layers, the mesosporium (M), the episporium (E) and an outer readily separable electron-opaque layer (inset, arrowed).
Figure 20.5c   *Aniptodera chesapeakensis*, single-layered ascus wall (AW).
Figure 20.5d   *Halosarpheia marina*, showing the single-layered ascus wall (AW) and the pole of an ascospore with a wall comprised of a bilaminate mesosporium (M) and an outer episporium (E).

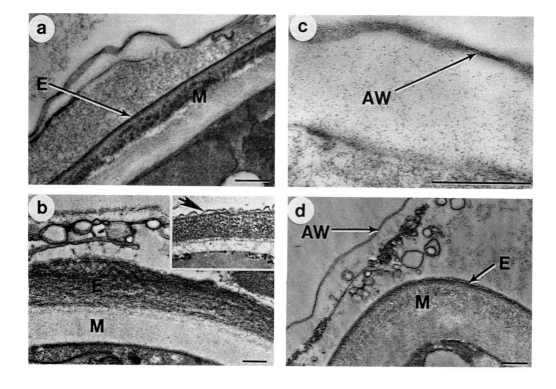

Johnson 1980).

All four species investigated possessed polarly attached, thread-like
appendages. However preliminary studies have indicated that appendage
ontogeny is not similar in all four species, for example in *Halosarpheia
marina* the appendage appears to be an extension of the mesosporium,
while in *H. trullifera* the appendage is not associated with a wall
layer and is formed prior to wall deposition. Further work is required
before any taxonomic conclusions based on appendage ontogeny may be
drawn.

The ascus wall in all four species studied was a 40-70 nm thick
electron-opaque layer which was continuous over the apical apparatus.
In *A. chesapeakensis*, *H. marina* and *Halosarpheia* sp. the basic structure
of the ascus apical apparatus comprised a ring-shaped thickening with a
central plug. In *A. chesapeakensis* and *H. marina* the thickening was
restricted to the apical region of the ascus but in *Halosarpheia* sp. it
extended subapically. In *A. chesapeakensis* and *H. marina* the structure
interpreted by previous workers as a subapical thickening of the ascus
wall was found to be a retraction of the plasma membrane away from a wall
which lacked modification. The plasma membrane in the apical region
remained closely adpressed to the apical apparatus except in the plug
region. In *Halosarpheia* sp. the ascus plasma membrane was closely
associated with both the apical ring and its subapical extension. In
this species no apparent subapical thickening similar to that observed
with the light microscope in *A. chesapeakensis* and *H. marina* was
present. *Halosarpheia trullifera* was distinguished from the other
species of *Halosarpheia* investigated in that it lacked an apical plug,
the plasma membrane was not retracted from the subapical region of the
ascus wall and the ascospore wall possessed an outer electron-opaque
layer. This concurs with previous observations of *H. trullifera*
(Kohlmeyer & Kohlmeyer 1979) at the light microscope level in which no
subapical thickening to the ascus wall was described. However Jones,
Moss & Cuomo (1983) described young asci of *H. trullifera* as thick-
walled below the apex and with an apical apparatus.

Development of the apical apparatus has not been followed for any
species. Correspondingly it has not been possible to determine whether
the apical modification develops from an existing inner ascus wall
layer which is lost subapically or is an additional layer. Further, the
function of the apical apparatus in the species examined requires
elucidation. In *H. marina* (Cribb & Cribb 1956) and *A. chesapeakensis*
(Shearer pers. com.) forcible ejection of the ascospores through an
apical pore has been recorded. However in other species of *Aniptodera*
and *Halosarpheia* which possess an apical ring ascospore release is
reported to occur by deliquescence of the ascus wall.

Kohlmeyer & Kohlmeyer (1977) described the ascus of the type species of
Halosarpheia, *H. fibrosa*, as subapically thickened but without an apical
apparatus. On the evidence presented in this paper of ascus structure
in *H. marina* and *A. chesapeakensis* it is speculated that *H. fibrosa* may
also possess an apical ring with subapical retraction of the plasma

membrane. Based on ascus structure it appears that there is some anomaly in the delineation of *Aniptodera* and *Halosarpheia* and ultrastructural examination of *H. fibrosa* is required. The character which presently unites all species of *Halosarpheia* is the presence of polar thread-like appendages on the ascospores. If the ascospore appendage is of major taxonomic importance at the generic level then *A. chesapeakensis* should be assigned to *Halosarpheia*. If, however, ascus structure is a more stable character and thus of greater taxonomic importance then *H. marina* and *Halosarpheia* sp. should be transferred to *Aniptodera*. Based on this evidence a re-evaluation of the taxonomic characters used to separate these two genera is required.

REFERENCES

Barghoorn, E.S. & Linder, D.H. (1944). Marine fungi. Their taxonomy and biology. Farlowia, 1, 395-467.

Beckett, A. (1981). The ascus with an apical pore: Development, composition and function. In Ascomycete Systematics. The Luttrelian Concept, ed. D.R. Reynolds, pp. 7-26. New York: Springer Verlag.

Cavaliere, A.R. & Johnson, T.W. (1966). Marine ascomycetes: Ascocarp morphology and its application to taxonomy. V. Evaluation. Nova Hedwigia, 10, 453-461.

Chadefaud, M. (1960). Les végétaux non vasculaires (Cryptogamie). In Traité de Botanique Systématique, Vol. I, eds. M. Chadefaud & L. Emberger. Paris: Masson.

Cribb, A.B. & Cribb, J.W. (1956). Marine fungi from Queensland. II. University of Queensland Papers. Dept. of Botany, 3, 95-105.

Johnson, R.G. (1980). Ultrastructure of ascospore appendages of marine ascomycetes. Botanica mar., 23, 501-527.

Johnson, R.G., Jones, E.B.G. & Moss, S.T. (1984). Taxonomic studies of the Halosphaeriaceae: *Remispora* Linder, *Marinospora* Cavaliere and *Carbosphaerella* Schmidt. Botanica mar., 27, 557-566.

Jones, E.B.G., Johnson, R.G. & Moss, S.T. (1983). Taxonomic studies of the Halosphaeriaceae: *Corollospora* Werdermann. J. Linn. Soc. (Bot), 87, 193-212.

Jones, E.B.G., Johnson, R.G. & Moss, S.T. (1984). Taxonomic studies of the Halosphaeriaceae: *Halosphaeria* Linder. Botanica mar., 27, 129-143.

Jones, E.B.G. & Moss, S.T. (1978). Ascospore appendages of marine ascomycetes. An evaluation of appendages as taxonomic criteria. Mar. Biol., 49, 11-26.

Jones, E.G.B., Moss, S.T. & Cuomo, V. (1983). Spore appendage development in the lignicolous marine pyrenomycetes *Chaetosphaeria chaetosa* and *Halosphaeria trullifera*. Trans. Brit. mycol. Soc., 80, 193-200.

Kohlmeyer, J. (1960). Wood inhabiting marine fungi from the Pacific North West and California. Nova Hedwigia, 2, 293-343.

Kohlmeyer, J. (1962). *Corollospora maritima* Werderm: Ein Ascomycet. Berichten der Deutschen botanischen. Gesellschaft, 74, 305-310.

Kohlmeyer, J. (1972). A revision of Halosphaeriaceae. Can. J. Bot., 50, 1951-1963.

Kohlmeyer, J. (1984). Tropical marine fungi. P.S.Z.N.I.: Mar. Ecol., 5, 329-378.

Kohlmeyer, J. & Kohlmeyer, E. (1964-1969). Icones fungarum maris. Vol. 1. Tabs. 1-90. Weinheim/Lehr., J. Cramer.

Kohlmeyer, J. & Kohlmeyer, E. (1977). Bermuda Marine Fungi. Trans. Brit. mycol. Soc., 68, 207-219.

Kohlmeyer, J. & Kohlmeyer, E. (1979). Marine Mycology. The Higher Fungi. New York: Academic Press.

Luttrell, E.S. (1951). Taxonomy of the Pyrenomycetes. University of Missouri Studies, 24, 1-120.

Patil, S.D. & Borse, B.D. (1982). Marine fungi from Maharashtra (India). I. The genus *Halosarpheia* J. & E. Kohlm. Indian bot. Reptr., 1, 102-106.

Shearer, C.A. & Crane, J.L. (1980). Fungi of the Chesapeake Bay and its tributaries. VII. Ascomycetes with unfurling appendages. Botanica mar., 23, 607-615.

Shearer, C.A. & Miller, M. (1977). Fungi of the Chesapeake Bay and its tributaries. V. *Aniptodera chesapeakensis* gen. et sp. nov. Mycologia, 69, 887-898.

A. Nakagiri

K. Tubaki

INTRODUCTION
       Species of the ascomycete genus *Corollospora* are dominant
and cosmopolitan members of the intertidal arenicolous marine
mycoflora. Their perithecia are black, carbonaceous, thick-walled and
form on sand grains, shell fragments and other hard materials. The
somatic mycelium ramifies between sand grains and utilizes organic
substrates including buried driftwood, seaweeds and seagrasses. The
genus contains five validly described species, namely *C. maritima*
Werderm., *C. intermedia* Schmidt, *C. lacera* (Linder) Kohlm., *C. pulchella*
Kohlm., Schmidt et Nair and *C. luteola* Nakagiri et Tubaki; a further six
species isolated from the Japanese coast await formal description.

Ascospores of *Corollospora* species occur commonly in sea foam following
their release from perithecia. Species may be differentiated on the
size, colour and septation of their ascospores but all are
characterized by equatorial and polar secondary appendages formed by
fragmentation of the exosporium. Primary spine-like appendages are
also present at each pole although in some species they are
inconspicuous. Appendage type and ontogeny have been considered to be
of fundamental importance at the generic level (Jones & Moss 1978,
1980; Johnson 1980; Jones *et al.* 1983, 1984; Nakagiri & Tubaki 1985).

The perithecial wall of *Corollospora* species has a basic two-layered
construction although we have found a third outermost layer in some
species. This chapter provides information on the structure of
ascocarp peridial walls and ascospores in species of *Corollospora* and
evaluates their taxonomic significance.

ASCOSPORE MORPHOLOGY
       The range of ascospore size, septation and colour of
Corollospora species based on our data and those of Kohlmeyer &
Kohlmeyer (1979) is presented in Table 21.1 and Figures 21.1a-k. The
ascospores of *Corollospora* sp. 6 are distinct from those of all other
species in their large size, possession of longitudinal in addition to
transverse septa (Figure 21.11), melanized walls and the presence of
longitudinally orientated ridges on the spore surface (Figure 21.1m).
Despite this diversity in ascospore morphology all the species are
assigned to *Corollospora* owing to their possession of polar and
equatorial secondary appendages formed by exosporial fragmentation.

Further support for their inclusion in a single genus is provided by
similar modes of conidiogenesis in those species where the anamorphic
state has been identified (Nakagiri & Tubaki 1985).

ASCOCARP WALL
        The ascocarp peridial wall structure of *Corollospora* species
is shown in Figures 21.2a-k. These micrographs show that the basic
peridial wall is two-layered, an inner layer composed of flattened
cells and an outer layer composed of spherical to polyhedral cells
(Figures 21.2a-h). However in *Corollospora* sp. 4, *Corollospora* sp. 5
and *Corollospora* sp. 6 additional outermost layers are present
(Figures 21.2i-1). In *Corollospora* sp. 4 and 5 this outermost layer is
composed of spherical paliform cells (Figures 21.2i,j) whereas in
Corollospora sp. 6 it is composed of prosenchymatous (Figure 21.2k)
cells. It is of interest that those species with a three-layered
peridial wall are also characterized by large ascospores and ascocarps.
A similar correlation between ascocarp and ascospore characters occurs
in species of *Lindra* and *Lulworthia*. The arenicolous species *Lindra*
obtusa Nakagiri et Tubaki, *Lulworthia lignoarenaria* Koch et Jones and
Lulworthia crassa Nakagiri all produce large ascospores within large
carbonaceous ascocarps whose peridial walls are thicker and more
complex than those of lignicolous species of the same genera (Nakagiri
& Tubaki 1983; Koch & Jones 1984; Nakagiri 1984). A thick, carbonaceous
peridial wall with an outer layer of large paliform cells is also
characteristic of the arenicolous species *Carbosphaerella*
leptosphaerioides Schmidt (Figure 21.21).

It is speculated that these modifications of the peridial wall of
arenicolous marine Ascomycotina may be a protective adaptation to
prevent mechanical damage by sand grains or other hard materials.
Further, within the genus *Corollospora* there may be an evolutionary
trend towards large ascospores produced within thick-walled ascocarps.

Figures 21.1a-m Released ascospores of *Corollospora* species showing
polar and equatorial appendages. Figures 21.1a-1, light micrographs;
Figure 21.1m, scanning electron micrograph. Bars = 10 µm.
Figure 21.1a *Corollospora maritima.*
Figure 21.1b *Corollospora* sp. 1.
Figure 21.1c *Corollospora intermedia.*
Figure 21.1d *Corollospora* sp. 2.
Figure 21.1e *Corollospora* sp. 3.
Figure 21.1f *Corollospora luteola.*
Figure 21.1g *Corollospora lacera.*
Figure 21.1h *Corollospora pulchella.*
Figure 21.1i *Corollospora* sp. 4.
Figure 21.1j *Corollospora* sp. 5.
Figures 21.1k,1 *Corollospora* sp. 6 showing transverse and longitudinal
septa.
Figure 21.1m *Corollospora* sp. 6 showing equatorial appendages and
longitudinal ridges on the spore wall.

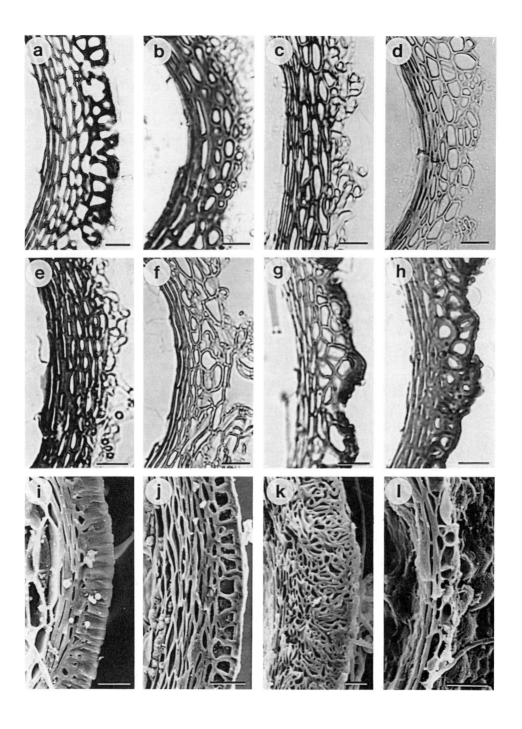

Certainly the inoculum potential of large multiseptate ascospores is considered greater than small few celled ascospores. Correspondingly, we consider that species characterized by large ascospores and thick-walled ascocarps are evolutionarily the most advanced.

TAXONOMIC EVALUATION

Despite the described variation in ascospore morphology and ascocarp wall structure between species of *Corollospora* we do not consider these characters of taxonomic significance at the generic level. However the similarity in appendage ontogeny and the basically two-layered ascocarp peridial wall in all species of *Corollospora* are considered of phyletic importance. Additional support for the assignment of all the eleven species investigated to the same genus is provided by the following teleomorph-anamorph connections. *Clavatospora bulbosa* (Anastasiou) Nakagiri et Tubaki, *Sigmoidea luteola* Nakagiri et Tubaki and *Varicosporina* sp., which are the anamorphs of *Corollospora pulchella*, *C. luteola* and *C. intermedia* respectively, and all show sympodial conidium development (Nakagiri & Tubaki 1985). The taxonomic significance of the teleomorph-anamorph relationship is also demonstrated by species of *Lulworthia* and *Lindra*. The arenicolous species *Lulworthia lignoarenaria* and *Lulworthia crassa* have septate or nonseptate filiform ascospores with mucous filled end chambers and ascocarps with thick peridial walls composed of two layers. Conversely, arenicolous *Lindra obtusa* has septate filiform ascospores without end chambers and with an ascocarp peridial wall unlike that of *Lulworthia* species (Nakagiri & Tubaki 1983). The anamorphic states support the separation of these two genera; the anamorph of *Lulworthia uniseptata* Nakagiri is *Zalerion maritimum* (Linder) Anastasiou which is characterized by terminal holoblastic conidia, while *Anguillospora marina* Nakagiri et Tubaki, the anamorph of *Lindra obtusa*, produces filiform conidia in percurrent proliferations (Nakagiri & Tubaki 1983; Nakagiri 1984).

Figures 21.2a-l Ascocarp peridial wall structures of *Corollospora* species and *Carbosphaerella leptosphaerioides*. Figures 21.2a-h, light micrographs of sectioned ascocarps; Figures 21.2i-l, scanning electron micrographs of fractured ascocarps. Bars = 10 μm.

Figure 21.2a *Corollospora maritima*.
Figure 21.2b *Corollospora* sp. 1.
Figure 21.2c *Corollospora intermedia*.
Figure 21.2d *Corollospora* sp. 2.
Figure 21.2e *Corollospora* sp. 3.
Figure 21.2f *Corollospora luteola*.
Figure 21.2g *Corollospora lacera*.
Figure 21.2h *Corollospora pulchella*.
Figure 21.2i *Corollospora* sp. 4.
Figure 21.2j *Corollospora* sp. 5.
Figure 21.2k *Corollospora* sp. 6.
Figure 21.2l *Carbosphaerella leptosphaerioides*.

Table 21.1   Ascospore morphology in species of *Corollospora*

Species	C. maritima	C. sp. 1	C. intermedia	C. sp. 2	C. sp. 3	C. luteola	C. lacera	C. pulchella	C. sp. 4	C. sp. 5	C. sp. 6
Ascospores											
Length (μm)[1]	20–37.5	34–45	25–34	37–57	41.5–59	50–85	39–60	52.5–102.5	77–108	87–120	63–220
Diameter (μm)[1]	6–11.5	3–5.5	7–11.5	3–7.5	7.5–9.8	4.8–7.5	10–16	7–12.5	13–25	5–8	20–38
Septa[2]	1	1	3	3	5	5	5	7	7	13	12–21

[1] excluding secondary appendages.
[2] characteristic number.

It is considered that the demonstrated correlations between ascospore appendage ontogeny, ascocarp peridial wall structure and the conidial ontogeny of the anamorphic state indicate that these characters are of taxonomic importance at the generic level in the genus *Corollospora* and other arenicolous marine fungi.

REFERENCES

Johnson, R.G. (1980). Ultrastructure of ascospore appendages of marine Ascomycetes. Botanica mar., 23, 501-527.

Jones, E.B.G. & Moss, S.T. (1978). Ascospore appendages of marine ascomycetes: an evaluation of appendages as taxonomic criteria. Mar. Biol., 49, 11-26.

Jones, E.B.G. & Moss, S.T. (1980). Further observations on the taxonomy of the Halosphaeriaceae. Botanica mar., 23, 483-500.

Jones, E.B.G., Johnson, R.G. & Moss, S.T. (1983). Taxonomic studies of the Halosphaeriaceae: *Corollospora* Werdermann. J. Linn. Soc. (Bot), 87, 193-212.

Jones, E.B.G., Johnson, R.G. & Moss, S.T. (1984). Taxonomic studies of the Halosphaeriaceae: *Halosphaeria* Linder. Botanica mar., 27, 129-143.

Koch, J. & Jones, E.B.G. (1984). *Lulworthia lignoarenaria*, a new marine pyrenomycete from coastal sands. Mycotaxon, 20, 389-395.

Kohlmeyer, J. & Kohlmeyer, E. (1979). Marine Mycology: The Higher Fungi. New York: Academic Press.

Nakagiri, A. (1984). Two new species of *Lulworthia* and evaluation of genera-delimiting characters between *Lulworthia* and *Lindra* (Halosphaeriaceae). Trans. mycol. Soc. Japan, 25, 377-388.

Nakagiri, A. & Tubaki, K. (1983). *Lindra obtusa*, a new marine ascomycete and its *Anguillospora* anamorph. Mycologia, 75, 487-497.

Nakagiri, A. & Tubaki, K. (1985). Teleomorph and anamorph relationships in marine ascomycetes (Halosphaeriaceae). Botanica mar., 28, 485-500.

22 THE SIGNIFICANCE OF TELEOMORPH/ANAMORPH CONNECTIONS IN THE CLASSIFICATION OF MARINE ASCOMYCOTINA

C.A. Shearer

Taxonomists are interested not only in characterizing and describing species, but also in demonstrating relationships between them. Generally, it is hoped that relationships between species will assist not only in developing classification schemes but also in understanding how fungi have evolved. Fungal species can be related to one another morphologically, physiologically, biochemically, ecologically and according to their life cycles. All of these types of data can and should be used to classify fungi. This chapter is concerned with the relationships between marine Ascomycotina based on the anamorphic states in their life cycles.

Anamorphic states have been reported for a number of marine Ascomycotina in several different classes (Table 22.1). These Ascomycotina can be grouped according to the type of anamorphic state they produce. Members of the Halosphaeriaceae produce hyphomycetous anamorphs and/or chains of dematiaceous, spherical cells. Hyphomycetous anamorphs are found also in *Heleococcum japonense* (Nectriaceae), *Zopfiella latipes* (Sordariaceae) and *Pleospora triglochinicola* (Pleosporaceae). Pycnidial anamorphs have been found for *Mycosphaerella suaedae-australis*, *Didymosphaeria danica*, *Leptosphaeria typhicola*, *Phaeosphaeria typharum* and *Pleospora spartinae*. Ascocarps of *Pharcidia balani*, *P. laminariicola*, *P. rhachiana*, *Leiophloea pelvetia*, *Didymella gloiopeltidis*, *Mycosphaerella staticicola* and *Leptosphaeria albopunctata* are found in association with pycnidial-like structures but it is not known whether the conidia produced in these pycnidia function as conidia or spermatia.

Although classification of the marine Ascomycotina is based on characteristics of the sexual state, the anamorphic state can be a very important taxonomic character, especially at the specific or generic level, where it may serve to delineate closely related species. For example, Mouzouras & Jones (1985), who recently connected *Nereiospora cristata* to *Monodictys pelagica*, found that anamorphic characteristics supported the recent removal of *N. cristata* from *Corollospora* by Jones *et al.* (1983).

The usefulness of the anamorphic state in indicating relationships at all taxonomic levels is limited at the present time for a number of reasons. Ascomycotina with proven anamorphic states represent a minor proportion (~10-15%) of all Ascomycotina found in marine habitats. With so few proven connections, it is difficult to generalize about

Table 22.1 Reported anamorphs of marine Ascomycotina

Teleomorph	Anamorph	Other structures	Author
PHYSOSPORELLACEAE			
Buergenerula spartinae Kohlm. et Gessner	unnamed (phialide-like, micro-conidial? or spermatial? state)		Kohlmeyer & Gessner 1976
HALOSPHAERIACEAE			
Ceriosporopsis circumvestita (Kohlm.) Kohlm.		chains of reddish or grayish brown, globose to ellipsoidal cells	Kohlmeyer & Kohlmeyer 1979 / Shearer unpublished
Ceriosporopsis halima Linder	[1]*Zalerion maritimum* (Linder) Anastasiou sub *Helicoma salinum*	[2]curved or coiled chains of reddish-brown, globose to ellipsoidal cells	[1]Barghoorn & Linder 1944 / [2]Kohlmeyer & Kohlmeyer 1979
Ceriosporopsis tubulifera (Kohlm.) Kirk		chains of spherical, brown cells	Shearer unpublished data
Corollospora intermedia Schmidt	*Varicosporina* sp.		Nakagiri & Tubaki 1983
Corollospora luteola Nakagiri et Tubaki	*Sigmoidea luteola* Nakagiri et Tubaki		Nakagiri & Tubaki 1982
Corollospora maritima Werdermann		chains of brown, spherical cells fused together to form hyphal ropes	Shearer unpublished data
Corollospora pulchella Kohlm., Schmidt et Nair	*Clavariopsis bulbosa* Anastasiou	chains of deep-olive brown to black, globose to ellipsoidal cells	Shearer & Crane 1971
Halosphaeria cucullata (Kohlm.) Kohlm.	*Periconia prolifica* Anastasiou		Kohlmeyer 1969 (Validated-Shearer unpublished data)
Halosphaeriopsis mediosetigera (Cribb et Cribb) Johnson	*Trichocladium achrasporum* (Meyers et Moore) Dixon ex Shearer et Crane)	chains of brown, globose to ellipsoidal cells	Shearer & Crane 1977
Lindra obtusa Nakagiri et Tubaki	*Anguillospora marina* Nakagiri et Tubaki		Nakagiri & Tubaki 1983

Table 22.1 continued

Teleomorph	Anamorph	Other structures	Author
Lulworthia fucicola Sutherland		chains of spherical, brown cells	Shearer unpublished data
Lulworthia uniseptata Nakagiri et Tubaki	*Zalerion maritimum* (Linder) Anastasiou		Nakagiri & Tubaki 1984
Nereiospora cristata E.B.G. Jones, R.G. Johnson et Moss	*Monodictys pelagica* (T.W. Johnson) E.B.G. Jones		Mouzouras & Jones 1985
HYPOCREACEAE			
Halonectria milfordensis E.B.G. Jones	pycnidia (or spermogonia?) conidia (or spermatia?) one-celled, hyaline, straight or curved		Jones 1965
Heleococcum japonense Tubaki	*Trichothecium*-type conidial anamorph		Tubaki 1967
SORDARIACEAE			
Zopfiella latipes (Lundqvist) Malloch et Cain	*Humicola*-like		Furuya & Udagawa 1973
VERRUCARIACEAE			
Pharcidia balani (Winter) Bauch	pycnidia (or spermogonia) conidia (spermatia?) ellipsoidal, one-celled, hyaline, catenulate		Kohlmeyer & Kohlmeyer 1979
Pharcidia laminariicola Kohlm.	pycnidia (or spermogonia?) conidia (spermatia?) ellipsoidal, one-celled, hyaline		Kohlmeyer 1973
Pharcidia rhachiana Kohlm.	pycnidia (or spermogonia?) conidia (or spermatia?) ellipsoidal, one-celled, hyaline		Kohlmeyer 1973
MYCOPORACEAE			
Leiophloea pelvetiae (Sutherland) Kohlm. et Kohlm.	pycnidia (or spermogonia?) conidia (or spermatia?) bacilliform, hyaline, holoblastic		Sutherland 1915

Table 22.1 continued

Teleomorph	Anamorph	Other structures	Author
Didymella gloiopeltidis (Miyabe et Tokida) Kohlm. et Kohlm.	pycnidia (or spermogonia?) conidia (or spermatia?) bacilliform, one-celled, hyaline		Miyabe & Tokida 1948
Mycosphaerella staticicola (Patouillard) S. Dias	pycnidia (or spermogonia?) conidia (or spermatia?) ellipsoidal to cylindrical, one-celled, hyaline		Kohlmeyer & Kohlmeyer 1979
Mycosphaerella suaedae-australis Hansford	pycnidia, conidia 3-septate, hyaline, filiform straight or curved		Hansford 1954
PLEOSPORACEAE			
Didymosphaeria danica (Berlese) Wilson et Knoyle	*Phoma marina* Lind		Lind 1913
Leptosphaeria albopunctata (Westendorp) Saccardo	pycnidia (or spermogonia?) conidia (or spermatia?) cylindrical to ellipsoidal, one-celled, sub-hyaline to yellowish brown		Lucas & Webster 1967
Leptosphaeria typhicola Karsten	pycnidia, conidia globose, ovoid or ellipsoidal, hyaline		Lucas & Webster 1967
Phaeosphaeria typharum (Desmaz.) Holm	*Hendersonia typhae* Oudemans		Webster 1955
Pleospora spartinae (Webster et Lucas) Apinis et Chesters	pycnidia, conidia cylindrical, 5-7 septate, yellow to pale brown		Webster & Lucas 1961
Pleospora triglochinicola Webster	*Stemphylium triglochinicola* Sutton et Pirozyn.		Webster 1969

relationships using anamorphic states. The small number of proven
connections may reflect, in part, the small number of anamorphs present
in marine habitats. Sea water apparently exerts a negative selective
effect on species of Deuteromycotina. Shearer (1972) found that the
percentage of Deuteromycotina species decreased while the percentage of
Ascomycotina species increased with increasing salinity. Data from
various marine mycotas indicate that the ratio of Ascomycotina to
Deuteromycotina species ranges from 2.71 to 5.00 (Table 22.2). This
suggests that there are insufficient anamorphs for all the teleomorphs
and that for at least half the Ascomycotina, connections to anamorphs
may not be made. Other possible alternatives are that enough anamorphs
for all the Ascomycotina will be found eventually or that different
Ascomycotina may share the same anamorphic state.

Another problem affecting the taxonomic usefulness of anamorphs is the
unpredictability of the connections. Within the Halosphaeriaceae,
several of the teleomorphs and anamorphs were well known and widely
collected prior to the establishment of their connections. A case in
point is *Halosphaeriopsis mediosetigera* and its anamorphic state,
Trichocladium achrasporum (Shearer & Crane 1977). Most isolates of this
teleomorph do not produce the anamorph and no isolate of the anamorph
has been known to produce the teleomorph. Secondly, thus far, knowing
the teleomorph or anamorph of one species of a genus has not had any
predictive value for the other species of the genus. The anamorphs of
three species of *Corollospora* are completely different and one anamorph,
Zalerion maritimum, has been connected to two different teleomorphs
(Table 22.1).

Although only nine connections have been reported for members of the
Halosphaeriaceae, these connections bring some interesting points to
mind. The largest number of connections (three) have been made for the
genus *Corollospora* and all of the reported anamorphs are different. The
anamorph of *C. intermedia*, *Varicosporina* sp., produces branched conidia,
while the anamorph of *C. luteola*, *Sigmoidea luteola*, produces sigmoidal
conidia. The third species, *C. pulchella*, has a tetraradiate anamorphic
state, *Clavariopsis bulbosa*. Although these three anamorphs are
morphologically distinct from one another, their conidia are all
produced holoblastically by sympodial proliferation of conidiogenous
cells and they are all similar morphologically to fresh water anamorphs
considered highly adapted to the aquatic environment (Webster & Descals
1981). *Varicosporina* and *Sigmoidea* are genera originally established
for fresh water anamorphs. Conidia of *Clavariopsis bulbosa*, although
superficially similar to those of *Clavariopsis aquatica* De Wilde., a
fresh water species, are formed holoblastically by sympodial
proliferation of the conidiogenous cell rather than through percurrent
proliferation of the apex of the conidiogenous cell as in *C. aquatica*.
Clavariopsis bulbosa is more similar, according to conidiogenesis and
morphology, to *Clavatospora longibrachiata* (Ingold) S. Nilsson ex Marv.
et S. Nilsson, a fresh water species characterized by tetraradiate
conidia produced holoblastically on sympodially proliferating
conidiogenous cells (Marvanova 1980). Nakagiri & Tubaki (1985) have
recently transferred *Clavariopsis bulbosa* to *Clavatospora*. Although it

Table 22.2   Ascomycete/deuteromycete ratios for species reported in various marine mycotas, and for all species listed by Kohlmeyer & Kohlmeyer 1979

Geographical location	Number of Ascomycotina	Number of Deuteromycotina	Ascomycete/deuteromycete ratios	Author
Sri Lanka	19	7	2.71	Koch 1982
Denmark	38	12	3.17	Koch & Jones 1983
Tropics	35	7	5.00	Kohlmeyer 1984
Seychelles	57	18	3.17	Hyde 1985
Chile	29	8	3.63	Shearer unpublished data
Worldwide	153	64	2.39	Kohlmeyer & Kohlmeyer 1974

might have been expected that the three similar *Corollospora* species
would produce three similar anamorphs, this is not the case. The same
type of discrepancy can be seen for three species of *Massarina*, an
ascomycete found in fresh water habitats. Three morphologically
different anamorphs, *Dactylella aquatica* (Ingold) Ranzoni, *Anguillospora
longissima* (Sacc. et Syd.) Ingold and *Clavariopsis aquatica*, have been
connected to three different species of *Massarina* (Webster & Descals
1979). As with the anamorphs of *Corollospora*, the anamorphs of
Massarina, although morphologically dissimilar, all produce conidia in a
similar manner, holoblastically by percurrent proliferation of
conidiogenous cells. It is difficult to understand why the anamorphs of
both *Corollospora* and *Massarina* have evolved differently while the
teleomorphs have not. There is the possibility that the teleomorphs
involved are not as closely related as previously thought; clearly,
closer examination of the species in both genera is needed.

Two different teleomorphs, *Ceriosporopsis halima* and *Lulworthia
uniseptata*, have been connected to a single anamorph, *Zalerion
maritimum*. Since six different taxa have been reduced to synonymy with
Z. maritimum (Kohlmeyer & Kohlmeyer 1979), a re-examination of these
taxa may reveal that more than one species exists within the complex.

By far, the most intriguing aspect of the anamorphic states in the
Halosphaeriaceae is that all of them have sibling species found in fresh
water habitats (Shearer 1972; Webster & Descals 1981; Shearer & Crane
1986). This suggests at least three possible theories regarding the
evolution of the Halosphaeriaceae: (1) marine species originated in
fresh water habitats and invaded marine habitats secondarily; (2) marine
species originated in sea water and invaded fresh water habitats
secondarily; (3) the aquatic environment has acted as a selecting force
on two different groups of Ascomycotina, one fresh water and one marine,
producing parallel evolutionary trends in the anamorphs. One piece of
evidence against theories one and two is that although similar anamorphs
occur in both habitats, the Ascomycotina to which they are connected are
quite different. No fresh water counterparts of *Ceriosporopsis*,
Corollospora, *Halosphaeria*, *Halosphaeriopsis*, *Lindra* and *Nereiospora*
have been reported. The only anamorphic genus connected to ascomycete
species in both habitats is *Anguillospora*. This genus, however, has
been connected to such diverse ascomycete genera as *Lindra*, in the
Halosphaeriaceae, *Massarina*, in the Loculoascomycetes, and *Mollisia*,
Pezoloma and *Orbilia*, in the Discomycetes (Nakagiri & Tubaki 1983).
Clearly *Anguillospora* is a heterogeneous genus comprised of species
whose anamorphs may have undergone convergent evolution. There are
species in the Halosphaeriaceae which bridge the gap between fresh water
and marine habitats. *Aniptodera chesapeakensis* Shearer et Miller,
Halosarpheia retorquens Shearer et Crane, and *Nais inornata* Kohlm. have
been reported from fresh water (Shearer & von Bodman 1983) and sea water
(Kohlmeyer & Kohlmeyer 1979). None of these species, however, has been
connected to an anamorphic state. Kohlmeyer & Kohlmeyer (1979)
postulated that the marine, cellulolytic, saprobic Sphaeriales (which
includes the Halosphaeriaceae) evolved originally from a fungal-algal
ancestor in the sea via a pathway involving species parasitic on marine

algae. In light of the types of teleomorph-anamorph connections
recently made, a reappraisal of evolutionary pathways of this group is
warranted.

Information regarding teleomorph-anamorph connections and their
significance is tantalizing but scant. A number of activities should be
carried out in the future to broaden the teleomorph-anamorph data base.
Investigators need to isolate routinely (using single spore techniques)
the fungi they collect. This is especially important since not all
strains of a holomorph will fruit in culture. Cultures from single
spore isolations need to be deposited in accessible culture collections.
These isolates are important for comparative study, experimental work
and as a genetic repository. Experimental studies to determine the
factors which effect the expression of teleomorphic and anamorphic
states are very much needed. Occurrences of both anamorphic and
teleomorphic states must be reported and voucher specimens deposited in
herbaria. More distributional and ecological studies are needed in
order to determine the importance of each state to the maintenance of
the species. Very little is known about a sizeable group of fungi, the
algicolous species. Functional studies are needed for this group to
resolve the pycnidial/spermogonial? structures. Attempts to isolate
algicolous species must be made so that their morphology and behaviour
in pure culture can be studied. Lastly, attempts must be made to move
away from the single-character taxonomy which has been used for
classification of the marine Ascomycotina. Although this type of
taxonomy is perhaps necessary in the beginning and certainly has been
useful, as more is known about the marine Ascomycotina, a more natural
classification scheme based on all known characteristics must be
developed.

With regard to anamorphic states and nomenclature of the marine fungi,
Article 59 of the International Code of Botanical Nomenclature (1975)
clearly states that for fungi with two or more states in the life cycle,
"the correct name of all states which are states of any one species is
the earliest legitimate name typified by the perfect state". The
imperfect state may or may not be given a name different from that of
the perfect state, depending on the choice of the author. Article 59
does not prevent the use of names of imperfect states in works referring
to those states. Separate names for perfect and imperfect states of the
same fungus are useful in distributional and ecological studies where
perfect and imperfect states may be widely separated in time and/or
space.

REFERENCES
Barghoorn, E.S. & Linder, D.H. (1944). Marine fungi: Their taxonomy and
        Biology. Farlowia, 1, 395-467.
Furuya, K. & Udagawa, S. (1973). Coprophilous Pyrenomycetes from Japan
        III. Trans. mycol. Soc. Japan, 14, 7-30.
Hansford, C.G. (1954). Australian fungi II. New records and revisions.
        Proc. Linn. Soc. N.S.W., 79, 97-141.

Hyde, K.D. (1985). Spore settlement and attachment in marine fungi.
        Ph.D. Thesis, CNAA, Portsmouth Polytechnic, U.K.
International Code of Botanical Nomenclature. (1978). Adopted by the
        XIIth Int. Bot. Congr. Leningrad, 1975. Regn. Veg., 97,
        1-77.
Jones, E.B.G. (1965). *Halonectria milfordensis* gen. et sp. nov., a
        marine Pyrenomycete on submerged wood. Trans. Br. mycol.
        Soc., 48, 287-290.
Jones, E.B.G., Johnson, R.G. & Moss, S.T. (1983). Taxonomic studies of
        the Halosphaeriaceae: *Corollospora* Werderman. J. Linn. Soc.
        (Bot.), 87, 193-212.
Koch, J. (1982). Some lignicolous marine fungi from Sri Lanka. Nord. J.
        Bot., 2, 163-169.
Koch, J. & Jones, E.B.G. (1983). Vedboende havsvampe fra danske kyster.
        Svampe, 8, 49-65.
Kohlmeyer, J. (1969). Marine fungi of Hawaii including the new genus
        Helicascus. Can. J. Bot., 47, 1469-1487.
Kohlmeyer, J. (1973). Fungi from marine algae. Botanica mar., 16, 201-205.
Kohlmeyer, J. (1984). Tropical marine fungi. P.S.Z.N.I. Mar. Ecol., 5,
        329-378.
Kohlmeyer, J. & Gessner, R.V. (1976). *Buergenerula spartinae* sp. nov.,
        an Ascomycete from saltmarsh cordgrass, *Spartina*
        alterniflora. Can. J. Bot., 54, 1759-1766.
Kohlmeyer, J. & Kohlmeyer, E. (1979). Marine Mycology. The Higher
        Fungi. New York: Academic Press.
Lind, J. (1913). Danish Fungi as Represented in the Herbarium of
        E. Rostrup. Copenhagen: Nordisk Forlag.
Lucas, M.T. & Webster, J. (1967). Conidial states of British species of
        Leptosphaeria. Trans. Br. mycol. Soc., 50, 85-121.
Marvanova, L. (1980). New or noteworthy aquatic hyphomycetes.
        Clavatospora, *Heliscella*, *Nawawia* and *Heliscina*. Trans. Br.
        mycol. Soc., 75, 221-231.
Miyabe, K. & Tokida, J. (1948). Black-dots disease of *Gloiopeltis*
        furcata Post. et Rupr. caused by a new ascomycetous fungus.
        Bot. Mag., Tokyo, 61, 116-118.
Mouzouras, R. & Jones, E.B.G. (1985). *Monodictys pelagica* (T.W. Johnson)
        E.B.G. Jones the anamorph of *Nereiospora cristata*
        E.B.G. Jones, R.G. Johnson and Moss. Can. J. Bot., 63,
        2444-2447.
Nakagiri, A. & Tubaki, K. (1982). A new marine Ascomycete and its
        anamorph from Japan. Trans. mycol. Soc. Japan, 23, 101-110.
Nakagiri, A. & Tubaki, K. (1983). *Lindra obtusa*, a new marine Ascomycete
        and its *Anguillospora* anamorph. Mycologia, 75, 487-497.
Nakagiri, A. & Tubaki, K. (1984). Two new species of *Lulworthia* and
        evaluation of genera-delimiting characters between
        Lulworthia and *Lindra* (Halosphaeriaceae). Trans. mycol. Soc.
        Japan, 25, 377-388.
Nakagiri, A. & Tubaki, K. (1985). Teleomorph and anamorph relationships
        in marine Ascomycetes (Halosphaeriaceae). Botanica mar.,
        28, 485-500.
Shearer, C.A. (1972). Fungi of the Chesapeake Bay and its tributaries.
        III. The distribution of wood-inhabiting Ascomycetes and

Fungi Imperfecti of the Patuxent River. Am. J. Bot., 59, 961-969.

Shearer, C.A. & Crane, J.L. (1971). Fungi of the Chesapeake Bay and its tributaries. I. Patuxent River. Mycologia, 63, 237-260.

Shearer, C.A. & Crane, J.L. (1977). Fungi of the Chesapeake Bay and its tributaries. VI. *Trichocladium achrasporum*, the imperfect state of *Halosphaeria mediosetigera*. Mycologia, 69, 1218-1223.

Shearer, C.A. & Crane, J.L. (1986). Illinois fungi XII. Fungi and Myxomycetes from wood and leaves submerged in southern Illinois swamps. Mycotaxon, in press.

Shearer, C.A. & von Bodman, S.B. (1983). Patterns of occurrence of Ascomycetes associated with decomposing twigs in a midwestern stream. Mycologia, 75, 518-530.

Sutherland, G.K. (1915). New marine Pyrenomycetes. Trans. Br. mycol. Soc., 5, 147-155.

Tubaki, K. (1967). An undescribed species of *Heleococcum* from Japan. Trans. mycol. Soc. Japan, 8, 5-10.

Webster, J. (1955). *Hendersonia typhae* the conidial state of *Leptosphaeria typharum*. Trans. Brit. mycol. Soc., 38, 405-408.

Webster, J. (1969). The *Pleospora* state of *Stemphylium triglochinicola*. Trans. Br. mycol. Soc., 53, 478-482.

Webster, J. & Lucas, M.T. (1961). Observations on British species of *Pleospora* II. Trans. Br. mycol. Soc., 44, 417-436.

Webster, J. & Descals, E. (1979). The teleomorphs of water-borne hyphomycetes from freshwater. In The Whole Fungus, Vol. II, ed. W.B. Kendrick, pp. 419-451. Ottawa, Canada: National Museum of Natural Sciences.

Webster, J. & Descals, E. (1981). Morphology, distribution and ecology of conidial fungi in fresh water habitats. In The Biology of Conidial Fungi, eds. G.T. Cole & W.B. Kendrick, pp. 295-355. New York and London: Academic Press.

P.W. Kirk, Jr.

INTRODUCTION

The Halosphaeriaceae is a large natural family of obligately marine Pyrenomycetes in which genera and species are based mostly on ascospore ornamentations. Although there is little evidence to indicate which of these ornamentations are primitive or derived, this question is to be addressed now, for two reasons. First, the Halosphaeriaceae occupies a pivotal position between opposing views on the origin of Ascomycotina, hence examining the direction of evolution in this family may throw light on the broader problem. Second, recent electron microscope studies have provided much new information on the ascospores that invites phyletic interpretation. After considering the origin and evolutionary processes of these fungi, I shall stress the phylogeny of ascospore appendages, and suggest new researches into the problem.

ORIGINS OF ASCOMYCOTINA AND HALOSPHAERIACEAE

The Ascomycotina may be derived either from marine Rhodophyta (Kohlmeyer & Kohlmeyer 1979) or from terrestrial *Taphrina*-like fungi (Barr 1983). According to the rhodophytan theory, the Halosphaeriaceae arose from ancestral Spathulosporales, marine Ascomycotina that parasitize red algae and in several respects resemble the host. From this viewpoint the primitive condition in the Halosphaeriaceae seems best represented by *Chadefaudia*, which also parasitizes red algae but lacks trichogynes and spermatia. The asci are deliquescent, the centrum pseudoparenchyma dissolves completely, and the ascospores bear a cap of mucilage at each end. Terrestrial Pyrenomycetes presumably arose from ancient Halosphaeriaceae.

Proponents of the taphrinian theory argue that the similarities between Rhodophyta and Spathulosporales represent convergent evolution, and support their position with evidence too voluminous to cite here. This includes morphology, cytology, biochemistry, host-parasite relations and palaeontology, as well as basic tenets of evolutionary biology. That the preponderance of Ascomycotina occurs on land strongly supports a terrestrial origin for the subdivision, and the low genus to species ratio in the Halosphaeriaceae (*ca* 1:3) suggests a recent radiation. Nevertheless, as a result of long isolation, the family is not readily accommodated by any order containing terrestrial Pyrenomycetes, and I have disposed it in Barr's (1983) subclass Parenchymatomycetidae. In Figure 23.1 it is suggested that the Halosphaeriaceae is distantly related to sordariaceous and diaporthaceous terrestrial Pyrenomycetes, and more closely to the obligate marine genus *Aniptodera*. The latter is often excluded from the Halosphaeriaceae because the ascus has an apical apparatus like that in some terrestrial fungi. However, species of *Aniptodera* also have three traits considered primitive in the Halosphaeriaceae: ostiolar periphyses; persistent asci; unspecialized interascal remnants of the centrum pseudoparenchyma termed catenophyses. All three traits occur in members of *Halosarpheia* and *Trichomaris*, and two each in *Lignincola*, *Nais*, *Haligena*, *Marinospora* and *Remispora* (Figure 23.1). The predominance of deliquescent asci without periphyses or catenophyses elsewhere in the family is considered a derived condition having adaptive value in the aquatic environment.

Based on the supposition that traits of the ascocarp and the ascospore have evolved together, it is expected that the above genera may embody features of the ascospore as it emerged from the adaptive threshold (constriction, Figure 23.1) into the new adaptive zone (PRE-HALOSPHAERIACEAE). In species of *Aniptodera*, *Haligena*, *Halosarpheia*, *Lignincola*, *Marinospora*, *Nais*, *Remispora* and *Trichomaris* most ascospores are ellipsoidal, hyaline and two celled. *Lignincola* spp. are not known to form appendages, while species of *Aniptodera* and *Nais* may produce mucilaginous ones only under certain conditions (Shearer & Crane 1980). Species of *Halosarpheia* exude mucilage through apical pores whereas in species of *Haligena* and *Trichomaris* mucilaginous ornamentations form at the ascospore poles which lack pores. In *Remispora maritima* Linder poorly differentiated apical appendages with fibrillar and amorphous components result from fragmentation of the outer layer, the exosporium, of the ascospore wall. In most other genera, the appendages are formed in part by the exosporium, which develops within a unique membranous structure termed the "membrane complex" (Johnson 1980). Therefore, the prototype of the Halosphaeriaceae may have produced ellipsoidal, hyalodidymous ascospores within a membrane complex, with a generalized capacity for internal and external synthesis of capsular material that under certain conditions was deposited amidst a sparse fibrillar skeleton containing polymerases at the spore poles.

The ascospore wall in the Halosphaeriaceae is multilayered (Kirk 1976), although the exosporium and perhaps the membrane complex are absent in some genera. Enveloped by these, and always present, are an episporium of chitin and protein, and to the inside of this a mesosporium of pliant glucans. In most genera a chitinous endosporium forms within germinating ascospores and is continuous with the wall of the germ tube.

EVOLUTIONARY PROCESSES IN HALOSPHAERIACEAE
    The land-sea interface is a dynamic zone, producing gradients of many parameters that serve as coincident isolating mechanisms. Under such rigorously selective conditions the pre-existence or appearance of subtle, even inefficient features can enable a population to cross an ecological barrier and radiate profusely in relative isolation. One feature that could have served to isolate ancestral Halosphaeriaceae from related terrestrial or semiaquatic fungi

is suggested by Jennings (1983). The facultative marine deuteromycete,
Dendryphiella salina (Sutherland) Pugh et Nicot, maintains a K:Na ratio
higher than in sea water by having a plasma membrane ATPase with a pH
optimum above that in terrestrial fungi. Thus, the alkaline pH and
sodium ions in sea water work in concert as a selective agent. In the
critical spore germination phase of the life cycle, these adverse
effects are exacerbated by organic, mycostatic factors in coastal sea
water, which as a rule inhibit terrestrial fungi more than marine fungi
(Kirk 1980).

Rees (1980) and Rees & Jones (1984), and many fresh water mycologists,

Figure 23.1 Schema of proposed evolutionary trends within
the Halosphaeriaceae and its relation to other fungi.
Numbers refer to clusters of genera thought to have a common
ancestry; the adjacent clusters also share fundamental
traits. (A = asci persistent, C = catenophyses, P =
periphyses in some species, -U = ultrastructure unobserved,
-S = SEM only observed).

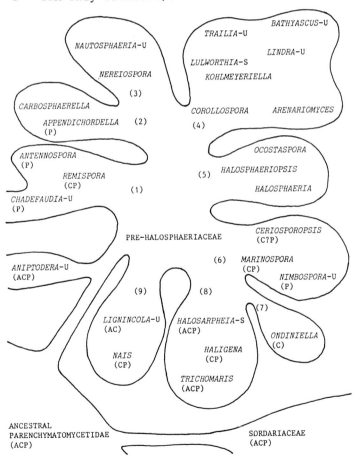

have shown that spore ornamentations are quite adaptive in aquatic fungi, they assist flotation and attachment. It is this adaptive value and the resultant convergent evolution that complicates the use of spore appendages as taxonomic traits at the generic level (Kohlmeyer & Kohlmeyer 1979). There is a tendency among disparate genera to form filamentous spore configurations with mucilaginous tips. In *Lulworthia* spp. this is achieved by filiform ascospores with polar mucin pouches, and in *Ceriosporopsis halima* Linder by ellipsoidal ascospores with long, polar mucilaginous appendages.

Species of the Halosphaeriaceae show much phenotypic plasticity in ascocarp morphology, spore dimensions and nutrition. It is pertinent to this discussion that the greatest variation occurs in *Ceriosporopsis halima* and *Lulworthia* spp., which are the most frequent pioneers in an apparent fungal succession on wood in sea water. Margalef (1968) noted that pioneers generally are adapted to changing conditions in time and space by phenotypic plasticity, a short life cycle, excellent dispersal and simple nutritional requirements. The two taxa above fruit readily on simple culture media. Such fungi are thought to be of recent origin and capable of rapid evolution. Fungi like *Remispora* spp., which are less variable, more fastidious in culture, and appear later in succession on wood, may be better integrated biochemically with their environment and phylogenetically older. It is also relevant that variation occurs in the pattern of ornamentation, as well as the size, septation and pigmentation of ascospores in the Halosphaeriaceae. In *Corollospora maritima* Werdermann I have observed abnormal spines, situated subpolarly on the ascospore and oriented toward the midseptum, which are suggestive of the immature spines in *Arenariomyces trifurcatus* Höhnk ex E.B.G. Jones, R.G. Johnson et Moss. These two species are related ecologically and in other morphological features, and until recently were considered congeneric (Jones *et al*. 1983 a). The length of ascospores or of polar appendages, or both, is often greatest in tropical, intermediate in cosmopolitan, and least in temperate species of the same genus, as in other plankters. These variations are not only the raw material for continued speciation, but also indicate latent genetic capacity, giving insight into the origin of normal characteristics in the same or other species.

EVOLUTIONARY TRENDS STRESSING ASCOSPORE APPENDAGES
        Phylogenetic pathways are constrained by prior genetic history, so that natural selection generally fashions new structures from pre-existing ones. However, because selection is most rigorous where a prototype crosses an adaptive threshold, few intermediate forms remain and close homologies are difficult to find. To this end extensive cytochemical, ultrastructural and morphological studies have been conducted (Kirk 1966, 1976; Lutley & Wilson 1972 a,b; Moss & Jones 1977; Kohlmeyer & Kohlmeyer 1979; Jones & Moss 1978, 1980; Johnson 1980; Kohlmeyer 1980; Shearer & Crane 1980; Porter 1982; Jones *et al*. 1983 a,b,c, 1984; Johnson *et al*. 1984; Koch & Jones 1984; Rees & Jones 1984; Zainal & Jones 1984; Hyde & Jones 1985). These workers have shown that most ascospore appendages in the Halosphaeriaceae arise

Figure 23.2 Schema of relationships among selected genera and species of Halosphaeriaceae. Clusters numbered as in Figure 23.1.

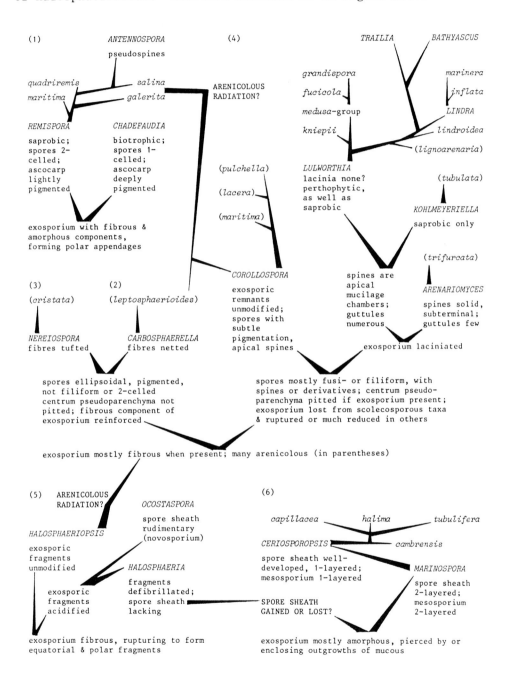

either as exudations or outgrowths from the spore, or as modifications
of the exosporium. Within each ontogenic category the appendages differ
in the proportions of fibrillar and amorphous components. Still, in
some cases homologies are difficult to interpret. What originated as a
chitinous exosporic fragment could have evolved to become an acidic
mucilaginous process that closely resembles an outgrowth. Having data
on the ascocarp and other features is essential. Therefore,
Figure 23.1, the focal point of this synthesis, does not stress single
traits as is often done in constructing phylogenetic trees. Close
relationships among genera are implied by placing the taxa together in
one of nine clusters, based on my intuitive impression of all available
data. The proximity of the clusters to one another is a further
indication of presumed interrelationships. Figure 23.2 provides more
detail on clusters 1-6 than can be shown in Figure 23.1. For brevity,
the above references on which the following discussion is largely based
will not be cited in the subsequent text.

Ascospores in the Halosphaeriaceae may have evolved along two complex
lines, represented by clusters 1-6 (Figures 23.1, 2) and 7-9
(Figure 23.1). In clusters 7-9 the exosporium is unknown, the spores
mostly hyalodidymous, and the ascocarps presumably primitive. In
species of *Lignincola* the asci function as dispersal units, and in *Nais
inornata* Kohlmeyer the ascospore may be coated at times with ephemeral
mucilage derived from the ascus cytoplasm (cluster 9). In *Ondiniella
torquata* (Kohlmeyer) E.B.G. Jones, R.G. Johnson et Moss (cluster 7) the
ascospore forms elongate polar processes as outgrowths of the
episporium, and a midseptal bulge by the synthesis of mesosporial mucins
that exert pressure beneath the episporium without rupturing it or
forming a pore. *O. torquata* resembles species of *Marinospora* (cluster
6) except the latter have sheaths in the exosporic position and form
midseptal processes rather than bulges. Comparative morphology alone
can not determine whether cluster 6 gave rise to 7 by the loss of the
sheath, or vice versa, if the two are related at all.

In *Halosarpheia* spp. (cluster 8) mucin synthesized in the ascospore
exudes through polar pores, often draping the episporium before
unfurling into elongate appendages. The ascus has a flattened tip as in
Aniptodera spp. Ancestral species of *Halosarpheia* may have evolved into
species of *Haligena* and *Trichomaris* by closure of the polar pores with a
continuation of external mucin synthesis. In *Trichomaris invadens*
Hibbits, Hughes et Sparks the mesosporium extends into the episporium at
intervals, in a manner suggestive of the midseptal bulge in *O. torquata*.
Perhaps clusters 6-8 contain derivatives of an ancestral trend from
internal to external mucin synthesis, as I can not perceive what would
select for the reverse trend. Perhaps in early sporogenesis in many
members of Halosphaeriaceae there is continuity between internal and
external synthesis of wall materials, that at maturity is manifest only
by the position of the germ pore. I have seen polar fragility where
ectodesmata occur in young *Halosphaeriopsis mediosetigera* (Cribb et
Cribb) T.W. Johnson (cluster 5) ascospores.

In clusters 1-6 an exosporium or its derivative occurs in all but the scolecosporous genera (cluster 4). The asci are deliquescent from the base; catenophyses occur only in clusters 1 and 6, and periphyses only in clusters 1, 2 and 6. Ancestral species of *Remispora* (cluster 1) are thought to have given rise to all other clusters, except possibly cluster 6, in this complex line of evolution. Little is known of *Chadefaudia*, tentatively placed in cluster 1, because it is periphysate and the one-celled spores at maturity bear polar appendages superficially resembling those in *Remispora* spp. (Cluster 1 is open to the left of *Chadefaudia*, showing the origin of the Halosphaeriaceae according to the rhodophytan theory). Ascospores of *R. maritima* prior to fragmentation of the exosporium resemble those of the hypothetical prototype of the family. From *R. maritima* to *R. salina* (Meyers) Kohlmeyer there is a trend toward the loss of catenophyses and periphyses, and greater differentiation of the exosporic fragments to resemble outgrowths from the ascospore wall.

Remispora salina resembles *Antennospora quadricornuta* (Cribb et Cribb) T.W. Johnson in many respects. The ascocarps are basically similar, both are tropical, and the immature hyalodidymous ascospores are enveloped by exosporic appendages that consist of parallel fibres in an amorphous matrix, and are attached subterminally to the mesosporium by a knob or depression. The mature appendages extend away from the spore, usually occurring in pairs in *A. quadricornuta* and 3's in *R. salina*, but the number is variable. *A. quadricornuta* appendages might be termed pseudospines, as they resemble the true spines in species of *Arenariomyces* (cluster 4) which are outgrowths of the wall and have an amorphous rather than fibrous core. Another reason which suggests that the genus *Antennospora* may be derived from *R. salina*, rather than vice versa, is that the former is a pioneer that fruits in culture, but not the latter.

A few traits of the genus *Remispora* suggest that its ancestors may have radiated to produce the arenicolous Halosphaeriaceae, which are adapted to sand and shell beaches. This ecological group includes the genera *Carbosphaerella* (cluster 2), *Nereiospora* (cluster 3), *Corollospora*, *Arenariomyces*, *Kohlmeyeriella* and *Lulworthia lignoarenaria* Koch et E.B.G. Jones (cluster 4). These fungi share some of the following traits to a greater or lesser degree: carbonaceous ascocarps with reduced necks that attach to sand and shell; multiseptate spores; dematiaceous spore pigments; polar spines; a fragmenting exosporium or derivative; pit-like connections between cells of the centrum pseudoparenchyma; drought resistance; sensitivity to mycostatic factors, and affinity for wood. *Remispora salina* (and *A. quadricornuta*) often occur on *Teredo* tubes, and the former has pitted centrum pseudoparenchyma. *Remispora galerita* Tubaki and *R. stellata* Kohlmeyer produce an exosporium of anastomosing fibres like that in species of *Carbosphaerella*. Figure 23.2 shows the suspected radiation from *Remispora*.

Species of *Carbosphaerella* (cluster 2) and *Nereiospora* (cluster 3) have phaeodictyous or phaeophragmous ascospores enveloped by stout exosporic fibres that form in predetermined patterns on the episporium. In

Carbosphaerella spp. the fibres are net-like at first, but later lose the cross connections. *Nereiospora* spp. have tufts of very thick fibres that resemble outgrowths at the poles and around the midseptum. Though much stouter, the equatorial fibres especially are suggestive of those in *Corollospora*, a genus with several phragmosporous species that often contain traces of pigment in the mesosporium and middle lamellae of septa. The genera *Appendichordella* and *Nautosphaeria* are placed here mainly because their appendages resemble those of *Carbosphaerella* and *Nereiospora*, respectively, but neither genus is arenicolous.

Cluster 4 includes a trend beginning with ancestral *Corollospora* spp. and culminating with several saprobic and perthophytic, scolecosporous genera that lack an exosporium. In species of *Corollospora* apical spines result from growth of the episporium and mesosporium, while fibrous exosporic fragments cling to the tips of spines and around the equator. In *C. maritima* and *C. intermedia* Schmidt the mesosporium is bipartite at the base of the spines, forming a cone-shaped cavity that resembles the apical mucilage pouches in scolecosporous *Kohlmeyeriella tubulata* (Kohlmeyer) E.B.G. Jones, R.G. Johnson et Moss ascospores. In *K. tubulata* and species of *Arenariomyces* the exosporium is mostly reduced to a fine laciniation, but vestigial exosporic lamellae also cover the base of spines in *Arenariomyces* spp. That the segregation of the arenicolous genera appears justifiable on morphological grounds, despite their having many features in common, is further evidence of recent radiation in the Halosphaeriaceae. Figure 23.2 indicates a pattern of speciation in the genus *Lulworthia*, leading to *Lindra*, which is very close to *Lulworthia lindroidea* Kohlmeyer except for having inflated spore tips. The ascospores of *Lulworthia* spp. are tipped with mucin filled, modified spines, derived possibly from species of *Corollospora* via *Kohlmeyeriella*. Little is known about *Trailia* and *Bathyascus*, placed here mainly because they are genera with scolecosporous ascospores.

Cluster 5 includes a trend, from the ancestral genus *Halosphaeriopsis*, through *Ocostaspora*, to *Halosphaeria*, in which chitinous exosporic fragments like those in *Corollospora* are transformed into more mucilaginous structures that resemble outgrowths from the spore apices and equator. Owing to the presence of fusiform ascospores with occasional extra septa, the genus *Halosphaeriopsis* is another possible source of the arenicolous radiation (Figure 23.2). However, this genus appears to be of much more recent origin than *Remispora*, because it fruits readily in culture and lacks the allegedly primitive ascocarpic features. In species of *Ocostaspora* and *Halosphaeria* the episporium is thin and the mesosporium thick where the appendages attach, but these may be traits only of mature ascospores with little phyletic significance. All genera in cluster 5 are hyalodidymosporous.

The genera in cluster 6 form mucilaginous processes at the poles and equator of hyalodidymous ascospores. That these processes are often appressed to the spore surface during development, and later rise up like exosporic fragments, suggests a continuation of the trend begun in cluster 5. However, in addition to these processes the ascospores in

cluster 6 are also enveloped by a mucilaginous sheath. Reports of
sulphated mucins in the sheath and processes are disputed owing to the
lack of demonstrated controls. The sheath may have originated in
ancestral *Ocostaspora apilongissima* E.B.G. Jones, R.G. Johnson et Moss,
in which exosporic fragments appear to be connected at first to the
episporium by thin mucin strands. Cytochemical evidence shows that the
processes are formed before the sheath during ascosporogenesis, hence
following Haeckel's biogenetic law, the processes are phylogenetically
older. The ascospore sheath in cluster 6 may have originated by the
selection of latent synthetic capacity in the episporium in either
cluster 5 or 7, and hence might more accurately be termed a novosporium.

The idea that cluster 6 and the novosporium may have originated in
cluster 5, is seriously challenged by the primitive ascocarpic features
in 6 that link it either to clusters 1 or 7 and 8. The genus
Marinospora (cluster 6) seems close to *Ondiniella* (cluster 7) in that
both have catenophyses, the ascospore surface is ridged, and the long
terminal appendages appear to be outgrowths from the episporium.
Although the midseptal bulge in *O. torquata* is unique, this at least
demonstrates some synthetic capacity in the same region as the
equatorial appendages in *Marinospora* spp.

SOURCES OF CORROBORATIVE EVIDENCE
Because the Halosphaeriaceae is a natural group with clearly
defined evolutionary trends in the spore appendages, its systematics
seems ripe for intensive investigation. Electron microscopy should
search for subtle ascospore ornamentations in species of *Aniptodera* and
cluster 9, for laciniation in the ascospore wall of *Lulworthia* spp.,
and homologies in young stages of sporogenesis in clusters 5-8.
Toluidine blue (0.1% in 3% acetic acid) should be adopted as the
standard for reporting acidity in mucins. The enzyme cytochemistry of
young appendages may show the direction of evolution better than the
constituent cytochemistry. The biochemistry of the mycelium should use
gel electrophoresis to show isozyme variation in species of *Lulworthia*,
serological and fluorescent antibody tests for antigenic relationships,
signature lipid analysis, and studies like Jennings' (1983) on
mechanisms of saline tolerance. Studies of the ascocarp should stress
cytogenetics. The doubled mesosporium could have resulted from
polyploidy, an event having little evolutionary significance in the
genera *Carbosphaerella* and *Marinospora*, but perhaps in ancestral
Corollospora making possible the evolution of mucin pouches in species
of *Kohlmeyeriella*. The search for fossil Halosphaeriaceae might
concentrate on deposits containing calcareous algae and *Teredo* spp. It
is possible that some Halosphaeriaceae depend upon species of *Teredo* to
provide habitat in the interior of wood while the mollusc does not
require the fungus at all, thus suggesting that *Teredo* spp. may be the
more ancient of the two.

Not all new contributions need be so sophisticated. It is important to
continue exploring the oceans in remote regions and unusual habitats to
find new taxa, like *O. apilongissima* and *Lulworthia* spp. with septate

spores, that fill important gaps in the evolutionary puzzle. The
continued conscientious reporting of variations in spore morphology
serves a similar end. Hyphal fusion studies could prove more useful
than DNA base sequence analysis to clarify relationships within the
family. Laboratory and field studies on the relation of ascospore
length to water temperature are needed to explain the pattern of
variation in *Lulworthia* and other genera. For example, in cold water do
the polar processes in *Ceriosporopsis halima* have a greater tendency to
remain appressed to the spore, thus resembling in form and function the
spore sheath in the temperate species *C. tubulifera* (Kohlmeyer) Kirk *in*
Kohlmeyer and *C. circumvestita* (Kohlmeyer) Kohlmeyer? The author is
currently investigating an inverse linear relation between salinity and
encystment in higher marine fungi, that could serve as an isolating
mechanism. The continuing efforts by several investigators to define
the conditions that promote fruiting in fastidious members of the
Halosphaeriaceae are of intrinsic value, and basic to cytogenetics and
ultrastructure.

REFERENCES

Barr, M.E. (1983). The ascomycete connection. Mycologia, 75, 1-13.
Hyde, K.D. & Jones, E.B.G. (1985). Marine fungi from Seychelles. I.
  Nimbospora effusa and *Nimbospora bipolaris* sp. nov. from
  driftwood. Can. J. Bot., 63, 611-615.
Jennings, D.H. (1983). Some aspects of the physiology and biochemistry
  of marine fungi. Biol. Rev., 58, 423-459.
Johnson, R.G. (1980). Ultrastructure of ascospore appendages of marine
  Ascomycetes. Botanica mar., 23, 501-527.
Johnson, R.G., Jones, E.B.G. & Moss, S.T. (1984). Taxonomic studies of
  the Halosphaeriaceae: *Remispora* Linder, *Marinospora*
  Cavaliere and *Carbosphaerella* Schmidt. Botanica mar., 27,
  551-566.
Jones, E.B.G. & Moss, S.T. (1978). Ascospore appendages of marine
  Ascomycetes: an evaluation of appendages as taxonomic
  criteria. Mar. Biol., 49, 11-26.
Jones, E.B.G. & Moss, S.T. (1980). Further observations on the taxonomy
  of the Halosphaeriaceae. Botanica mar., 23, 483-500.
Jones, E.B.G., Johnson, R.G. & Moss, S.T. (1983 a). Taxonomic studies
  of the Halosphaeriaceae: *Corollospora* Werdermann. J. Linn.
  Soc. (Bot), 87, 193-212.
Jones, E.B.G., Johnson, R.G. & Moss, S.T. (1983 b). *Ocostaspora*
  apilongissima gen. et sp. nov: a new marine pyrenomycete
  from wood. Botanica mar., 26, 353-360.
Jones, E.B.G., Moss, S.T. & Cuomo, V. (1983 c). Spore appendage
  development in the lignicolous marine Pyrenomycetes
  Chaetosphaeria chaetosa and *Halosphaeria trullifera*. Trans.
  Br. mycol. Soc., 80, 193-200.
Jones, E.B.G., Johnson, R.G. & Moss, S.T. (1984). Taxonomic studies of
  the Halosphaeriaceae: *Halosphaeria* Linder. Botanica mar.,
  27, 129-143.
Kirk, P.W. Jr. (1966). Morphogenesis and microscopic cytochemistry of
  marine pyrenomycete ascospores. Nova Hedwigia, 22, 1-128.

Kirk, P.W. Jr. (1976). Cytochemistry of marine fungal spores. In Recent
    Advances in Aquatic Mycology, ed. E.B. Gareth Jones,
    pp. 177-192. London: Elek Science.
Kirk, P.W. Jr. (1980). The mycostatic effect of sea water on spores of
    terrestrial and marine higher fungi. Botanica mar., 23,
    233-238.
Koch, J. & Jones, E.B.G. (1984). *Lulworthia lignoarenaria*, a new marine
    pyrenomycete from coastal sands. Mycotaxon, 20, 389-395.
Kohlmeyer, J. (1980). Tropical and subtropical filamentous fungi of the
    western Atlantic Ocean. Botanica mar., 23, 529-540.
Kohlmeyer, J. & Kohlmeyer, E. (1979). Marine Mycology: the Higher Fungi.
    New York: Academic Press.
Lutley, M. & Wilson, I.M. (1972 a). Development and fine structure of
    ascospores in the marine fungus *Ceriosporopsis halima*.
    Trans. Br. mycol. Soc., 58, 393-402.
Lutley, M. & Wilson, I.M. (1972 b). Observations on the fine structure
    of ascospores of marine fungi: *Halosphaeria appendiculata*,
    Torpedospora radiata and *Corollospora maritima*. Trans. Br.
    mycol. Soc., 59, 219-227.
Margalef, R. (1968). Perspectives in Ecological Theory. Chicago: Univ.
    Chicago Press.
Moss, S.T. & Jones, E.B.G. (1977). Ascospore appendages of marine
    Ascomycetes: *Halosphaeria mediosetigera*. Trans. Br. mycol.
    Soc., 69, 313-315.
Porter, D. (1982). The appendaged ascospore of *Trichomaris invadens*
    (Halosphaeriaceae), a marine ascomycete parasite of the
    tanner crab, *Chionoecetes bairdi*. Mycologia, 74, 363-375.
Rees, G. (1980). Factors affecting the sedimentation rate of marine
    fungal spores. Botanica mar., 23, 375-385.
Rees, G. & Jones, E.B.G. (1984). Observations on the attachment of
    spores of marine fungi. Botanica mar., 27, 145-160.
Shearer, C.A. & Crane, J.L. (1980). Fungi of the Chesapeake Bay and its
    tributaries. VIII. Ascomycetes with unfurling appendages.
    Botanica mar., 23, 607-615.
Zainal, A. & Jones, E.B.G. (1984). Observations on some lignicolous
    marine fungi from Kuwait. Nova Hedwigia, 39, 569-583.

G.C. Hughes

"... there will always be puzzles that do not easily fit any hypothesis. Of course, this may be in part because we do not have all the information. But it will be a sad world when we know everything and there is nothing left to surprise or puzzle us".

J.W. Hedgpeth 1979

INTRODUCTION

Biogeography is a perplexing science. Studies of the geographical distributions of living organisms and their various relationships with environments, past and present, have appealed to biologists, geologists, and geographers since the days of Linnaeus, Lamarck, Humboldt, and A.P. de Candolle; yet, there are no institutes of biogeography, no departments of biogeography, nor indeed, not many professors of biogeography (Nelson 1978). In those halcyon early days few credentials, aside from interest in the subject and experience travelling to distant places, were required for setting oneself up as a practising "biogeographer". However, our changing, expanding knowledge of earth history has led to a ferment of change in biogeography over the past two or three decades. This biogeographical renaissance has led to the development of new principles and new methodologies and to unprecedented growth in the literature and conceptual framework of the discipline. Numerous textbooks have been written (e.g. Pears 1978; Pielou 1979; Simmons 1979; Brown & Gibson 1983; Cox & Moore 1985); a new journal has been founded (*Journal of Biogeography* in 1974); symposia have been held (N.F. Hughes 1973; Davidse 1975; Society of Systematic Zoology 1975; Gray & Boucot 1979; New Zealand D.S.I.R. 1979; Ballance 1980; Nelson & Rosen 1981; Patterson 1981; Endler 1982; Pirozynski & Walker 1983); and a number of reviews have been published (e.g. Sauer 1969; Cracraft 1974; Crovello 1981; Taylor 1984) covering various aspects of this rapidly growing field.

Which organisms and how many organisms are to be found in a particular place is determined by "... past patterns of speciation; dispersal ability and history [historical biogeography]; and current ecology [ecological biogeography]" (Endler 1984). The development of new concepts, principles, and methodologies in biogeography has produced its share of problems, particularly in historical biogeography. Some of these problems that might affect fungal biogeography are mentioned in the discussions below.

Historical biogeography
In the dispersalist tradition that has dominated historical biogeography for approximately a hundred years, species and genera are considered to have originated in a particular locality in the past (centre of origin) and to have achieved their present distributions, including disjunctions, as a result of separate, independent dispersals of individual taxa from the centre of origin. Paramount in this early concept of biogeography was the notion of fixed continents and permanent ocean basins (Dana 1856). This is the view of biogeography presented by Darwin (1859) in the *Origin of Species* and by Wallace (1876) in his *Geographical Distribution of Animals*, volumes that some authors (e.g. Patterson 1981; Mayr 1982 re Darwin) consider the starting points for biogeography as a field of scientific endeavour. The most complete exposition of Darwin-Wallace dispersalism ("Wallace's synthesis" *fide* Nelson 1981) can be found in the first ten chapters of Wallace's, *Island Life*, published in 1880, twenty-one years after publication of the *Origin of Species*. The influence of the Darwin-Wallace dispersalist view of historical biogeography has extended well into the twentieth century as can be seen in the books of Briggs (1974), Darlington (1957, 1965), Mayr (1963, 1970, 1982), Simpson (1965, 1980 a, 1980 b), and others.

The revival, widespread acceptance, and expansion of Wegener's (1915, 1929) ideas of continental drift in the 1960's (see Bishop 1981) and general acceptance of the mechanics of plate tectonics (see Valentine & Moores 1974) have led to new and different explanations of distributions that are based primarily on abiotic vicariance events rather than dispersal events. Vicariists contend that, on relatively fewer occasions throughout history, simultaneous disjunctions have arisen in all taxa of a region as a result of vicariance events. The general causal principle of vicariance is considered to be geological (Croizat *et al.* 1974; Rosen 1975), especially events associated with the ruptures and separations of continental tectonic plates, which carry the organisms along as passive passengers. However, subtler events such as sea level changes, climatic fluctuations, elevation changes associated with mountain building, erosion, or glaciations may also function to isolate parts of a biota (Croizat *et al.* 1974; Pielou 1981). Dispersalists consider that disjunctions (commonly seen in marine fungal distributions) arise as taxa disperse across preexisting barriers whereas vicariists consider that disjunctions arise as barriers form within previously established ranges of taxa. Many of the conceptual and, especially, methodological bases of the vicariance approach to biogeography originated in a series of books in which Leon Croizat (1952, 1958, 1961, 1964) developed his ideas of panbiogeography, which provided a world view of biotic interrelationships as they relate to vicariance events.

More recently Croizat's original ideas of panbiogeography have been "hybridized" (Croizat's term, 1982) with Hennig's (1966) phylogenetic systematics (cladistics) to produce a new "vicariance biogeography" *sensu* Platnick & Nelson (1978), Cracraft (1980), and Nelson & Platnick (1981). Croizat made it clear in 1982 that he was opposed to such interpretations of his work and rejected as unwarranted any notion that his panbiogeography and Hennig's phylogenetic systematics were "substantially similar" contributions as suggested by Nelson & Platnick (1981). Nonetheless, a notable school of vicariance biogeography has grown from this "hybridization" and along with it considerable confusion as to what vicariance biogeography really is and how it relates to the more traditional dispersal biogeography. I think the confusions have been augmented by adherents of vicariance biogeography who have taken such an antagonistic view of dispersalism and those who accept dispersalist explanations. This has in turn led to extended and often heated debates between vicariance biogeographers and the dispersalists. The extent and seriousness of the disagreements can be realized by perusing issues of the last ten or twelve volumes of *Systematic Zoology*, a major platform for the views of the new vicariance biogeography.

Biogeography is a synthetic science relying on botany, zoology, ecology, population biology, geological sciences, climatology, evolutionary biology, systematics, geography, biometrics, oceanography, and related sciences for its data. I agree with McDowall (1980) that the dispersalist and vicariance approaches to biogeography need not necessarily be incompatible or mutually exclusive and that explanations of present-day distribution patterns might be found in various combinations between these two possibilities. As noted by McDowall, biogeographers who live on islands and continually see plants and animals arrive on their shores from different distant places find it illogical, even impossible, to dismiss dispersal as a factor in explaining geographical distributions. Likewise, marine mycologists, who have found species like *Ceriosporopsis halima* Linder or *Monodictys pelagica* (Johnson) E.B.G. Jones virtually everywhere marine fungi have been sought, also have reason to favour dispersalist views of biogeography. Edmunds' (1981) suggestion that a test for vicariance versus dispersal be applied in explanations of disjunctive distribution patterns has merit. He argued that associated and coevolved organisms would be expected to share a disjunct pattern if the disjunction were caused by vicariance but not to share such a pattern if the disjunction were caused by dispersal. This kind of approach to the questions of historical biogeography will certainly produce more significant results than vituperative arguments about whose methods are the most "scientific", particularly when so many of the present arguments appear to be based on prior beliefs or opinions of the disputants. Few arguments can be resolved simply because "... one believes that the results of others are inherently incorrect because of the method of analysis or the technique utilized" (Cracraft 1985).

Ecological biogeography
        The environmental parameters determining the diversity and the number of coexisting species in local habitats and the interactions (competition, predation, and symbiotic associations) that

produce their distribution patterns are the primary interests of the
ecological biogeographer. Most studies of ecological biogeography have
emphasized the role or roles of abiotic environmental factors in
limiting distributions of individual species and in the determination of
the diversity of species in different regions. Brown & Gibson (1983)
stated that, "Important advances in technique and instrumentation since
the mid-1960s have permitted a flurry of activity..." in ecological
biogeography. Among the most notable of these advances has been an
increase in the application of the analytical methods of community
ecology to biogeographical problems. Wicklow & Carroll (1981) have
provided an overall introduction to the roles these methods now play in
fungal ecology. Of particular significance has been the use of
ordination techniques (Whittaker 1978 a), classification techniques
(Whittaker 1978 b), and multivariate analyses (Gauch 1982; Pielou 1984))
in finding solutions to problems in ecological biogeography. As
biogeographical data bases grow larger, the value of these analytical
methods in interpreting them can only become more important to the
biogeographer. Fortunately the advent of the desk-top microcomputer
makes it possible for almost any worker to carry out these numerical
analyses with relative ease.

FUNGI AND BIOGEOGRAPHY

The fungi have not been the object of serious
biogeographical studies until quite recently (see Pirozynski & Walker
1983). Early workers concentrated primarily on fungal distributions,
producing long lists of fungi collected from particular localities or
regions around the globe. Although abundant, accurate distribution data
are prerequisite to biogeographical analyses, such listings of fungi
from particular geographic localities are not in themselves
biogeographical studies. They become so only when the distributions are
analysed either from a historical perspective or in terms of the
ecological parameters which determine the observed patterns.

The earliest study to consider fungal distributions in any real
ecological sense is that of Elias Petrus Fries (1857). This important
paper by Elias Magnus Fries' second son (who died in 1858 at the age of
24) was based on his father's collections and dealt primarily with the
worldwide distribution of the higher fungi. Fries considered "heat and
humidity" the determiners of global distribution patterns and suggested
that differences in fungal distributions are caused by variations in the
amount of atmospheric moisture and rain. He also discussed the division
of the globe into temperature-determined zones and how these zonal
classifications might apply to fungal distributions. He did not refer
to any specific classification scheme but could well have been
familiar with the early zonal classifications of Humboldt (1807) or
de Candolle (1820). At any rate, Fries concluded that in considerations
of the geographical distributions of the higher fungi such temperature
zones "... have very little practical or useful application..."
(English translation 1862). He also noted that ... "It is enough to
admit of the existence of two zones, peculiar in their fungaceous
growths -- namely, a temperate and a tropical zone; for the frigid zone

of geographers produces no peculiar types different from those of the
temperate zone; it is merely poorer in species. As to the tropical and
subtropical zones, no essential distinction can be pointed out in the
present state of our information respecting their fungaceous
inhabitants". Subsequent to Fries' paper mycologists have not
completely neglected studies of biogeographical importance but most have
continued to stress distribution rather than biogeography. In general,
it has been concluded that distributions of hosts or substrata are more
significant than climate in determining fungal distributions (Bisby
1933, 1943; Diehl 1937) or that studies of coevolved fungi and their
hosts or associated organisms are likely to provide the most fruitful
biogeographic insights (Pirozynski & Weresub 1979; Horak 1983;
Pirozynski 1983). Wolf & Wolf (1947) concluded that, "... any
consideration of the geographical distribution at this stage of
mycological development has a limited usefulness, partly because to date
this phase of inquiry has received little attention". This or something
very like it is the general apologia given from the time of Fries to the
present day by almost every worker who has considered our lack of
explanations for fungal distribution patterns (see Pirozynski 1968).

It is obvious to most mycologists interested in biogeography that we
neither have as much biogeographical data as we would like to have for
the fungi nor, even more importantly, are we ever likely to have that
much biogeographical data for most fungal taxa. The ideal data base
depends on extensive distributional data for taxa throughout their
geographical ranges and the distributional data, in turn, depend on
application of stable phylogenetic classifications of the organisms in
question, classifications the vicariance biogeographers would expect to
be phylogenetic, in the sense of Hennigian cladistics. Mycologists will
have to accept biogeographical data bases that are less than ideal if we
are to consider fungal biogeography at all. Limited distribution data
and unreliable phylogenies should not deter us from analysing the fungal
distribution data we do have in biogeographically acceptable ways.

Marine fungi

Pirozynski (1983) suggested that the absence of a more
nearly phylogenetic systematics and an incomplete geographical
distribution record offer the fungal biogeographer little more "... than
mapping available occurrences of a given taxon to ascertain whether or
not they fall within broadly defined biogeographical groupings. This is
true of many nutritionally less specialized fungi, and *particularly of
the marine saprotrophs*..." (italics mine). An examination of some of
the published accounts of marine fungal distributions (e.g. Hughes 1974;
Boyd & Kohlmeyer 1982; Kohlmeyer 1983) shows more than a simple mapping
of available occurrences; all of these authors used maps that were
divided into biogeographic regions based on sea-surface water
temperatures. There is, of course, much we can do to increase and
improve the ecological and biogeographical information conveyed by our
maps. The systematics of the marine Ascomycotina, especially the family
Halosphaeriaceae, has recently stabilized considerably, largely as a
consequence of an outstanding series of papers by Gareth Jones and his

co-workers in Portsmouth. The resulting classification may not be a
phyletic one but much can be learned if biogeographical analyses of the
distributions of these taxa proceed as if it were!

There is no reason to expect definitive biogeographical statements
regarding either historical biogeography or ecological biogeography of
any group of marine fungi. It will be enough for the present if trends
can be recognized, explanations suggested, and, if the results indicate
where additional data and what kinds of additional data are needed, to
enable acceptable biogeographical explanations of the observed patterns
as we study these fungi around the world. I agree with Pirozynski
(1983) that studies of associated or coevolved organisms will serve to
test more rigorously the mechanisms responsible for disjunct
distributions. This is borne out by Horak's (1983) treatment of agaric
species ectomycorrhizal with species of *Nothofagus*, *Eucalyptus*, and
Leptospermum in the South Pacific area and by Korf's (1983)
comprehensive treatment of the *Cyttaria-Nothofagus* association. Marine
mycologists might profitably direct their attention to fungi associated,
perhaps coevolved, with mangroves or sea grasses. Certainly the
palaeobiogeography, evolution, and modern biogeography of both of these
groups have received enough attention recently (Brasier 1975; McCoy &
Heck 1976; Heck & McCoy 1979; Bunt *et al*. 1982; Mepham 1983) that one
might expect to reap significant biogeographic returns from studies of
fungi associated with them. However, this does not mean that marine
mycologists should neglect the saprobic fungi for which we have more
distributional data.

Marine biogeography has a rich history and a large literature which
marine mycologists should make a part of their biogeographical studies
of the marine fungi. Two of the most notable early works in the field
are J.D. Dana's (1853) isocrymal oceanic chart which illustrated the
geographical distribution of marine animals, a paper that provided the
general outlines for marine zoogeographic mapping for many years, and
Forbes & Godwin-Austen's book, *The Natural History of the European Seas*
(1859). This small volume is largely a summary of extant knowledge
about marine distributions which attests to the breadth of knowledge
that was available to the authors even at that early date.
Subsequently, a number of other workers also published summaries of
progress in marine zoogeography, notably Ortmann (1896), who was
especially interested in crustacean distributions, and Ekman (1935,
1953) whose comprehensive summary has been the starting place for many
present-day students of marine zoogeography and biogeography. The
literature of marine zoogeography will be richly rewarding to
mycologists who consult it for information which is applicable to
studies of the ecological biogeography of marine fungi. For example,
the multitude of papers which deal with the animals of marine muds and
sands (see Eltringham 1971 for introduction) is certain to provide
mycologists with valuable information on the environmental parameters of
these ubiquitous marine habitats; habitats for fungi as well as for
molluscs and worms.

There are also a number of analytical methods and new mapping techniques

that can be applied profitably to marine fungal distribution data.
These, along with my views about some of the methods and techniques
we have used, conclude this chapter.

Coefficients of similarity

Coefficients of similarity and difference based on binary
data have been available to biologists since Jaccard introduced his
Coefficient of Community in 1901. Their use in both taxonomic and
bioassociational analyses (including biogeography) has increased
considerably over the last two decades. Coefficients of similarity are
not only of interest for comparing samples, sites, habitat types, etc.,
but their calculation is a necessary first step in applying the
techniques of multivariate analysis to biogeographical data bases (see
Booth & Kenkel Chapter 25). The available coefficients have been
assessed in considerable detail by Peters (1968), Cheetham & Hazel
(1969), and Goodall (1975). Cheetham & Hazel reviewed and compared
twenty-two different coefficients in their paper, noting especially the
ones that emphasize either the differences or the similarities between
the sample pairs being compared. Peters (1968) provided a guide for
programming a microcomputer to calculate the various coefficients and
Goodall (1975) gave the best explanations of the rationales and
mathematical niceties of the various coefficients.

I have calculated three of the most frequently used similarity
coefficients for the same set of data and illustrated them in
Figure 24.1. These 2 x 2 data matrices and "Trellis diagrams" (*sensu*
Gage 1972) show numerically and graphically the per cent similarities
(based on presence-absence data) among the sets of *Penicillium* species
that have been reported in the literature from muds, sands, or sediments
in each of seven marine or marine dominated habitats: Supratidal Beaches
and Sands (SBS); Intertidal Beaches and Sands (IBS); Subtidal Muds and
Sands, Estuarine and Coastal (SMS); the Water Column, Estuarine and
Coastal (WC); Salt Marshes (SM); Mud Flats (MF); Mangrove Muds (MM). In
addition, *Penicillium* spp. reported from two terrestrial localities in
Wisconsin (TER-2, TER-3) and one in Long Island, New York (TER-1), are
also included in Figure 24.1 where they are compared with each other and
with the species reported from the marine habitat types. Since marine
mycologists have generally treated *Penicillium* as a terrestrial genus
whose species have been isolated fortuitously from marine habitats,
similarity coefficients may indicate just how similar Penicillia from
terrestrial and marine habitats actually are. Are marine Penicillia
where they are as a result of conidia or other propagules blowing,
floating, or otherwise being transported to marine habitats from
terrestrial communities? Or are there marine species or strains of
Penicillium that are autochthonous components of the marine mycota? The
coefficients of similarity compared in Figure 24.1 include the Jaccard
Coefficient (Figure 24.1a), the Sorensen Index (Figure 24.1b), and the
Otsuka (Ochiai) Coefficient (Figure 24.1c). Cheetham & Hazel (1969)
observed that the Jaccard Coefficient emphasizes the differences between
the groups of taxa being compared (with the resulting lower per cent
similarities), whereas both the Sorensen and Otsuka Coefficients

emphasize the similarities between the samples or taxa being compared and result in higher per cent similarities (cf. Figures 24.1a & 1b). Furthermore, Gage (1972) cautioned that the Sorensen Index is markedly dependent on sample size for good results, an aspect that should be considered before choosing this index, or the Otsuka Coefficient, over others that are available (see discussion and comparisons in Goodall 1975).

Another index which is of value in ecological and biogeographical studies is the dissimilarity index; the reciprocal of the similarity index. Figures 24.1d and 1e show dissimilarity indices plotted from the per cent values for the Jaccard Coefficient (Figure 24.1a) and the Sorensen Index (Figure 24.1b). The value of dissimilarity indices to bioassociational studies seems to have been first noted in 1965 by Huheey, who called them "faunistic divergence factors" in his biogeographical analysis of the herpetofauna of Illinois.

Jaccard's "coefficient of community", is the oldest and one of the most widely used of the similarity coefficients. I have found no instances where this coefficient has been used in comparisons of fungal samples or populations but it was used with success by Murray & Littler (1981) in their analyses of marine algal floras in southern California. The formulation of the Jaccard Coefficient is as follows:

$$\frac{C}{N_1 + N_2 - C} \qquad (1)$$

where: C = Number of taxa present in both samples being compared.

N_1 = Total number of taxa present in first, smaller (or equal) of the two samples compared, but absent in the second.

N_2 = Total number of taxa present in second, larger (or equal) of the two samples compared, but absent in the first. (see Cheetham & Hazel 1969).

One of the most popular coefficients is Sorensen's (1948) "quotient of similarity". It can be represented as:

$$\frac{2C}{N_1 + N_2} \qquad (2)$$

Sorensen's index has been used frequently in assessing similarities between populations or samples of soil fungi (e.g. Gochenaur 1984 [TER-1]; Gochenaur & Backus 1967 [TER-2]; Christensen *et al.* 1962 [TER-3]) but has not been used as much in comparing populations or samples of aquatic fungi. Exceptions are Dick's (1971) use of Sorensen's index in his study of water mould ecology in lakes and its use by Wood-Eggenschwiler & Barlocher (1985) in their study of the

Figure 24.1 Similarity and dissimilarity matrices along with percentages and trellis diagrams to show degrees of similarity and difference between all possible pairs of three terrestrial populations of *Penicillium* spp. and Penicillia reported from seven marine habitats, calculated using three coefficients of similarity: the Jaccard Coefficient; the Sorensen Index; the Otsuka Coefficient.

a

	N	TER-1	TER-2	TER-3	SBS	IBS	SMS	WC	SM	MF	MM
TER-1	31		25	33	13	12	23	21	24	10	16
TER-2	24	•		43	27	14	24	26	22	18	16
TER-3	29	•	•		19	13	28	31	17	16	19
SBS	14	•	•	•		43	23	21	21	16	20
IBS	16	•	•	•	•		18	28	27	15	13
SMS	44	•	•	•	•	•		26	23	28	28
WC	39	•	•	•	•	•	•		28	29	20
SM	26	•	•	•	•	•	•	•		14	18
MF	15	•	•	•	•	•	•	•	•		24
MM	21	•	•	•	•	•	•	•	•	•	

b

	N	TER-1	TER-2	TER-3	SBS	IBS	SMS	WC	SM	MF	MM
TER-1	31		40	50	22	21	37	34	31	17	27
TER-2	24	•		60	42	25	38	41	36	31	31
TER-3	29	•	•		33	22	44	47	29	27	32
SBS	14	•	•	•		60	38	34	35	28	33
IBS	16	•	•	•	•		30	44	43	26	38
SMS	44	•	•	•	•	•		41	37	31	31
WC	39	•	•	•	•	•	•		43	46	33
SM	26	•	•	•	•	•	•	•		24	38
MF	15	•	•	•	•	•	•	•	•		39
MM	21	•	•	•	•	•	•	•	•	•	

c

	N	TER-1	TER-2	TER-3	SBS	IBS	SMS	WC	SM	MF	MM
TER-1	31		40	50	24	23	38	35	39	19	27
TER-2	24	•		61	44	26	40	43	36	32	31
TER-3	29	•	•		35	23	45	48	29	29	32
SBS	14	•	•	•		60	44	39	37	28	34
IBS	16	•	•	•	•		34	48	44	26	38
SMS	44	•	•	•	•	•		41	38	36	33
WC	39	•	•	•	•	•	•		44	41	35
SM	26	•	•	•	•	•	•	•		25	39
MF	15	•	•	•	•	•	•	•	•		39
MM	21	•	•	•	•	•	•	•	•	•	

d

	N	TER-1	TER-2	TER-3	SBS	IBS	SMS	WC	SM	MF	MM
TER-1	31		75	67	87	88	77	79	76	90	84
TER-2	24	•		57	73	86	76	74	78	82	84
TER-3	29	•	•		81	87	72	69	83	84	81
SBS	14	•	•	•		57	77	79	79	84	80
IBS	16	•	•	•	•		82	72	73	85	77
SMS	44	•	•	•	•	•		74	77	82	82
WC	39	•	•	•	•	•	•		73	71	80
SM	26	•	•	•	•	•	•	•		86	82
MF	15	•	•	•	•	•	•	•	•		76
MM	21	•	•	•	•	•	•	•	•	•	

e

	N	TER-1	TER-2	TER-3	SBS	IBS	SMS	WC	SM	MF	MM
TER-1	31		60	50	78	79	63	66	69	83	73
TER-2	24	•		40	58	75	62	59	64	69	69
TER-3	29	•	•		67	78	56	53	71	73	68
SBS	14	•	•	•		40	62	66	65	72	67
IBS	16	•	•	•	•		70	56	57	74	62
SMS	44	•	•	•	•	•		59	63	69	69
WC	39	•	•	•	•	•	•		57	54	67
SM	26	•	•	•	•	•	•	•		76	62
MF	15	•	•	•	•	•	•	•	•		61
MM	21	•	•	•	•	•	•	•	•	•	

KEY:

a) Per cent Similarity: Jaccard Coefficient X 100
b) Per cent Similarity: Sorensen Index X 100
c) Per cent Similarity: Otsuka Coefficient X 100
d) Per cent Dissimilarity: 1 - Jaccard Coefficient X 100
e) Per cent Dissimilarity: 1 - Sorensen Index X 100

SYMBOLS

Similarity	Dissimilarity
10-19.9%	40-49.9%
20-29.9%	50-59.9%
30-39.9%	60-69.9%
40-49.9%	70-79.9%
50-59.9%	80-89.9%
60-69.9%	90-99.9%

geographical distribution of Ingoldian fungi.

Still another coefficient which has been of considerable interest to ecologists and the only one that has been used in biogeographical analyses of marine fungal distributions (Booth & Kenkel Chapter 25), is the Otsuka or Ochiai Coefficient. Although this measure is often called the Ochiai Coefficient, Ochiai (1957) attributed it to Otsuka in his widely cited paper. This coefficient can be stated:

$$\frac{C}{\sqrt{N_1 \times N_2}} \qquad (3)$$

The Otsuka Coefficient would probably have been more popular had the calculation not involved the square root of $N_1 \times N_2$, an inconvenience which has been effectively eliminated by the general availability of microcomputers.

Figure 24.1 shows clearly the coefficient differences produced by the same set of data. In comparing habitats, the results generally show more similarities among the three terrestrial habitats than between the terrestrial habitats and any of the seven marine habitat types. The marine habitats most similar to the terrestrial sites are subtidal muds and sands (SMS) (30-45%, Figures 24.1b,c) and the water column (WC) (34-47%, Figures 24.1b,c) with salt marshes (SM) (29-44%, Figures 24.1b,c) third. Regardless of the index used, the similarity between *Penicillium* species of intertidal beach sands (IBS) and supratidal beach sands (SBS) is high; 40% for the Jaccard Coefficient and 60% for the Sorensen and Otsuka Coefficients. This, and the relatively high degree of similarity (> 40%) between these two habitat types and the water column (WC), is to be expected. What was not expected was the relatively lower degree of similarity (generally < 40%) for the Penicillia of mud flats (MF) and mangrove muds (MM) as compared with the three terrestrial populations and the other five marine habitats. Salt marshes (SM) fall into an intermediate position (29-44%, Sorensen) on this scale. It seems possible that MF, MM, and SM are sufficiently dissimilar (see Figures 24.1c,d) that they can be expected to support autochthonous species of *Penicillium* in addition to species shared with other marine habitats. A total of 100 *Penicillium* species has been recorded from the marine habitats and the three terrestrial areas used in these comparisons, thirteen of them only from MF, SM, and MM, and another thirty species restricted to SBS, IBS, SMS, and WC. Although the results presented in Figure 24.1 are preliminary, they provide a clear indication that we should no longer dismiss *Penicillium* simply as a terrestrial genus whose species fortuitously get into our samples from marine localities.

Multivariate analysis
The methods of multivariate analysis should be applied to marine fungal distribution data whenever possible. Booth & Kenkel

(Chapter 25) have shown the objective values these techniques bring to our understanding of marine fungal biogeography. As the size of our data bases increases, these methods will become necessary for interpreting large files of distributional data in biogeographically meaningful ways. Every fungal biogeographer should be on "speaking terms" with a book like Pielou's splendid primer, *The interpretation of ecological data* (1984).

Maps and mapping
        Maps have been and will continue to be major components of our treatments of marine fungal distributions and biogeography. In the past our maps have not conveyed sufficient ecological information nor have we considered seriously the biogeographical possibilities afforded by mapping marine fungal distributions on palaeogeographic, palaeoclimatic, or palaeoceanographic maps. Such maps give us useful insights into the historical relationships between present-day distributions and ancient marine habitats, especially as they relate to considerations of vicariance biogeography and the marine fungi. Mycologists have previously given little attention to such possibilities. A notable exception is Demoulin's (1973) treatment of the biogeography of the genus *Lycoperdon* in terms of continental drift and the opening of the Atlantic Ocean (late Cretaceous-middle Eocene). A plot of the present distribution of the marine ascomycete, *Antennospora quadricornuta* (Cribb et Cribb) T.W. Johnson, on a map of the Tethys Sea area during the middle Cretaceous (Figure 24.2) shows some of the values in such maps. For example, Figure 24.2 indicates that the present-day distribution of *A. quadricornuta* could have resulted from dispersal of the fungus during the Cretaceous when the Tethys Sea was open and ocean currents were such as to transport fungal spores east to west over the relatively shorter distances from an origin in the islands of southeast Asia. Maps based on palaeoenvironmental data, therefore, have considerable value in directing our thinking toward possible routes of dispersal or toward possible ways that vicariance events might have determined present-day distributions.

Conventional distribution maps will be of greater biogeographical value if they convey a maximum amount of ecological information. Figure 24.3 shows a series of maps of the same region of the eastern Australian coastline which illustrates the kinds of maps we have used in the past and some others we should consider using in the future. The first map (Figure 24.3a) is a simple distribution map showing distributions of the marine ascomycete, *Antennospora quadricornuta* (open circles) and the marine hyphomycete, *Trichocladium achrasporum* (Meyers et Moore) Dixon (closed circles). The first is a species whose distribution is greatly affected by temperature; the second has a worldwide distribution that is apparently independent of sea water temperatures (Hughes 1974). It is impossible to determine either of these facts from Figure 24.3a. Although the information content of distribution maps is fairly minimal, they are more informative than a simple listing of geographic localities where the fungi have been found. Lists of this kind are found for each

of the species treated by Kohlmeyer & Kohlmeyer (1979), e.g. the "Range"
of *Asteromyces cruciatus* Moreau et Moreau ex Hennebert is given as:
"Atlantic Ocean – France, Germany (Helgoland), Great Britain (England,
Wales), United States (Massachusetts, New Jersey); Baltic Sea – Germany
(GDR); Pacific Ocean – United States (California)". The map in
Figure 24.3a becomes more meaningful if one also finds a Station Map
(Figure 24.3b) showing the location of each sampling station where wood
samples were collected along this portion of the Australian coast.
Station maps should always accompany reports of distribution or
occurrence of marine fungi; when compared with occurrence maps they make
it possible to see locations where a species was not found and also
permit rough calculations of species frequencies for those that were
found. The station map can be replaced by a detailed station list
(Hughes & Chamut 1971) but both map and list are desirable if

Figure 24.2 Present-day distribution of *Antennospora quadricornuta*
plotted on a map of the Tethys Sea area, showing major ocean currents
and relative positions of continents in the middle Cretaceous. (Map
from Mepham 1983, currents adapted from Gordon 1973 and van Andel 1979).

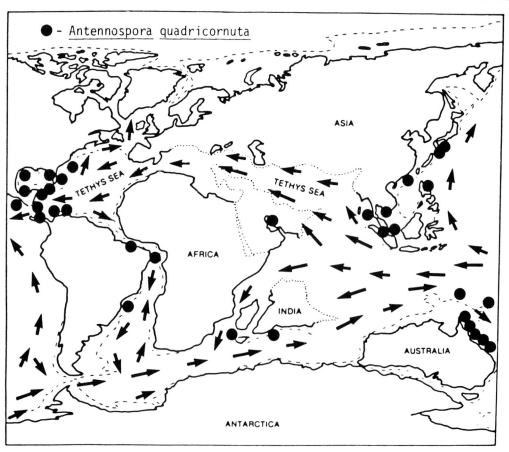

occurrence and distribution data are to be most useful ecologically or
biogeographically. Figure 24.3c shows the map of temperature-determined
biogeographical zones that I proposed in 1974; the tropical zone (TR) is
determined by the 20° C surface water isocryme for the coldest calendar
month (February in the north, August in the south), a sub-tropical zone
(ST) is found between the 17-20° C isocrymes, and a temperate zone (TM)
between the 17°C isocryme and the 10° C isothere for the warmest calendar
month (August in the north, February in the south). This map is a
direct descendant of similar zonal maps published by Dana (1856),
Ortmann (1896) and Hedgpeth (1957, based on data from Ekman 1935).

A major drawback of my 1974 zonal map and most other maps of its type
is that temperature zones are based on single isocrymal or isotheral
lines that are fixed at certain points along coastlines, e.g. the 20° C
isocryme that determines the boundary of the tropical zone in
Figure 24.3c. Isotherms are not, in fact, fixed points as can be seen
in Figure 24.3d, a map showing the position of the 24° C isotherm on the
Australian east coast throughout the course of a calendar year (months
are numbered 1 to 12). This isotherm moves from its southernmost
position near Newcastle, NSW, in February to its northernmost position
at Cape Melville, Qld. in August (approximately 2400 km) and then back
to its southernmost position by the next February. What this means
biogeographically is that any particular site on the coast may well
experience several temperature regimes over the year, a situation not
obvious in Figure 24.3c. For example, surface water temperatures at
Cape Hatteras, North Carolina (Lat. 35.14 N) approach 27°C in August, by
October they are nearer 21°C, in December they are approximately 15° C,
and by February, the coldest month of the year, the sea-surface
temperature is 10°C, giving an annual range of 16-17° C (Hutchins 1947).
Hutchins also observed that summer and winter sea-surface temperatures
have wider ranges in temperate regions than polar or tropical regions
and that more extreme temperature combinations are found on western
shores of oceans than on eastern shores. He proposed a polyzonal
treatment of temperature's role in determining geographic distributions
of marine organisms; advocating that the zonal boundaries of organisms
should be interpreted in terms of both summer maximum and winter
minimum temperatures, since as he put it, "Distributional boundaries are
extremes, and it is reasonable to study them in terms of the coinciding
extremes of environmental conditions". Hutchins further observed that
most organisms have a given range of temperature over which survival is
possible, and within the survival limits a somewhat narrower range of
temperature conditions over which reproduction and repopulation can be
completed. Both survival and repopulation are limited at the extremes
by temperature. It is therefore most meaningful if our distribution
maps show both of the sea water temperature extremes that characterize
our sampling sites. Such a map is shown in Figure 24.3e for the
eastern Australian coast with the distribution of *Antennospora*
quadricornuta (open circles) indicated. This fungus has been found most
frequently along the Australian coast in regions where water
temperatures range from 24° to 28° C throughout the year. The
southernmost record (arrow, Figure 24.3e) is from a region where
temperatures range from 21° to 24°C. It is of interest, however, that

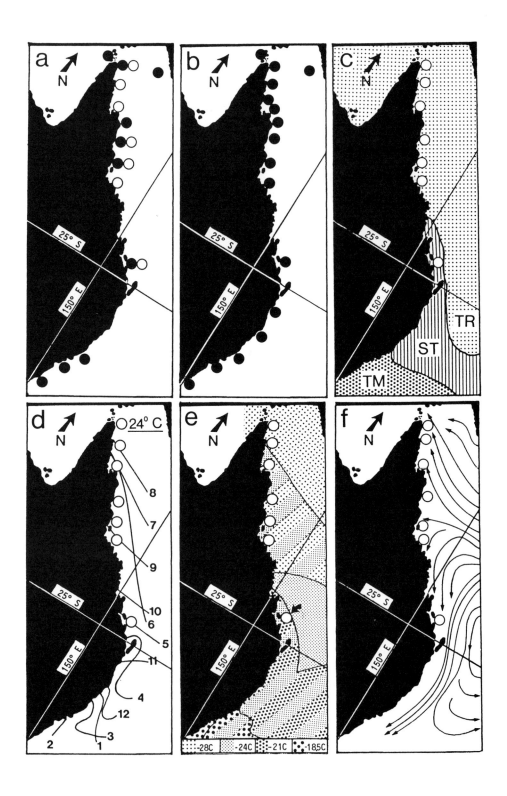

two of the three times *A. quadricornuta* has been found in this southern
area have been times (see Figure 24.3d) when the surface water
temperatures were 24°C, i.e. in January and March (Cribb & Cribb 1956).
The third record dates from an October collection. There is no
question that maps which show seasonal extremes of temperature will be
more useful in understanding temperature effects on occurrence and
distribution than maps that do not.

Furthermore, since the occurrence of any lignicolous marine fungus is
dependent on our finding reproductive structures, the plotting of both
occurrence and dates of collection on such maps (Figure 24.3e) may be
of great assistance in correlating our laboratory findings concerning
temperature effects on growth and reproduction (Boyd & Kohlmeyer 1982)
with temperatures and reproductive events in nature. We may be failing
to recognize the lignicolous fungi present at temperature extremes of
their distributional ranges since survival extremes may exceed the
narrower temperature ranges which limit their reproduction.

Finally, Figure 24.3f shows a map of the major currents along the
Australian east coast in July. If, as is so often suggested, marine
fungi owe their present distribution patterns to long-range dispersals,
it seems reasonable to assume that ocean currents are the main forces
behind such dispersals. Marine mycologists have never, however,
included current maps in their attempts to explain geographic
distributions. Maps of oceanic or coastal currents (past or present,
whichever is appropriate) should be included routinely in biogeographic
studies of marine fungi or, at very least, major coastal currents
should be incorporated into our general distribution maps or
temperature zone maps.

ACKNOWLEDGEMENTS
          Preparation of this paper was supported by the Natural
Sciences and Engineering Research Council of Canada through
Grant A-2561. My thanks to Gareth Jones, Geoff Pugh, and especially to
Steve Moss for their hard work that made it possible for me to share my
ideas about biogeography and marine fungi with those assembled for the
4th International Marine Mycology Symposium in Portsmouth, August 1985.

Figures 24.3a-f Mapping marine fungal distribution data, Australian
east coast. 3a. Distribution map of *Antennospora quadricornuta* and
Trichocladium achrasporum. 3b. Station map for the east coast area.
3c. Temperature-based zonal map (Hughes 1974). 3d. Seasonal positions
of 24°C isotherm along Australian east coast. 3e. Map based on Hutchins'
(1947) polyzonal concepts; seasonal temperature extremes plotted.
3f. Map of major coastal currents, July, Australian east coast
(Australia Pilot 1962). The distribution of *A. quadricornuta* is plotted
on all maps except Fig. 3b.

REFERENCES

Australia Pilot (1962). Vol. 4, comprising the eastern coast of
	Queensland from Sandy Cape to Cape York, ... including
	Great Barrier Reefs. London: Hydrographic Dept., The
	Admiralty.

Ballance, F., ed. (1980). Plate tectonics and biogeography in the
	southwest Pacific: The last 100 million years. Palaeogeog.
	Palaeoclimatol. Palaeoecol., 31, 101-372.

Bisby, G.R. (1933). The distribution of fungi as compared with that of
	phanerogams. Am. J. Bot., 20, 246-254.

Bisby, G.R. (1943). Geographical distribution of fungi. Bot. Rev., 9,
	466-482.

Bishop, A.C. (1981). The development of the concept of continental
	drift. In The Evolving Earth, ed. L.R.M. Cocks, pp. 89-101.
	London: British Museum (Nat. Hist.) & Cambridge: Cambridge
	University Press.

Boyd, P.E. & Kohlmeyer, J. (1982). The influence of temperature on the
	seasonal and geographic distribution of three marine fungi.
	Mycologia, 74, 894-902.

Brasier, M.D. (1975). An outline history of seagrass communities.
	Palaeontology, 18, 681-702.

Briggs, J.C. (1974). Marine Zoogeography. New York: McGraw-Hill Book
	Co.

Brown, J.H. & Gibson, A.C. (1983). Biogeography. St. Louis, Mo.:
	C.V. Mosby Co.

Bunt, J.S., Williams, W.T. & Duke, N.C. (1982). Mangrove distributions
	in north-east Australia. J. Biogeogr., 9, 111-120.

Candolle, A.P. de (1820). Géographie botanique. In Dictionnaire des
	Sciences Naturelles, ed. F.C. Levrault, vol. 18,
	pp. 359-436. Paris: Levrault. (Issued separately as Essai
	Elémentaire de Géographie Botanique, Paris & Strasbourg,
	1820).

Cheetham, A.H. & Hazel, J.E. (1969). Binary (presence-absence)
	similarity coefficients. J. Paleontol., 43, 1130-1136.

Christensen, M., Whittingham, W.F. & Novak, R.O. (1962). The soil
	microfungi of wet-mesic forests in southern Wisconsin.
	Mycologia, 54, 374-388.

Cox, C.B. & Moore, P.D. (1985). Biogeography. An Ecological and
	Evolutionary Approach. 4th ed. Oxford: Blackwell Scientific
	Publications.

Cracraft, J. (1974). Continental drift and vertebrate distribution.
	A. Rev. Ecol. Syst., 5, 215-261.

Cracraft, J. (1980). Biogeographic patterns of terrestrial vertebrates
	in the southwest Pacific. Palaeogeogr., Palaeoclimatol.,
	Palaeoecol., 31, 353-369.

Cracraft, J. (1985). Monophyly and phylogenetic relationships of the
	Pelecaniformes: A numerical cladistic analysis. Auk, 102,
	834-853.

Cribb, A.B. & Cribb, J.W. (1956). Marine fungi from Queensland. II.
	Pap. Dept. Bot. Univ. Qd, 3, 97-105.

Croizat, L. (1952). Manual of Phytogeography. The Hague: W. Junk BV.

Croizat, L. (1958). Panbiogeography. 3 vols. Caracas: Published by the
	author.

Croizat, L. (1961). Principia Botanica. 2 vols. Caracas: Published by
    the author.
Croizat, L. (1964). Space, Time, Form: the Biological Synthesis.
    Caracas: Published by the author.
Croizat (-Chaley), L. (1982). Vicariance/vicariism, panbiogeography,
    "vicariance biogeography", etc.: A clarification. Syst.
    Zool., 31, 291-304.
Croizat, L., Nelson, G. & Rosen, D.E. (1974). Centers of origin and
    related concepts. Syst. Zool., 23, 265-287.
Crovello, T.J. (1981). Quantitative biogeography: an overview. Taxon,
    30, 563-575.
Dana, J.D. (1853). On an isothermal oceanic chart, illustrating the
    geographical distribution of marine animals. Am. J. Sci.,
    2nd ser., 16, 153-167, 314-327.
Dana, J.D. (1856). On the plan of development in the geologic history
    of North America. Am. J. Sci., 2nd ser., 22, 335-349.
Darlington, P.J., Jr. (1957). Zoogeography: the Geographical
    Distribution of Animals. New York: John Wiley & Sons.
Darlington, P.J., Jr. (1965). Biogeography of the southern end of the
    world. Cambridge, Mass.: Harvard University Press.
Darwin, C. (1859). On the Origin of Species by Means of Natural
    Selection, or the Preservation of Favoured Races in the
    Struggle for Life. London: John Murray. (Facsimile edition,
    ed. E. Mayr, 1964, Cambridge, Mass.: Harvard University
    Press).
Davidse, G., ed. (1975). Biogeography: The twenty-first systematics
    symposium. Ann. Mo. bot. Gdn., 62, 225-385.
Demoulin, V. (1973). Phytogeography of the fungal genus Lycoperdon in
    relation to the opening of the Atlantic. Nature, 242,
    123-125.
Dick, M.W. (1971). The ecology of Saprolegniaceae in lentic and
    littoral muds with a general theory of fungi in the lake
    ecosystem. J. gen. Microbiol., 65, 325-337.
Diehl, W.W. (1937). A basis for mycogeography. J. Wash. Acad. Sci., 27,
    244-254.
Edmunds, G.F., Jr. (1981). Discussion of 'Vicarious plant distributions
    and paleogeography of the Pacific Region' by R. Melville.
    In Vicariance Biogeography: A Critique, eds. G. Nelson &
    D.E. Rosen, pp. 287-297. New York: Columbia University
    Press.
Ekman, S. (1935). Tiergeographie des Meeres. Leipzig: Akad. Verlagsges.
Ekman, S. (1953). Zoogeography of the sea. London: Sidgwick & Jackson
    Ltd. (English transl. by Elizabeth Palmer).
Eltringham, S.K. (1971). Life in Mud and Sand. London: English
    Universities Press.
Endler, J.A. (1982). Alternative hypotheses in biogeography:
    Introduction and synopsis of the symposium. Am. Zool., 22,
    349-354.
Endler, J.A. (1984). Review of Biogeography by J.H. Brown & A.C. Gibson,
    1983. St. Louis: C.V. Mosby Co. Syst. Zool., 33, 249-250.
Forbes, E. & Godwin-Austen, R. (1859). The Natural History of the
    European Seas. London: John Van Voorst. (pp. 1-126 by
    Forbes).

Fries, E.P. (1857). Anteckningar öfver Svamparnes geografiska
        Utbredning. Akad. Afhandling, Upsala (Leffler). (English
        transl. by J.T. Arlidge, Observations on the geographical
        distribution of fungi. Ann. Mag. nat. Hist., 9, 269-288,
        1862).
Gage, J. (1972). A preliminary survey of the benthic macrofauna and
        sediments in Lochs Etive and Creran, sea-lochs along the
        west coast of Scotland. J. mar. biol. Ass. U.K., 52,
        237-276.
Gauch, H.G., Jr. (1982). Multivariate Analysis in Community Ecology.
        Cambridge: Cambridge University Press.
Gochenaur, S.E. (1984). Fungi of a Long Island oak-birch forest II.
        Population dynamics and hydrolase patterns for the soil
        Penicillia. Mycologia, 76, 218-236.
Gochenaur, S. & Backus, M.P. (1967). Mycoecology of willow and
        cottonwood lowland communities in southern Wisconsin.
        II. Soil microfungi in the sand-bar willow stands.
        Mycologia, 59, 893-909.
Goodall, D.W. (1975). Sample similarity and species correlation. In
        Ordination of Plants, ed. R.H. Whittaker, pp. 101-149. The
        Hague: W. Junk BV.
Gordon, W.A. (1973). Marine life and ocean surface currents in the
        Cretaceous. J. Geol., 81, 269-284.
Gray, J. & Boucot, A.J., eds. (1979). Historical Biogeography, Plate
        Tectonics, and the Changing Environment. Proc. 37th Biology
        Colloquium. Corvallis: Oregon State University Press.
Heck, K.L. & McCoy, E.D. (1979). Biogeography of seagrasses: evidence
        from associated organisms. In Proceedings of the
        International Symposium on Marine Biogeography and Evolution
        in the Southern Hemisphere, vol. 1, pp. 109-127. Wellington,
        New Zealand: Science Information Division, D.S.I.R.
Hedgpeth, J.W. (1957). Marine biogeography. In Treatise on Marine
        Ecology and Paleoecology, vol. 1, ed. J.W. Hedgpeth,
        pp. 359-382, New York: Geological Society of America,
        Memoir 67.
Hedgpeth, J.W. (1979). Prologue: At sea with provinces and plates. In
        Historical Biogeography, Plate Tectonics, and the Changing
        Environment. Proc. 37th Biology Colloquium, eds. J. Gray &
        A.J. Boucot, pp. 1-7. Corvallis: Oregon State University
        Press.
Hennig, W. (1966). Phylogenetic Systematics. 2nd ed. (Transl. by
        D.D. Davis & R. Zanderl). Urbana, Ill.: University of
        Illinois Press.
Horak, E. (1983). Mycogeography of the South Pacific region: Agaricales,
        Boletales. Aust. J. Bot., Suppl. Ser., 10, 1-41.
Hughes, G.C. (1974). Geographical distribution of the higher marine
        fungi. Veröff. Inst. Meeresforsch. Bremerh., Suppl., 5,
        419-441.
Hughes, G.C. & Chamut, P.S. (1971). Lignicolous marine fungi from
        southern Chile, including a review of distributions in the
        southern hemisphere. Can. J. Bot., 49, 1-11.

Hughes, N.F., ed. (1973). Organisms and continents through time. Spec. Pap. Palaeontol. No. 12. London: The Palaeontological Association.

Huheey, J.E. (1965). A mathematical method of analyzing biogeographical data. I. Herpetofauna of Illinois. Am. Midl. Nat., 73, 490-500.

Humboldt, F.H.A. von (1807). Essai sur la géographie des plantes: ... accompagne d'un tableau physique des regions equinoxiales. Paris: Levrault, Schoell & Co.

Hutchins, L.W. (1947). The bases for temperature zonation in geographical distribution. Ecol. Monogr., 17, 325-335.

Jaccard, P. (1901). Distribution de la flore alpine dans le Bassin des Dranses et dans quelques regions voisines. Bull. Soc. vaud. Sci. nat., 37, 241-272.

Kohlmeyer, J. & Kohlmeyer, E. (1979). Marine Mycology: The Higher Fungi. New York: Academic Press.

Kohlmeyer, J. (1983). Geography of marine fungi. Aust. J. Bot., Suppl. Ser., 10, 67-76.

Korf, R.P. (1983). *Cyttaria* (Cyttariales): coevolution with *Nothofagus*, and evolutionary relationship to the Boedinjpezizeae (Pezizales, Sarcoscyphaceae). Aust. J. Bot., Suppl. Ser., 10, 77-87.

McCoy, E.D. & Heck, K.L., Jr. (1976). Biogeography of corals, seagrasses, and mangroves: An alternative to the center of origin concept. Syst. Zool., 25, 201-210.

McDowall, R.M. (1980). Freshwater fishes and plate tectonics in the southwest Pacific. Palaeogeogr., Palaeoclimat., Palaeoecol., 31, 337-351.

Mayr, E. (1963). Animal Species and Evolution. Cambridge, Mass.: Harvard University Press.

Mayr, E. (1970). Populations, Species, and Evolution. Oxford & London: Oxford University Press.

Mayr, E. (1982). The Growth of Biological Thought. Diversity, Evolution, and Inheritance. Cambridge, Mass.: Belknap Press of Harvard University Press.

Mepham, R.H. (1983). Mangrove floras of the southern continents. Part 1. The geographical origin of Indo-Pacific mangrove genera and the development and present status of the Australian mangroves. S. Afr. Tydskr. Plantk., 2, 1-8.

Murray, S.N. & Littler, M.M. (1981). Biogeographical analysis of intertidal macrophyte floras of southern California. J. Biogeog., 8, 339-351.

Nelson, G. (1978). From Candolle to Croizat: Comments on the history of biogeography. J. Hist. Biol., 11, 269-305.

Nelson, G. (1981). Summary. In Vicariance Biogeography: A Critique, eds. G. Nelson & D.E. Rosen, pp. 524-537. New York: Columbia University Press.

Nelson, G. & Platnick, N.I. (1981). Systematics and Biogeography. Cladistics and Vicariance. New York: Columbia University Press.

Nelson, G. & Rosen, D.E., eds. (1981). Vicariance Biogeography: A Critique. New York: Columbia University Press.

New Zealand D.S.I.R. (1979). Proceedings of the International Symposium on Marine Biogeography and Evolution in the Southern Hemisphere, Auckland, N.Z., 17-20 July 1978, 2 vols. Wellington, N.Z.: Science Information Division, Department of Scientific and Industrial Research.

Ochiai, A. (1957). Zoogeographic studies of the Solenoid fishes found in Japan and its neighbouring regions. Bull. Jap. Soc. scient. Fish., 22, 526-530.

Ortmann, A.E. (1896). Grundzuge der marinen Tiergeographie. Jena: Gustav Fischer.

Patterson, C. (1981). Biogeography: In search of principles. (Review of symposium, Time and Space in the Emergence of the Biosphere, British Museum of Natural History, 1981). Nature, 291, 612-613.

Pears, N. (1978). Basic Biogeography. London: Longman Group Ltd.

Peters, J.A. (1968). A computer program for calculating degree of biogeographic resemblance between areas. Syst. Zool., 17, 64-69.

Pielou, E.C. (1979). Biogeography. New York: John Wiley & Sons.

Pielou, E.C. (1981). Crosscurrents in biogeography. (Review of Vicariance Biogeography: A critique, eds. G. Nelson & N. Rosen, 1981. New York: Columbia University Press). Science, 213, 324-325.

Pielou, E.C. (1984). The Interpretation of Ecological Data: A Primer on Classification and Ordination. New York: John Wiley & Sons.

Pirozynski, K.A. (1968). Geographical distribution of fungi. In The Fungi: An Advanced Treatise, Vol. III, The Fungal Population, eds. G.C. Ainsworth & A.S. Sussman, pp. 487-504. New York: Academic Press.

Pirozynski, K.A. (1983). Pacific mycogeography: an appraisal. Aust. J. Bot., Suppl. Ser., 10, 137-159.

Pirozynski, K.A. & Walker, J., eds. (1983). Pacific mycogeography: A preliminary approach. Aust. J. Bot., Suppl. Ser., 10, 1-172.

Pirozynski, K.A. & Weresub, L.K. (1979). A biogeographic view of the history of Ascomycetes and the development of their pleomorphism. In The Whole Fungus, ed. W.B. Kendrick, Vol. 1, pp. 93-123. Ottawa: National Museums of Canada.

Platnick, N.I. & Nelson, G. (1978). A method of analysis for historical biogeography. Syst. Zool., 27, 1-16.

Rosen, D.E. (1975). A vicariance model of Caribbean biogeography. Syst. Zool., 24, 431-464.

Sauer, J.D. (1969). Oceanic islands and biogeographical theory: a review. Geogr. Rev., 59, 582-593.

Simmons, I.G. (1979). Biogeography: Natural and Cultural. London: Edward Arnold Publishers Ltd.

Simpson, G.G. (1965). The Geography of Evolution. Collected Essays. Philadelphia & New York: Chilton.

Simpson, G.G. (1980 a). Why and How. Some Problems and Methods in Historical Biology. Oxford: Pergamon Press.

Simpson, G.G. (1980 b). Splendid Isolation; the Curious History of Mammals in South America. New Haven, Conn.: Yale University Press.

Society of Systematic Zoology. (1975). Symposium: World Perspectives in
    Biogeography. Syst. Zool., 24, 407-488.
Sorensen, T. (1948). A method of establishing groups of equal amplitude
    in plant sociology based on similarity of species content.
    K. dansks Vidensk. Selsk. Biol. Skr., 5, 1-34.
Taylor, J.A., ed. (1984). Themes in Biogeography. London: Croom Helm
    Ltd.
Valentine, J.W. & Moores, E.M. (1974). Plate tectonics and the history
    of life in the oceans. Scient. Amer., 230, 80-87.
van Andel, T.H. (1979). An eclectic overview of plate tectonics,
    paleogeography, and paleoceanography. In Historical
    Biogeography, Plate Tectonics, and the Changing Environment.
    Proc. 37th Biology Colloquium, eds. J. Gray & A.J. Boucot,
    pp. 9-25. Corvallis: Oregon State University Press.
Wallace, A.R. (1876). The Geographical Distribution of Animals. 2 vols.
    London: Macmillan & Co.
Wallace, A.R. (1880). Island Life. London: Macmillan & Co.
Wegener, A. (1915). Die Entstehung der Kontinente und Ozeane.
    Braunschweig: Vieweg.
Wegener, A. (1929). The Origins of Continents and Oceans. 4th ed.
    (Transl. 1966, by J. Biram, New York: Dover Publications).
Whittaker, R.H., ed. (1978 a). Classification of Plant Communities. The
    Hague: W. Junk BV.
Whittaker, R.H., ed. (1978 b). Ordination of Plant Communities. The
    Hague: W. Junk BV.
Wicklow, D. & Carroll, G.C., eds. (1981). The Fungal Community. New
    York: Marcel Dekker Inc.
Wood-Eggenschwiler, S. & Barlocher, F. (1985). Geographical distribution
    of Ingoldian fungi. Verh. Int. Verein. Limnol., 22,
    2780-2785.
Wolf, F.A. & Wolf, F.T. (1947). The Fungi, vol. 2. New York: John Wiley
    & Sons.

ECOLOGICAL STUDIES OF LIGNICOLOUS MARINE FUNGI: A
DISTRIBUTION MODEL BASED ON ORDINATION AND CLASSIFICATION

T. Booth

N. Kenkel

Several factors are said to influence the distribution of
lignicolous marine fungi. Principal among these variables are carbon
level, competition, hydrogen ion activity, hydrostatic pressure,
inhibitory factors, ionic levels and ratios, light, osmotic response,
oxygen availability, pollutants, propagule numbers, salinity, substrate
types and condition, temperature and tidal exposure. Although it is
recognized that a combination of factors affect fungal occurrence
(Ritchie 1957; Brooks 1972; Kirk 1972; Shearer 1972; Byrne & Jones
1974, 1975; Henningsson 1974; Hughes 1975; Shearer & Crane 1978;
Kohlmeyer & Kohlmeyer 1979; Vrijmoed *et al.* 1982; Kohlmeyer 1983),
temperature has been ascribed the central role in the formulation of
the currently applied and accepted distribution model (Hughes 1974,
1975). This model, as applied by various workers (Hughes 1974; Booth
1979; Kirk & Brandt 1980; Boyd & Kohlmeyer 1982; Vrijmoed *et al.* 1982;
Kohlmeyer 1983), has resulted in the assignment of some fungal taxa to
the groups: 1) Arctic - Antarctic; 2) Temperate; 3) Sub-Tropical;
4) Tropical.

The limitation of temperature as a determinant of marine fungal
distribution is indicated by the temperature independence of various
species (Hughes 1974, 1975; Kohlmeyer & Kohlmeyer 1979) and the large
number of taxa which are cosmopolitan with respect to temperature
(Kohlmeyer & Kohlmeyer 1979; Kohlmeyer 1983). Therefore, it may be that
temperature alone is not entirely adequate in defining the distribution
of lignicolous fungi. Among the other variables used to describe
occurrences, salinity is the best known. Prior to proposal and
acceptance of the temperature based distribution model, several workers
(Ritchie 1957; Hughes 1969; Shearer 1972; Byrne & Jones 1974;
Henningsson 1974) considered salinity to play a major role. More
recent publications (Byrne & Jones 1975; Henningsson 1978; Aleem 1980;
Kirk & Brandt 1980; Kirk & Schatz 1980; Vrijmoed *et al.* 1982) have also
reported on occurrence and salinity levels. The evidence suggests that
salinity may have an important role in affecting the distribution of at
least some marine fungi.

Although temperature and salinity are considered to be major factors
controlling the distribution of marine fungi, the occurrence of
organisms in natural ecosystems is a function of several synergistically
interacting environmental factors (Mueller-Dumbois & Ellenberg 1974;
Whittaker 1975; Krebs 1978). Assuming that this is applicable to fungi,

a number of factors may influence the distribution of lignicolous organisms. The degree of factor influence depends on the nature of the variable and on the organism in question (Krebs 1978). Generally, however, taxa occur along complex environmental gradients, and factors are perhaps best expressed as levels of variability (Whittaker 1975).

The aim of this chapter is to consider some of the factors which may affect the worldwide distribution of lignicolous marine fungi, and to describe a model of their distribution based on multivariate methods of ordination and classification. Ordination methods are used to directly or indirectly generate a low dimensional space (Orlóci 1966, 1978; Hill 1973; Pielou 1979), with species or sites represented as points in this space. The relative placement of points is indicative of interspecific relationships along axes which are generally interpretable in an environmental context (Gauch 1982). Sites and species are classified using objective techniques of cluster analysis (Anderberg 1973). Concentration analysis (Feoli & Orlóci 1979) is used to examine the relationships between the site and species groupings imposed by the clusterings.

Data for the analyses were obtained from a concomitant study (Booth unpublished) as the presence or absence of 68 species over 31 sites. The sites represent coastal marine habitats which are widely distributed throughout the world (Figure 25.1). Booth (unpublished) has characterized each site by climate, temperature and salinity (when possible) based on previous reports of lignicolous fungi. The occurrence of species found in at least five of the sites was recorded over all 31 locations. Species nomenclature is based on Kohlmeyer & Kohlmeyer (1979). The failure to incorporate more recent nomenclatural changes is a necessary expedient and not a rejection of the new names.

Ordination of the sites using the Ochiai coefficient (Ochiai 1957; Orlóci 1978) in nonmetric multidimensional scaling (Kruskal 1964 a,b) resulted in a two-dimensional scattergram (Figure 25.2) with a stress of 16.3%. The horizontal axis represents a definite trend in temperature variation, with tropical sites to the right and temperate regions to the left. Sum of squares agglomerative cluster analysis (Ward 1963; Orlóci 1967) of the sites led to the recognition of three major site groups which reflect temperature differences (Figure 25.3): A = temperate; B = subtropical – temperate; C = tropical. As reflected by the tight clustering of these groups (Figures 25.2, 3) temperature is the dominant factor influencing the worldwide distribution of marine fungi.

Perusal of the vertical ordination axis suggests that it reflects temporal variation in temperature and salinity. Within the temperate group (A) the lower portion is represented by sites showing temporal variation in salinity while the upper portion is characterized by sites showing strong changes in temperature. The central portion includes sites showing changes in both temperature and salinity. The upper sites of group B are cooler than those lower in the group. This lower portion encompasses sites with limited seasonal temperature changes. Diurnal

Figure 25.1   Locations of the collecting sites.
[1 = Antilles (Ant); 2 = Arctic (Arc); 3 = Atlantic Scandinavia (Atscan); 4 = Atlantic Canada (Atcan);
5 = Atlantic Mexico (Atmex); 6 = Atlantic Scandinavia (Atscan); 7 = Austro-Melanesia (Aumel);
8 = Austro-Zealand (Auzea); 9 = Bahamas (Bah); 10 = Baltic (Balt); 11 = Bermuda (Ber); 12 = Britain
(Brit); 13 = California (Calif); 14 = Carolina-Mexico (Carmex); 15 = Chesapeake (Ches); 16 = Chile
(Chile); 17 = Columbia-Panama (Copan); 18 = Franco-Belgium (Frabel); 19 = Hawaii (Haw); 20 = Iberia
(Iber); 21 = Iceland (Ice); 22 = Japan (Jpn); 23 = Mediterranean (Med); 24 = New England (NEng);
25 = Olympia (Oly); 26 = Pacific Canada (Pacan); 27 = Pacific Mexico (Pacmex); 28 = Southern Brazil
(Sobra); 29 = Spanish Canary (Spancan); 30 = Venezuela (Ven); 31 = Western Indian (West)].

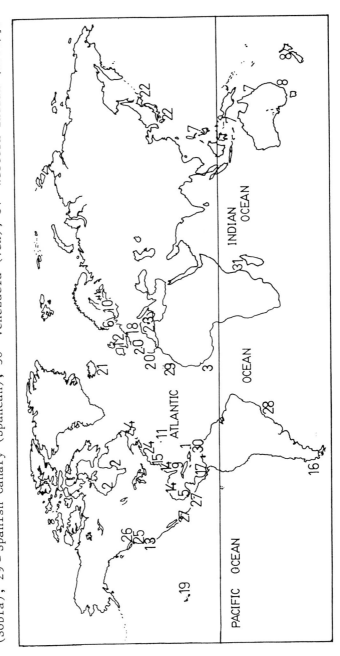

Figure 25.2 Nonmetric multidimensional scaling of sites with three major groups, A-C. (Site abbreviations as in Figure 25.1. Britain not shown as it is located between NEng and Frabel).

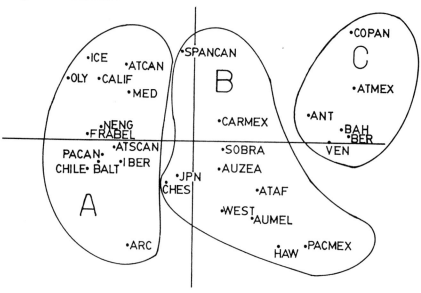

Figure 25.3 Cluster analysis of the sites, and the three major groups A-C. (Site numbers as in Figure 25.1).

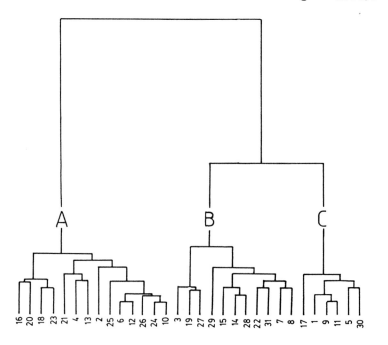

and seasonal changes in salinity are a prevalent feature of group C. Given these trends, we suggest a further subdivision of the major temperate, subtropical - temperate, and tropical groups A-C based on: 1) neighbouring sea water maximum temperature <17°C as cool water sites and >17°C as warm water sites; 2) summer maximum minus winter maximum (or, where applicable, a literature based temperature range) of >5°C as eurythermic and <5°C as homeothermic; 3) salinity changes >5⁰/oo as euryhaline and <5⁰/oo as homeohaline. This results in a number of recognizable groups (Figure 25.2). Cool water euryhalothermic (ETH) sites (derived from group A) include: Atlantic Scandinavia (Atscan), Baltic (Balt), Britain (Brit), Franco-Belgium (Frabel), Iberia (Iber), New England (NEng), and Pacific Canada (Pacan). Warm water euryhalothermic (EHT) sites (derived from group B) encompass: Austro-Zealand (Auzea); Carolina-Mexico (Carmex); Chesapeake (Ches); Japan (Jpn); and Southern Brazil (Sobra). From group (A) Atlantic Canada (Atcan), California (Calif), Iceland (Ice), Mediterranean (Med), and Olympia (Oly) are cool water eurythermic homeohaline (ETHH). Spanish Canary (Spancan) of group (B) is a homeohalothermic (HHT) region. Cool water homeothermic euryhaline (HTEH) sites (derived from group A) include Arctic (Arc) and Chile (Chile), while the warm water tropical counterparts are Antilles (Ant); Atlantic Mexico (Atmex); Bahamas (Bah); Bermuda (Ber); Columbia - Panama (Copan) and Venezuela (Ven). A subtropical homeothermic euryhaline (HTEH) site series (derived from group B) encompasses Atlantic Africa (Ataf); Austro - Melanesia (Aumel); Hawaii (Haw); Pacific Mexico (Pacmex) and Western Indian (West).

The species ordination, using the Ochiai coefficient in nonmetric multidimensional scaling and specifying a two-dimensional solution (stress = 18.3%), is shown in Figure 25.4. The six major species groups (I-VI) suggested by cluster analysis (Figure 25.5) are also indicated. Those species which appear in all three of the major site groups are circled on the scattergram and are considered to be widespread. Various trends can be recognized in the species after consultation of the temperature-salinity regime of each species from the literature (Table 25.1). Group I includes species from tropical and subtropical waters which are largely restricted to mangroves (Kohlmeyer & Kohlmeyer 1979). Taxa of group II, which also frequently occur on mangroves, are found mainly in warm water temperate through tropical sites. Groups III through V represent species from cool waters. These groups may be distinguished by considering environmental homeostasis of temperature and salinity. The entirety of group III and a portion of group IV includes species of euryhalothermic sites. Species of group V show preference for homeothermic euryhaline waters. Taxa in group VI are mostly widespread.

Relationships between site groups A-C and species groups I-VI were examined using concentration analysis (Figure 25.6). The canonical correlations (R_1 = .69 and R_2 = .29) of the concentration analysis axes indicate that the first axis, representing a temperature gradient, is dominant when considering worldwide occurrence of marine fungi. The second axis, which reflects temperature and salinity variation, is only

Table 25.1 Species number, name, ordination group and temperature-salinity regime

Species number	Species name	Group in ordination	Temperature-salinity regime* of sites with literature sources†
1	Amylocarpus encephaloides Currey	IV	4 (ETHH), 6 (EHT)
2	Asteromyces cruciatus Moreau et Moreau ex Hennebert	IV	2 (HTEH)
3	Ceriosporopsis calyptrata Kohlm.	IV	10 (EHT)
4	Ceriosporopsis cambrensis Wilson	III	
5	Ceriosporopsis circumvestita (Kohlm.) Kohlm.	V	2 (HTEH)
6	Ceriosporopsis halima Linder	VI	
7	Ceriosporopsis tubulifera (Kohlm.) Kirk	IV	4 (ETHH)
8	Cirrenalia macrocephala (Kohlm.) Meyers et Moore	VI	
9	Corollospora comata (Kohlm.) Kohlm.	IV	
10	Corollospora cristata (Kohlm.) Kohlm.	IV	4 (ETHH)
11	Corollospora lacera Linder	VI	4 (ETHH), 6 (ETHH + EHT)
12	Corollospora maritima Werdermann	VI	
13	Corollospora pulchella Kohlm., Schmidt et Nair	II	
14	Corollospora trifurcata (Höhnk) Kohlm.	VI	
15	Cremasteria cymatilis Meyers et Moore	V	6, 10, 11 (all HTEH)
16	Cytospora rhizophorae Kohlm. et Kohlm.	I	1
17	Dendryphiella salina (Sutherland) Pugh et Nicot	V	2, 4, 9 (all HTEH)
18	Dictyosporium pelagicum (Linder) G.C. Hughes	VI	10 (ETHH)
19	Didymosphaeria enalia Kohlm.	II	1 ETH
20	Didymosphaeria rhizophora Kohlm. et Kohlm.	I	
21	Digitatispora marina Doguet	IV	5, 12 (both ETHH)
22	Gnomonia longirostris Cribb et Cribb	VI	
23	Haligena amicta (Kohlm.) Kohlm. et Kohlm.	V	
24	Haligena elaterophora Kohlm.	IV	4 (ETHH), 10 (EHT)
25	Haligena spartinae E.B.G. Jones	IV	4, 6, 13 (all EHT)
26	Halosarpheia viscidula (Kohlm.) Shearer et Crane	II	1, 13, 14 (all EHT)
27	Halonectria milfordensis E.B.G. Jones	IV	
28	Halosphaeria appendiculata Linder	VI	2, 4, 5, 6, 10, 13 (all EHT)
29	Halosphaeria cucullata (Kohlm.) Kohlm.	II	
30	Halosphaeria galerita (Tubaki) I. Schmidt	V	
31	Halosphaeria hamata (Höhnk) Kohlm.	IV	
32	Halosphaeria maritima (Linder) Kohlm.	IV	2, 4, 6 (all EHT)
33	Halosphaeria pilleata (Kohlm.) Kohlm.	IV	4 (ETHH)
34	Halosphaeria quadricornuta Cribb et Cribb	II	14 (EHT)
35	Halosphaeria quadriremis (Höhnk) Kohlm.	VI	
36	Halosphaeria salina (Meyers) Kohlm.	VI	
37	Halosphaeria stellata (Kohlm.) Kohlm.	IV	4, 6 (both EHT)
38	Halosphaeria torquata Kohlm.	IV	6 (ETHH)
39	Humicola alopallonella Meyers et Moore	VI	
40	Hydronectria tethys Kohlm. et Kohlm.	I	

Table 25.1 continued

Species number	Species name	Group in ordination	Temperature-salinity regime* of sites with literature sources†
41	*Keissleriella blepharospora* Kohlm. et Kohlm.	I	
42	*Dactylospora haliotrepha* (Kohlm. et Kohlm.) Hafellner	I	
43	*Leptosphaeria albopunctata* (Westend.) Sacc.	III	
44	*Leptosphaeria australiensis* (Cribb et Cribb) G.C. Hughes	II	1, 6 (both EHT)
45	*Leptosphaeria avicenniae* Kohlm. et Kohlm.	I	1 (HTEH)
46	*Leptosphaeria marina* Ellis et Everh	III	5, 6 (both ETHH)
47	*Leptosphaeria obiones* (Crouan et Crouan) Sacc.	III	10 (ETHH)
48	*Leptosphaeria oraemaris* Linder	IV	4, 10, 13 (all EHT)
49	*Lignincola laevis* Höhnk	VI	
50	*Lindra inflata* Wilson	III	6 (ETHH)
51	*Lulworthia* sp.	I	
52	*Lulworthia fucicola* Sutherland	IV	4, 6, 8, 10 (all EHT)
53	*Lulworthia grandispora* Meyers	II	10 (EHT)
54	*Microthelia linderi* Kohlm.	IV	1, 6, 10, 11 (all EHT)
55	*Monodictys pelagica* (Johnson) E.B.G. Jones	VI	
56	*Mycosphaerella pneumatophorae* Kohlm.	I	
57	*Nais inornata* Kohlm.	IV	3, 10, 11, 14 (all EHT)
58	*Nautosphaeria cristaminuta* E.B.G. Jones	V	1, 8 (both HTEH)
59	*Nia vibrissa* Moore et Meyers	VI	
60	*Orbimyces spectabilis* Linder	IV	
61	*Papulospora halima* Anastasiou	VI	4, 6 (both ETHH)
62	*Phoma* sp.	VI	
63	*Pleospora gaudefroyi* Patouillard	V	
64	*Sphaerulina oraemaris* Linder	III	
65	*Torpedospora radiata* Meyers	VI	
66	*Trichocladium achrasporum* (Meyers et Moore) Dixon	II	
67	*Zalerion maritimum* (Linder) Anastasiou	VI	
68	*Zalerion varium* (Anastasiou) Anastasiou	VI	

† Literature: 1 = Aleem 1980; 2 = Byrne & Jones 1974; 3 = Davidson 1974; 4 = Henningsson 1974; 5 = Hughes 1968; 6 = Hughes 1969; 7 = Hughes 1974; 8 = Jones 1968; 9 = Jones 1971; 10 = Kirk & Brandt 1980; 11 = Kirk & Schatz 1980; 12 = Kohlmeyer 1983; 13 = Shearer 1972; 14 = Vrijmoed *et al.* 1982.

* Regime legends: EHT = euryhalothermic; ETHH = eurythermic homeohaline; HTEH = homeothermic euryhaline.

secondarily important. A close association of species group I (warm water HTEH) with site group C (warm water HTEH) is indicated. Species group II (warm water EHT) is related to the site group B (warm water EHT and subtropical HTEH). Species groups III (cool water ETHH), IV (cool water EHT) and V (cool water HTEH) show relationships with major site group A, which includes subgroups with concomitant water characteristics for the species group. Species group VI is associated with both site groups A and B, which suggests a composition of eurytolerant organisms. Thus, by directly applying the site characteristics to species groups (Figures 25.2, 3) and using literature based designations, the various species groups can be assigned: a) group I = warm water HTEH; b) group II = warm water EHT; c) group III and species 1, 7, 9, 10, 21, 33, 38, 60 of group IV = cool water ETHH; d) taxa 18, 28, 47 of group VI = cool water EHT; e) group V and species 27 = cool water HTEH.

The clear associations among the site and species groupings indicated by the concentration analysis were utilized to develop a distribution model

Figure 25.4 Ordination of species by number. (See Table 25.1 for number and corresponding species name).

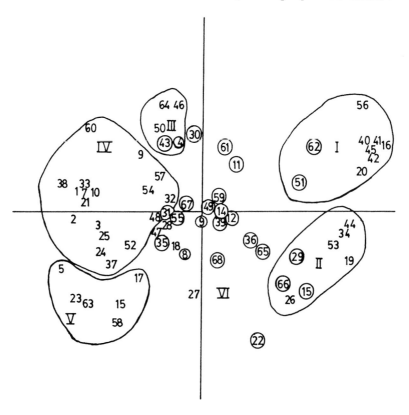

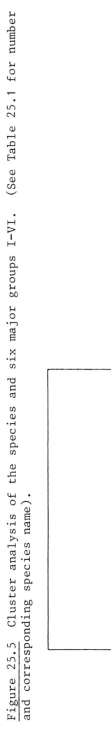
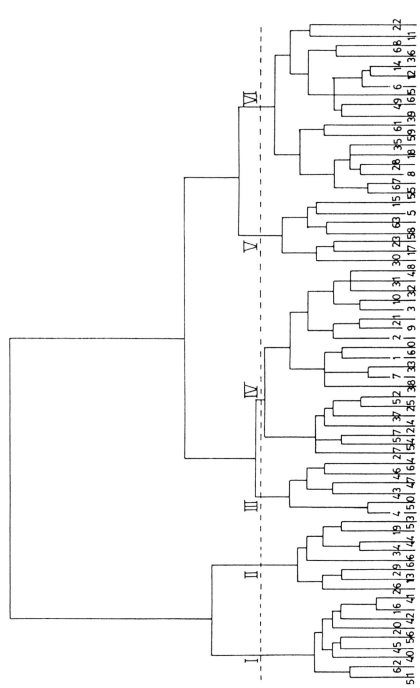

Figure 25.5 Cluster analysis of the species and six major groups I-VI. (See Table 25.1 for number and corresponding species name).

for lignicolous fungal species (Figure 25.7). The model, which is based
on temperature and salinity, is structured according to group placement
in the ordination. Three of the proposed conditions, shown with dashed
lines, are hypothetical but are included in order to complete all
possible combinations of temperature and salinity. The model represents
a gradient of temperature along the horizontal axis and temporal
variation in temperature and salinity along the vertical axis, in
confirmation with the axes of the ordinations and the concentration
analysis.

A chief limitation of the proposed model is the use of meagre and
imprecise temperature and salinity values in the data base. If temporal
variation in temperature and salinity are accepted as important
determinants of fungal distributions, quantitative measurement of
seasonal and diurnal changes may be critical to the production of a more
robust model. Other limitations reflect further inadequacies in the
data base. Firstly, the study sites in the original data (Booth
unpublished) are based on variable dimensions. Secondly, some of the
sites are associated with a number of water masses, indicating strong
site heterogeneity. Thirdly, sampling intensity at the sites varies;
the data base includes both sites with many literature sources and those
for which only a single source is available. Finally, the original data

Figure 25.6   Concentration analysis of species groups
(I-VI) and site arrays (A-C).

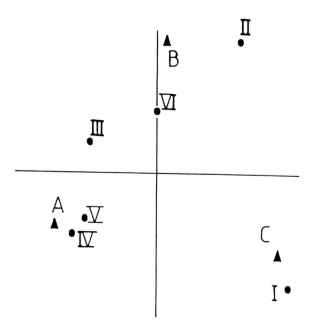

Figure 25.7   A distribution model for lignicolous marine fungi.

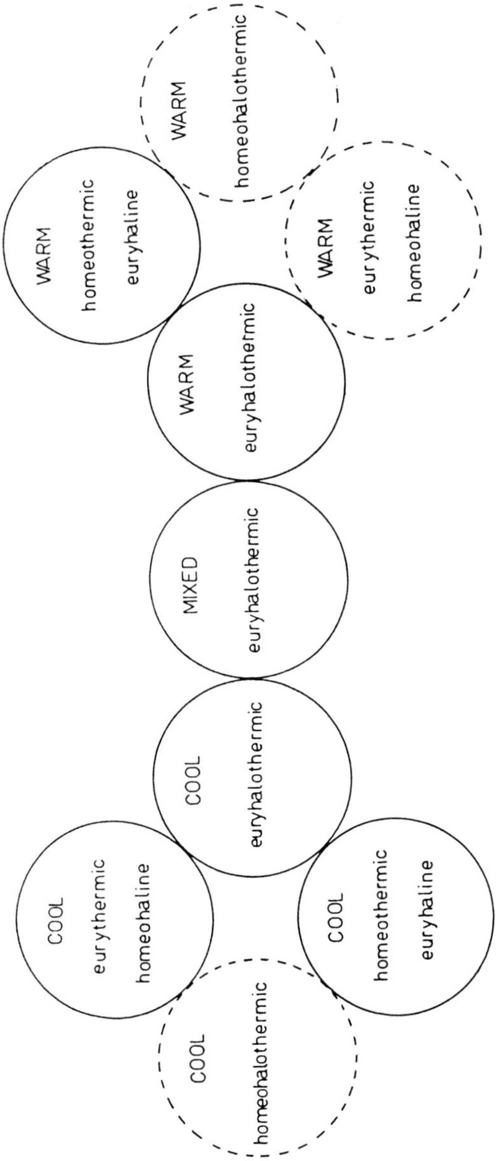

set is based on presence/absence rather than abundance measures.

Although the model is based on temperature and salinity regimes, it seems likely that other factors should be incorporated. One of the distinct advantages of the ordination - classification process is that the interspecific relationships which affect placement of sites in species space, and species in site space, are based on a large number of interacting environmental factors. The results of this study therefore remain open to further interpretation using factors such as substrate general habitat, oceanic currents and fungal interactions. Finally, the proposed model is not intended to represent a definitive, and therefore static, approach to predicting fungal distribution in the sea. Its aim is to elicit discussion and thought on the role of various factors controlling the presence and success of fungi in marine ecosystems.

REFERENCES

Aleem, A.A. (1980). Distribution and ecology of marine fungi in Sierra Leone (Tropical West Africa). Botanica mar., 23, 679-688.

Anderberg, M.R. (1973). Cluster Analysis for Applications. New York: Academic Press.

Booth, T. (1979). Strategies for study of fungi in marine and marine influenced ecosystems. Rev. Microbiol., (S. Paulo), 10, 123-138.

Boyd, P.E. & Kohlmeyer, J. (1982). The influence of temperature on the seasonal and geographic distribution of three marine fungi. Mycologia, 74, 894-902.

Brooks, R.D. (1972). Occurrence and distribution of wood-inhabiting marine fungi from Point Judith Pond. M.S. thesis, Univ. Rhode Island, U.S.A.

Byrne, P.J. & Jones, E.B.G. (1974). Lignicolous marine fungi. Veröff Inst. Meeresforsch. Bremerh., Suppl. 5, 301-320.

Byrne, P.J. & Jones, E.B.G. (1975). Effect of salinity on the reproduction of terrestrial and marine fungi. Trans. Br. myc. Soc., 65, 185-200.

Davidson, D.E. (1974). Wood-inhabiting and marine fungi from a saline lake in Wyoming. Trans. Br. myc. Soc., 63, 143-149.

Feoli, E. & Orlóci, L. (1979). Analysis of concentration and detection of underlying factors in structured tables. Vegetatio, 40, 49-54.

Gauch, H.G. (1982). Multivariate Analysis in Community Ecology. Cambridge studies in ecology: 1. Cambridge: Cambridge University Press.

Henningsson, M. (1974). Aquatic lignicolous fungi in the Baltic and along the west coast of Sweden. Svensk bot. Tidskr., 68, 401-425.

Henningsson, M. (1978). Physiology of aquatic lignicolous fungi from Swedish coastal waters. Mat. Org., 13, 129-168.

Hill, M.O. (1973). Reciprocal averaging: an eigenvector method of ordination. J. Ecol., 61, 237-249.

Hughes, G.C. (1968). Intertidal lignicolous fungi from Newfoundland. Can. J. Bot., 46, 1409-1417.

Hughes, G.C. (1969). Marine fungi from British Columbia: occurrence and
      distribution of lignicolous species. Syesis, 2, 121-140.
Hughes, G.C. (1974). Geographical distribution of the higher marine
      fungi. Veröff. Inst. Meeresforsch. Bremerh., Suppl., 5,
      419-441.
Hughes, G.C. (1975). Studies of fungi in oceans and estuaries since
      1961. I. Lignicolous, caulicolous and foliicolous species.
      Oceanogr. Mar. Biol., Ann. Rev., 13, 69-180.
Jones, E.B.G. (1968). The distribution of marine fungi on wood submerged
      in the sea. In Biodeterioration of Materials, eds.
      A.H. Walters & J.J. Elphick, pp. 460-485. Amsterdam:
      Elsevier.
Jones, E.B.G. (1971). The ecology and rotting ability of marine fungi.
      In Marine Borers, fungi and fouling organisms of wood, eds.
      E.B.G. Jones & S.K. Eltringham, pp. 237-248. Paris: O.E.C.D.
Kirk, P.W. (1972). Seasonal distribution of marine lignicolous fungi in
      the lower Chesapeake Bay. Am. J. Bot., 59, 667.
Kirk, P.W. & Brandt, J.M. (1980). Seasonal distribution of lignicolous
      marine fungi in the lower Chesapeake Bay. Botanica mar.,
      23, 657-668.
Kirk, P.W. & Schatz, S. (1980). Higher fungi affected by declining
      salinity and seasonal factors in a coastal embayment.
      Botanica mar., 23, 629-638.
Kohlmeyer, J. (1983). Geography of marine fungi. Aust. J. Bot., Suppl.
      Ser., 10, 67-76.
Kohlmeyer, J. & Kohlmeyer, E. (1979). Marine Mycology. The Higher Fungi.
      New York: Academic Press.
Krebs, C.J. (1978). Ecology, The Experimental Analysis of Distribution
      and Abundance (2nd ed.). New York: Harper & Row.
Kruskal, J.B. (1964 a). Multidimensional scaling by optimizing goodness
      of fit to a nonmetric hypothesis. Psychometrika, 29, 1-27.
Kruskal, J.B. (1964 b). Nonmetric multidimensional scaling: a numerical
      method. Psychometrika, 29, 115-129.
Mueller-Dombois, D. & Ellenberg, H. (1974). Aims and Methods of
      Vegetation Ecology. New York: Wiley.
Ochiai, A. (1957). Zoogeographic studies of the solenoid fishes found in
      Japan and its neighbouring regions. Bull. Jap. Soc.
      Scient. Fish., 22, 526-530.
Orlóci, L. (1966). Geometric models in ecology I. The theory and
      application of some ordination methods. J. Ecol., 54,
      193-215.
Orlóci, L. (1967). An agglomerative method for classification of plant
      communities. J. Ecol., 55, 193-206.
Orlóci, L. (1978). Multivariate Analysis in Vegetation Research. The
      Hague: Junk.
Pielou, E.C. (1979). Mathematical Ecology (2nd ed.). New York: Wiley.
Ritchie, D. (1957). Salinity optima for marine fungi affected by
      temperature. Am. J. Bot., 44, 870-874.
Shearer, C.A. (1972). Fungi of the Chesapeake Bay and its tributaries
      III. The distribution of wood inhabiting ascomycetes and
      fungi imperfecti of the Patuxent River. Am. J. Bot., 59,
      961-969.

Shearer, C.A. & Crane, J.L. (1978). The distribution of *Nais inornata*,
        a facultative marine Ascomycete. Mycotaxon, 7, 443-452.
Vrijmoed, L.L.P., Hodgkiss, I.J. & Thrower, L.B. (1982). Factors
        affecting the distribution of lignicolous marine fungi in
        Hong Kong. Hydrobiologia, 87, 143-160.
Ward, J.H. Jr. (1963). Hierarchical grouping to optimize an objective
        function. J. Am. Stat. Assoc., 58, 236-244.
Whittaker, R.H. (1975). Communities and Ecosystems. New York: Macmillan.

26 FREQUENCY OF OCCURRENCE OF LIGNICOLOUS MARINE FUNGI IN THE
    TROPICS

        K.D. Hyde

INTRODUCTION
        Most studies of lignicolous marine fungi have been conducted
in temperate zones (Hughes 1974), while studies of subtropical and
tropical marine fungi are mostly from the Atlantic Ocean. Many of
these studies present lists of marine fungi occurring in the locations
investigated and provide little indication of their frequency of
occurrence. The most detailed studies are those of Vrijmoed *et al.*
(1986 a,b) and Zainal & Jones (1986) who reported *Periconia prolifica*
and *Halosphaeria quadricornuta* Cribb et Cribb as the most frequently
recorded fungi in the tropics. In a recent investigation of the marine
fungi of the Seychelles, quantitative data on the frequency of
occurrence of marine fungi from various habitats were recorded. The
results are presented in this chapter and are compared with other
publications on tropical marine fungi. All the species recorded are new
records for the Seychelles and the data provide valuable information of
the marine fungi of the Indian Ocean, for which data are lacking.
Driftwood, roots and mangrove material from the various collecting
regions were sealed in large plastic bags for transport to the
laboratory. Storage facilities in the laboratory were lacking, so
material was subjected to microscopic examination as soon as possible.
Material was meanwhile stored in sealed plastic bags. The number of
collections of each species and the number of wood samples which
supported sporulating marine fungi were recorded for each habitat.

        RESULTS
        Table 26.1 lists the marine fungi which were identified
developing on driftwood, mangrove roots and branches, seagrasses,
angiosperm leaves and algae. A total of 77 fungi were identified:
57 Ascomycotina; 2 Basidiomycotina; 18 Deuteromycotina. Seven hundred
and fifty six samples of the driftwood and mangrove material examined
yielded marine fungi and the number of collections and percentage
occurrence of each species is given in Table 26.1. The percentage
occurrence was calculated from:

$$\frac{\text{number of collections of a particular species}}{\text{number of samples supporting sporulating marine fungi}} \times 100$$

The most common species collected in the Seychelles was *Antennospora
quadricornuta* which was found on 177 occasions (23.4% of samples).

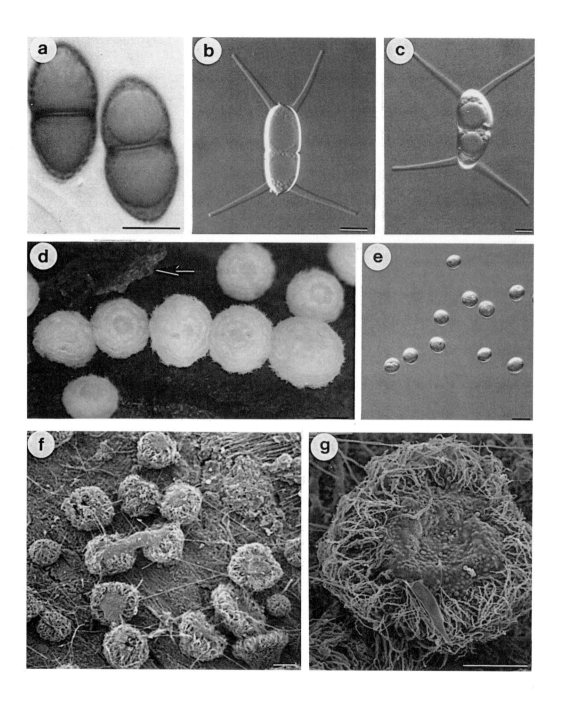

Halocyphina villosa (Basidiomycotina) from mangrove material, was the next most collected species (9.7%), while *Leptosphaeria australiensis*, *Lulworthia grandispora*, *Torpedospora radiata* and Ascomycete sp. (4) were frequently identified (above 5% of samples). Less frequently identified species (2-5%) included *Halosarpheia marina*, *Corollospora maritima*, *Dactylospora haliotrepha*, *Didymosphaeria enalia*, *Halosarpheia* sp., *Halosphaeria salina*, *Hydronectria tethys*, *Lulworthia floridana*-like, *L. medusa*-like, *Dictyosporium pelagicum* and *Humicola alopallonella*. All other species (60) were recorded on less than 2% of samples or on seagrasses, leaves and algae. Of the species identified 45 are new to the Indian Ocean.

Most of the fungi recorded could be identified to species level; four were similar to previously described forms, but differed in a number of respects (i.e. *Halosphaeria appendiculata*-like, *Mycosphaerella salicornae*-like, *Leptosphaeria neomaritima*-like, *Orcadia ascophylli*-like), while 24 could not be identified, or were identified to generic level only and some of these are new to science (i.e. Ascomycete sp. (1), *Massarina* sp. (Figure 26.2e).

The results presented in Table 26.1 are compiled from observations on material collected from beach and mangrove sites and provide an indication of the marine fungi of the Seychelles. Results from dead roots and driftwood collected from the beach sites showed that the most common fungus was *A. quadricornuta* (29.8% of samples), while *Corollospora maritima*, *Didymosphaeria enalia*, *Halosphaeria salina*, *Leptosphaeria australiensis*, *Lulworthia floridana*-like, *Dictyosporium pelagicum* and *Humicola alopallonella* were frequently collected (5.1-17.1%). The most common species from mangrove material was *Halocyphina villosa* (20.3% of samples), while *Halosarpheia marina*, *Antennospora quadricornuta*, Ascomycete sp. (4), *Halosarpheia* sp. and *Lulworthia grandispora* were frequently collected (5.2-13.5%). Several species frequently collected at the beach sites were also present within the mangrove ecosystem (i.e. *Antennospora quadricornuta*, *Didymosphaeria enalia*, *Halosarpheia salina*).

DISCUSSION

Quantitative data on the occurrence of tropical marine fungi have been published by Raghu Kumar (1973), Koch (1982), Kohlmeyer (1984), Zainal & Jones (1984, 1986) and Vrijmoed *et al.* (1986 a,b). However, in each of these studies, the number of collections was low and care must be exercised in the interpretation of the frequency data.

Figure 26.1a *Didymosphaeria enalia*, ascospores. L.M. Bar = 10 μm.
Figures 26.1b,c *Antennospora quadricornuta*, ascospores with subapical appendages. L.M. Bar = 10 μm.
Figure 26.1d *Halocyphina villosa*, basidiocarps on mangrove wood, ruptured bark arrowed. L.M. Bar = 1 mm.
Figure 26.1e *Halocyphina villosa*, hyaline basidiospores without appendages. L.M. Bar = 10 μm.
Figures 26.1f,g *Halocyphina villosa*, stalked basidiocarps at various stages of development. S.E.M. Bars = 100 μm.

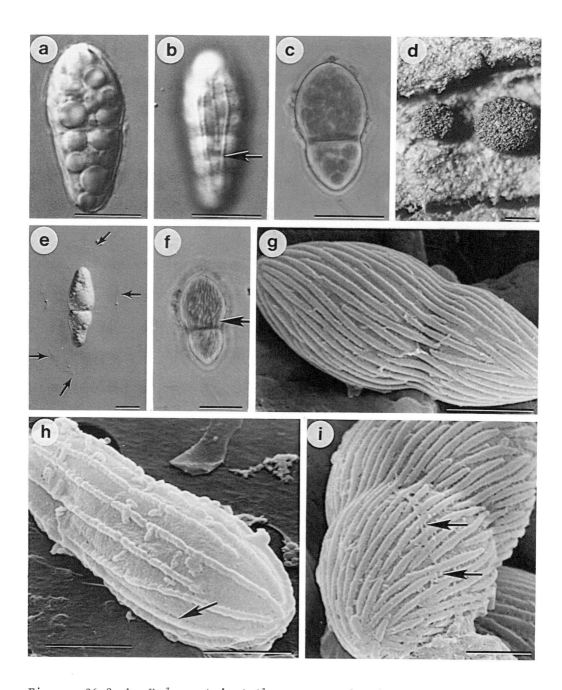

Figures 26.2a,b *Hydronectria tethys*, unappendaged ascospores, guttulate, one septate and with longitudinal ridges (arrowed). L.M. Bars = 10 µm.

Figures 26.2c,f *Dactylospora haliotrepha*, one septate ascospores with wall striations (arrowed). L.M. Bars = 10 µm.

Figure 26.2d *Dactylospora haliotrepha*, black apothecia. L.M. Bar = 200 µm.

The most common species to be collected in the Seychelles was
Antennospora quadricornuta (Figures 26.1b,c). *A. quadricornuta* has been
reported from tropical and subtropical waters on numerous occasions
(Kohlmeyer & Kohlmeyer 1979; Koch 1982; Booth 1983; Kohlmeyer 1984):
frequently as the most common species (Raghu Kumar 1973; Zainal & Jones
1986; Vrijmoed *et al.* 1986 a,b). Evidence in this study, and from other
workers, indicates that this is the most successful tropical lignicolous
marine fungus. It was found developing on driftwood in shallow lagoons
in the Seychelles where the temperature daily reached 37 °C. The
ability to grow under such high temperatures may be an important factor
in the success of this species. *Halocyphina villosa* (Figures 26.1d-g)
was the next most frequently collected species. While this species has
been previously recorded from the tropics, no data are available on its
frequency of occurrence. In this study it was only collected on
mangrove wood and it may be substrate specific as it was not collected
on submerged test blocks by either Vrijmoed *et al.* (1986 a,b) or Zainal
& Jones (1986). It has also been found to be abundant on mangrove wood
in Brunei (Hyde unpublished). *Leptosphaeria australiensis, Lulworthia
grandispora, Torpedospora radiata* and Ascomycete sp. (4) were commonly
identified. Of these *L. australiensis* and *T. radiata* have been reported
from tropical waters on numerous occasions (Kohlmeyer & Kohlmeyer 1979;
Koch 1982; Booth 1983; Kohlmeyer 1984).

Although Zainal & Jones (1986) and Vrijmoed *et al.* (1986 a,b) found
A. quadricornuta to be the most common species in their studies, the
sub- or co-dominant species were different. In this study *Halocyphina
villosa, Leptosphaeria australiensis, Lulworthia grandispora* and
Torpedospora radiata were the next most common, while in Kuwait,
*Ceriosporopsis halima, Halosphaeria salina, Corollospora maritima,
Lulworthia grandispora* and *Mycosphaerella* sp. were frequent (Jones &
Zainal 1986). Vrijmoed *et al.* (1986 a,b) noted that *Periconia
prolifica* (75% occurrence) was frequently co-dominant with
A. quadricornuta with the following fungi classed as frequent:
C. halima (44.23%), *Cirrenalia macrocephala* (51.9%), *Lulworthia* sp.
(25%), *Monodictys pelagica* (48%) and *Trichocladium achrasporum* (36%).

Frequently occurring species in this study (2-5% of samples) include
9 fungi belonging to the Ascomycotina and 2 belonging to the
Deuteromycotina (Table 26.1). All these species with the exception of
Halosarpheia marina have previously been reported from tropical and
subtropical waters on many occasions. *Halosphaeria salina* was the most
common species found in Kuwait waters (Zainal & Jones 1984), while

Figure 26.2e *Massarina* sp., one septate ascospore surrounded by a
gelatinous sheath (arrowed). L.M. Bar = 20 μm.
Figures 26.2g,i *Dactylospora haliotrepha*, ascospores with longitudinal
oblique wall striations, mucilage (arrowed) is present between
striations. S.E.M. Bar = 5 μm.
Figure 26.2h *Hydronectria tethys*, ascospore with longitudinally
orientated ridges, some do not extend the full length of the ascospore,
and associated mucilage (arrowed). S.E.M. Bar = 5 μm.

Table 26.1 Marine fungi collected from Seychelles - a complete list

A. Fungi from 756 samples of driftwood and mangrove material

Fungi	Number of collections	Percentage occurrence	New to Indian Ocean
Antennospora quadricornuta (Cribb et Cribb) T.W. Johnson (T)	177	23.4	
Halocyphina villosa Kohlmeyer et Kohlmeyer (T)	73	9.7	
Leptosphaeria australiensis (Cribb et Cribb) G.C. Hughes (T)	72	9.5	
Lulworthia grandispora T.W. Johnson (T)	50	6.6	✓
Ascomycete sp. (4)	50	6.6	✓
Torpedospora radiata Meyers (C)	49	6.5	
Halosphaeria salina (Meyers) Kohlmeyer (T)	37	4.9	
Humicola alopallonella Meyers et Moore (C)	33	4.4	
Halosphaeria marina (Cribb et Cribb) Kohlmeyer (T)	32	4.2	
Didymosphaeria enalia Kohlmeyer	32	4.2	
Dactylospora haliotrepha (Kohlmeyer et Kohlmeyer) Hafellner (T)	30	4.0	
Corollospora maritima Werdermann (C)	28	3.7	
Dictyosporium pelagicum (Linder) G.C. Hughes ex Johnson et Sparrow	27	3.6	✓
Lulworthia floridana-like	23	3.0	
Lulworthia medusa-like	21	2.8	
Hydronectria tethys Kohlmeyer et Kohlmeyer (T)	19	2.5	
Halosphaeria sp.	18	2.4	✓
Cirrenalia tropicalis Kohlmeyer (T)	14	1.9	✓
Monodictys pelagica (T.W. Johnson) Jones (C)	13	1.7	
Lulworthia opaca-like	12	1.6	✓
Pleospora sp. (1)	12	1.6	✓
Arenariomyces trifurcatus (Cribb et Cribb) T.W. Johnson (C)	11	1.5	✓
Caryosporella rhizophorae Kohlmeyer	10	1.3	✓
Clavariopsis bulbosa Anastasiou (T)	10	1.3	
Massarina sp.	9	1.2	✓
Aniptodera chesapeakensis Shearer et Miller	8	1.1	✓
Ascomycete sp. (5)	8	1.1	
Cirrenalia pygmea Kohlmeyer (T)	8	1.1	✓
Nimbospora bipolaris Hyde et Jones	7	0.93	✓
Cirrenalia pseudomacrocephala Kohlmeyer (T)	7	0.93	
Corollospora pulchella Kohlmeyer, I. Schmidt et Nair (C)	6	0.8	
Halosphaeria viscosa (Schmidt) Shearer et Crane	6	0.8	✓
Trematosphaeria lignatilis Kohlmeyer (C)	6	0.8	✓
Halosphaeria fibrosa Kohlmeyer et Kohlmeyer (T)	5	0.66	
Lignincola laevis Höhnk (C)	5	0.66	
Periconia prolifica Anastasiou (T)	5	0.66	

Table 26.1 continued

Fungi	Number of collections	Percentage occurrence	New to Indian Ocean
Sporodesmium ellipticum Moore	5	0.66	✓
Halosphaeria appendiculata-like	4	0.53	✓
Leptosphaeria sp.	4	0.53	✓
Lindra sp.	4	0.53	✓
Mycosphaerella pneumatophorae Kohlmeyer (T)	4	0.53	
Ocostaspora apilongissima Jones, R.G. Johnson et Moss (C)	4	0.53	✓
Ascomycete sp. (6)	3	0.4	
Ceriosporopsis halima Linder (C)	3	0.4	✓
Halonectria sp.	3	0.4	✓
Mycosphaerella salicorniae-like	3	0.4	✓
Pleospora sp. (2)	3	0.4	
Remispora crispa Kohlmeyer (T)	3	0.4	✓
Savoryella paucispora (Cribb et Cribb) Koch (T)	3	0.4	✓
Nia vibrissa Moore et Meyers (C)	3	0.4	✓
Chrysosporium sp.	3	0.4	✓
Curvularia sp.	3	0.4	✓
Diplodia sp.	3	0.4	✓
Keissleriella blepharospora Kohlmeyer et Kohlmeyer (T)	2	0.26	✓
Orcadia ascophylli-like	2	0.26	✓
Savoryella lignicola Jones et Eaton	2	0.26	✓
Zalerion varium Anastasiou (C)	2	0.26	
Halosarpheia cincinnatula Shearer et Crane	1	0.13	✓
Halosarpheia ratnagiriensis Borse	1	0.13	
Halosarpheia abonnis Kohlmeyer	1	0.13	✓
Halosarpheia-like sp. (3)	1	0.13	✓
Lanspora coronata Hyde et Jones	1	0.13	✓
Leptosphaeria neomaritima-like	1	0.13	
Nimbospora effusa Koch	1	0.13	✓
Remispora galerita Tubaki (C)	1	0.13	✓
Ascomycete sp. (1)	1	0.13	✓
Ascomycete sp. (2)	1	0.13	✓
Ascomycete sp. (3)	1	0.13	✓
Dictyosporium toruloides (Corda) Gueguen	1	0.13	✓
Trichocladium achrasporum (Meyers et Moore) Dixon (C)	1	0.13	
Trichocladium linderi Crane et Shearer	1	0.13	✓
Helicomyces sp.	1	0.13	✓
Hyphomycete sp. (1)	1	0.13	✓

Table 26.1 continued

Fungi	Number of collections	Percentage occurrence	New to Indian Ocean
B. Fungi from seagrasses, leaves or algae			
Emericellopsis stolkiae Davidson et Christensen			
Lindra marinera Meyers (T)	–	–	✓
Lindra thalassiae Orpurt, Meyers, Boral et Simms (T)	–	–	✓
Varicosporina ramulosa Meyers et Kohlmeyer (T)	–	–	✓

(T) = tropical species.
(C) = cosmopolitan species.

Kohlmeyer & Kohlmeyer (1979) regarded *Corollospora maritima* as the most
common tropical arenicolous marine fungus. The remaining 56 marine
fungi were identified occasionally, each having been collected on less
than 1.9% of driftwood samples. Of these, *Arenariomyces trifurcatus*,
Corollospora pulchella, *Keissleriella blepharospora*, *Cirrenalia pygmea*,
C. tropicalis, *Clavariopsis bulbosa* and *Periconia prolifica* have been
commonly recorded in the tropics, whereas other species have few records
(Kohlmeyer & Kohlmeyer 1979; Koch 1982; Booth 1983; Kohlmeyer 1984;
Hyde 1985).

The most common species from the beach sites (i.e. *A. quadricornuta*,
Corollospora maritima, *Didymosphaeria enalia* (Figure 26.1a),
Halosphaeria salina, *Leptosphaeria australiensis*, *Lulworthia floridana*-
like, *Dictyosporium pelagicum*, *Humicola alopallonella*) have all been
recorded on numerous occasions from tropical beach sites (Kohlmeyer &
Kohlmeyer 1979; Koch 1982; Booth 1983; Kohlmeyer 1984). Many of these
species were also common in the mangrove ecosystem (i.e.
A. quadricornuta, *D. enalia*), although generally they were not as common
as on beach sites; this may suggest less adaptation for the mesohaline
mangrove habitat. The common mangrove species, however, have not been
frequently reported from such habitats: *Halocyphina villosa* (4
occasions), *Aniptodera marina* (2 occasions), *A. quadricornuta* (6
occasions), *Lulworthia grandispora* (4 occasions). The low number of
published records for these fungi indicates that further studies are
required in order to establish the extent of the mycoflora of mangrove
habitats.

Prior to this study, a total of 35 Ascomycotina, 14 Deuteromycotina and
1 Basidiomycotina had been reported from the Indian Ocean (Koch 1982;
Zainal & Jones 1984, 1986). In this study 45 previously unrecorded
species have been found on driftwood, mangrove material, seagrasses,
angiosperm leaves and algal material. Notable new collections include
Nia vibrissa, *Lulworthia grandispora*, *Lindra marinera*, *L. thalassiae* and
Varicosporina ramulosa. Previous studies of fungi from tropical waters
by Koch (1982) and Booth (1983) have yielded only 27 and 32 species
respectively, while species diversity in the present study was high.
Island studies in particular have yielded few species: Hawaii, 27
species (Kohlmeyer 1969); Bermuda, 22 species (Kohlmeyer & Kohlmeyer
1977). The number of collections made in this study may have been
greater than for the previous investigations and may account for the
differences reported. Unidentified species are also included but this
has not always been the case in previous investigations.

Common tropical marine fungi (*Antennospora quadricornuta*, *Didymosphaeria
enalia*, *Halocyphina villosa*) are illustrated in Figures 26.1a-g.
H. villosa has previously been poorly documented, particularly at the
scanning electron microscope level. Basidiocarps are superficial,
white when young and yellowing with age. The young basidiocarp is
spherical, composed of closely packed tapering hyphae which radiate
towards the central depressed region (Figure 26.1f). Later in
development basidiocarps become stalked and napiform to turbinate, the
central region begins to separate and fill with basidiospores and the
hyphae become irregular (Figure 26.1g). Basidiospores are spherical and

Table 26.2   The frequency of occurrence of marine fungi in
the tropics

Common

    Antennospora quadricornuta (19)*
    Arenariomyces trifurcatus (13)
    Corollospora maritima (17)
    Didymosphaeria enalia (13)
    Halocyphina villosa (6)
    Halosphaeria salina (12)
    Keissleriella blepharospora (12)
    Leptosphaeria australiensis (14)*
    Lignincola laevis (14)
    Lulworthia sp. (23)
    Lulworthia grandispora (6)*
    Torpedospora radiata (18)*

Frequent

    Ceriosporopsis halima (10)
    Corollospora pulchella (10)
    Dactylospora haliotrepha (10)*
    Halosarpheia marina (3)*
    Halosarpheia sp. *
    Hydronectria tethys (8)*
    Leptosphaeria avicenniae   (10)
    Lindra thalassiae (8)
    Nia vibrissa (8)
    Cirrenalia pygmea (7)
    Clavariopsis bulbosa (8)
    Dictyosporium pelagicum (1)*
    Humicola alopallonella (7)*
    Periconia prolifica (11)
    Phoma spp. (8)
    Trichocladium achrasporum (11)
    Zalerion varium (10)

Occasional

    All the remaining 65 recorded species

Figures in brackets are the numbers of previously published
records of tropical marine fungi.

* These categories apply to this study.

smooth walled (Figure 26.1e).

Dactylospora haliotrepha and *Hydronectria tethys* are frequently
collected tropical marine fungi but have been poorly documented. Both
species have interesting wall ornamentations and these are illustrated
at the light microscope (Figures 26.2a-d,f) and scanning electron
microscope (Figures 26.2g-i) levels. The ornamentations in
D. haliotrepha are composed of short thickened striations which
protrude from the surface of the spore wall (Figure 26.2g). The
striations are positioned obliquely around the spore, extend from pole
to pole (Figure 26.2i), measure 2-25 x 0.3-0.6 µm, are rigid,
cylindrical, smooth and taper at each end. They appear to adhere to the
spore wall by mucilaginous material (Figure 26.2i, arrowed). Ascospores
of *H. tethys* are ornamented with ridges that extend longitudinally the
full length of the spore (Figures 26.2b,h). The ridges appear to be
coated with mucilaginous material (Figure 26.2h, arrowed).

It is clear from this study, and previous publications, that
Antennospora quadricornuta is one of the most successful and abundant
tropical marine fungi and that *Halocyphina villosa* is the most common
species on mangrove wood. Other species can be regarded as common,
frequent and occasional and are listed in Table 26.2.

ACKNOWLEDGEMENTS
I am grateful to Professor E.B. Gareth Jones and
Dr. S.T. Moss for their continued support, P. Crook for assistance with
field work in the Seychelles, G. Bremer, C. Derrick and the late
E. Hawton for photographic assistance and Ciba-Geigy for financial
support for the field work.

REFERENCES
Booth, T. (1983). Lignicolous marine fungi from São Paulo, Brazil. Can.
    J. Bot., 61, 488-506.
Hughes, G.C. (1974). Geographical distribution of the higher marine
    fungi. Veröff. Inst. Meeresforsch. Bremerh., Suppl. 5,
    419-441.
Hyde, K. (1985). Spore settlement and attachment in marine fungi. Ph.D.
    Thesis. CNAA, Portsmouth Polytechnic, U.K.
Koch, J. (1982). Some lignicolous marine fungi from Sri Lanka. Nord. J.
    Bot., 2, 163-169.
Kohlmeyer, J. (1969). Marine fungi from Hawaii including a new genus
    Heliascus. Can. J. Bot., 47, 1469-1487.
Kohlmeyer, J. (1984). Tropical marine fungi. Mar. Ecol., 5, 329-378.
Kohlmeyer, J. & Kohlmeyer, E. (1977). Bermuda marine fungi. Trans. Br.
    mycol. Soc., 68, 207-219.
Kohlmeyer, J. & Kohlmeyer, E. (1979). Marine mycology: The higher fungi.
    New York: Academic Press.
Raghu Kumar, S. (1973). Marine lignicolous fungi from India. Kavaka, 1,
    73-85.

Vrijmoed, L.L.P., Hodgkiss, I.J. & Thrower, L.B.   (1986 a). Factors
          affecting colonisation of submerged timber by lignicolous
          marine fungi, with special reference to the effects of
          fouling organisms. Hydrobiologia, in press.
Vrijmoed, L.L.P., Hodgkiss, I.J. & Thrower, L.B. (1986 b). Effects of
          different wood species on the colonisation of lignicolous
          marine fungi in Hong Kong coastal waters. Hydrobiologia, in
          press.
Zainal, A. & Jones, E.B.G. (1984). Observations on some lignicolous
          marine fungi from Kuwait. Nova Hedwigia, 39, 569-583.
Zainal, A. & Jones, E.B.G. (1986). Occurrence and distribution of
          lignicolous marine fungi in Kuwait coastal waters. Proc. 6th
          International Biodeterioration Symposium, Washington D.C.,
          in press.

G.J.F. Pugh

E.B.G. Jones

INTRODUCTION

Studies on micro-fungi in the Antarctic have been undertaken on an *ad hoc* basis of collections of materials as occasion has allowed (e.g. Tubaki 1961; Heal *et al.* 1967; Latter & Heal 1971). During the last twenty five years it has become possible to undertake work at the bases which are manned and operated by the British Antarctic Survey (BAS) particularly on Signy Island, South Orkney Islands, and at Grytviken on South Georgia. Such studies have been carried out by BAS personnel and visitors to the bases, and have largely been concerned with terrestrial bacteria and fungi which are involved with decomposition of plant remains. Headland (1982) has compiled a bibliography of publications including non-vascular plants on South Georgia up to 1981. More recent accounts of fungal decomposer activity include: Bailey & Wynn Williams (1982); Pugh & Allsopp (1982); Hurst *et al.* (1983). Elsewhere in the Antarctic, Kerry (1979) studied fungi in soils on Sabrina Island.

Investigations on fungi in Antarctic waters have been even less frequent. Tubaki & Asano (1965) isolated a number of species from material collected during the Japanese Expedition to the Antarctic 1956-62.

Casual collection of sea borne materials in the Antarctic is made difficult by two major factors: there is a dearth of drift wood on the beaches during the summer and macroalgae are restricted to sheltered bays where ice abrasion is limited; and during the winter the sea ice freezes to the land, so algae are normally absent from the intertidal zone. In the areas where the whaling stations were built, remains of wooden jetties could provide a source of materials which would be well worth examining. There are at least five remnants of such stations on South Georgia, and one on Signy Island.

The current work was undertaken during a short visit in January-March, 1980 to study terrestrial fungi; however other collections were made whenever possible, so as to expand our knowledge of fungi which occur in the South Atlantic. While there it was possible to immerse wood blocks in the sea at Factory Cove, Signy Island, South Orkney Islands (lat. 60.4°S, long. 45°W) and at Grytviken, Cumberland East Bay, South Georgia (lat. 54-55°S, long. 36-38°W). Because of limitations on the possible length of stay at Signy, blocks could only be immersed for about three months (Jan-March); at Grytviken they remained in the sea for a longer period (February-May).

This is a preliminary account of the fungi which have been recorded on the wood blocks which have been examined. Other blocks are awaiting examination. In view of the lack of knowledge of marine fungi in circumpolar waters (Hughes 1974, 1975; Kohlmeyer & Kohlmeyer 1979) this account provides information from a little-studied region.

MATERIALS AND METHODS

Wood blocks, approximately 40 mm x 40 mm x 8 mm were cut from planks of balsa (*Ochroma lagopus* L.), used as a readily perishable wood; from deal (*Pinus sylvestris* L.), which has commonly been used as a bait for lignicolous marine fungi; and from Ramin, a perishable tropical hardwood (*Gonystylus* sp.) to compare the fungal colonizers of different wood types. A hole was drilled through the centre of each block, and the blocks were threaded onto lengths of Cauline line. They were held apart by knots above and below each block, so that strings of 10 blocks of the same wood were formed.

In the Antarctic, the strings of blocks were attached to suitable weights. On Signy Island strings of each of the types of wood were located in deep water approximately 0.5 km north of the jetty in Factory Cove. The jetty itself was stone built, with a wooden slipway, the 'plan'. This was a remnant of an old whaling station which was used between 1912-1923. At the time of immersion of the blocks (January-March, 1980), the sea temperature was about 0.5° C. There was no obvious drift wood on the beaches in Factory Cove, and only occasional pieces of drift brown algae. While there is no growth of brown algae in the intertidal zone, because of ice scour, algae do grow abundantly in deeper water. Accumulations occur in the nearby Shallow Bay.

On South Georgia, strings of blocks were immersed from February to May 1980 off the jetty of the old Grytviken whaling station, in deep water. The jetty was built of greenheart (*Ocotea rodiaei* (Schomb.) Mez.), near a slipway, the 'plan' which was used for hauling whales from the water. A second wooden jetty is present at King Edward Point (about 1 km away), and there is also the hulk of an abandoned wooden sailing ship beached about 1 km away. There was little evidence of drift wood on the local beaches, but kelp grows abundantly around King Edward Point. There were drift brown algae, and many relics of whaling, mainly in the form of bones.

Strings of blocks of each wood were removed from the sea at Signy after 25, 35 and 39 days. At Grytviken, strings of each wood were recovered after 6 and 12 weeks. After recovery, the blocks were cut off their strings and wrapped individually: at Signy each block was wrapped in paper towelling moistened with sea water, and placed in polythene bags. At Grytviken, the blocks were placed separately in polythene bags which were heatsealed so as to keep the blocks separate from each other. All the blocks were transported by sea to the U.K. at about 4° C, and then stored at 5° C first at Aston University and subsequently at Portsmouth Polytechnic, until they could be examined. Single spore isolations were made of representative species. These were grown on cornmeal sea water agar.

RESULTS

A total of 56 balsa, 20 deal and 20 Ramin blocks have been examined from the Signy immersions. On the balsa blocks *Monodictys pelagica* was found on 55 blocks (98%) and *Ceriosporopsis tubulifera* was present on 26 blocks (46%). A sterile mycelium was present on 17 of the deal blocks and on all 20 Ramin blocks, while *M. pelagica* was found on 12 of the Ramin blocks. Diatoms were abundant on the deal blocks. These may have limited fungal colonization, which only occurred at 0.8 fungi per deal block, compared with 1.4 per block and 1.6 per block for balsa and Ramin respectively.

At Grytviken, from a total of 39 blocks examined (Table 27.1) 99 fungi were recovered, representing 9 identified and 4 unidentified species.

Table 27.1   Fungi collected on test blocks submerged in sea water at Grytviken, South Georgia

Wood type Length of immersion (days)	Balsa 42	Balsa 84	Deal 42	Ramin 42	Total No. of collections
Ceriosporopsis circumvestita (Kohlm.) Kohlm.	–	–	–	1	1
C. halima Linder	–	6	–	4	10
C. tubulifera (Kohlm.) Kirk *in* Kohlmeyer	8	13	2	14	37
Corollospora maritima Werdermann	–	1	–	–	1
Remispora maritima Linder	–	9	1	3	13
R. stellata Kohlm.	–	5	–	2	7
Humicola alopallonella Meyers et Moore	1	–	–	–	1
Monodictys pelagica (T.W. Johnson) E.B.G. Jones	4	4	–	6	14
Zalerion maritimum (Linder) Anastasiou	3	–	1	–	4
Hyaline sterile mycelium	–	–	–	5	5
Stysanus-like sp.	–	–	–	1	1
Unidentified sp.	1	–	–	–	1
White yeast	–	4	–	–	4
Total No. of collections	17	42	4	36	99
Number of blocks	10	13	2	14	39

C. tubulifera was the most frequent isolate, representing 37% of the total and occurred on all three types of wood (95% of the blocks). *M. pelagica* represented 14% of the total isolates (36% of the blocks) but was not recovered from the Ramin blocks. *Remispora maritima* occurred on all three wood types, representing 13% of the total isolates (33% of the blocks). It was followed in abundance by *C. halima* (10% of the total isolates; present on 26% of the blocks).

The length of time of immersion at Signy (25-39 days) did not appear to influence the degree of colonization of the blocks and this is illustrated by reference to Table 27.2.

However, at Grytviken over longer immersion times of 42 and 84 days, the rate of isolations rose overall from 2.2 per block to 3.2 per block after 84 days. On balsa there were 17 isolations (1.7 per block) at 42 days and 42 (3.2 per block) at 84 days. The increase on all the three timbers was caused by more frequent isolations of the identified fungi present at 42 days; only *Corollospora maritima* occurred (on balsa) after 84 days and not at 42 days.

DISCUSSION

The recovery of *Monodictys pelagica* from 98% of the balsa blocks and of *Ceriosporopsis tubulifera* on 46% of the balsa blocks so far examined from Signy shows that these two species are common in the very cold waters of the South Atlantic. Both species were able to colonize wood blocks immersed for 25 days, which implies a readily available source of inoculum in the local water.

The relative abundance of these two species was reversed at Grytviken, where *M. pelagica* yielded 14 isolates on a total of 39 blocks (38%), while *C. tubulifera*, with 37 isolates occurred on 95% of the blocks. Of the other seven identified species, *Remispora maritima* (33% of the blocks), *Ceriosporopsis halima* (26%), *R. stellata* (18%) and *Zalerion maritimum* (10%) occurred in descending frequency.

The larger number of fungi present at Grytviken than at Signy could

Table 27.2 Rate of colonization by fungi of different types of wood blocks during various lengths of immersion in sea water at Signy Island

Wood	Immersion time (days)	No. of fungal isolations	No. of blocks	Mean No. of fungal isolations per block
Balsa	25	11	6	1.8
Balsa	35	50	30	1.7
Deal	39	27	20	1.4
Ramin	39	32	20	1.6

reflect at least three possible factors:
1. The sea temperature. Although South Georgia lies south of the
Antarctic convergence, it has a mean sea temperature of 5° C compared
with about 1° C at Signy Island. Icebergs do occur, but are relatively
rare compared with Signy, where in the summer icebergs are almost always
visible, and in winter the sea freezes.
2. The quantity and time scale of wood present in the sea. Grytviken
was a base for whaling, and a large whale processing station was in use
for about 60 years from 1904. During this time the wooden slipway,
jetties and pilings, the visits by wooden ships and the continuing
presence of an old wooden hulk, and the presence of some algae could all
have provided substrates upon which marine fungi could grow to provide a
ready inoculum to colonize newly submerged wood blocks.
3. The circumpolar winds and currents. These funnel through the Drake
Passage between South America and the Antarctic Peninsula as the West
Wind Drift (Figure 27.1). Signy is situated where the ice from the

Table 27.3 Marine fungi reported from Antarctica or the Arctic region

Fungi	Reference
THRAUSTOCHYTRIALES	
Thraustochytrium antarcticum Bahnweg et Sparrow	Bahnweg & Sparrow 1972, 1974
T. kerguelensis Bahnweg et Sparrow	
T. rossii Bahnweg et Sparrow	
ASCOMYCOTINA	
Spathulospora antarctica Kohlm.	Kohlmeyer 1973
DEUTEROMYCOTINA	
Dendryphiella salina (Suth.) Pugh et Nicot	Tubaki & Asano 1965
Monodictys austrina Tubaki	
HEMIASCOMYCETES and BASIDIOMYCOTINA	
Cryptococcus lactativorus Fell et Phaff	Fell 1976
Leucosporidium antarcticum Fell, Statzell, Hunter et Phaff	
Rhodosporium capitatum Fell, Hunter et Tallman	
Rh. dacryoidum Fell, Hunter et Tallman	
Rh. malvinellum Fell et Hunter	
Rh. sphaerocarpum Newell et Fell	
Torulopsis austomarina Fell et Hunter	

Weddell Sea moves eastwards. The waters around the island are greatly influenced by such a movement. On the other hand, although South Georgia is within the Antarctic Convergence, it is some 650 km to the north, and is not subject to such extremes of pack ice and sea freezing. The northern eddies of the circumpolar currents can have a South American influence, providing sources of wood and possibly of fungal inocula. This has been illustrated by Smith (1985), who has catalogued stranded trees on the islands of the Southern Ocean. He has shown that *Nothofagus* spp. (southern beech) were the most common, with most records being of *N. pumilo* (Popp. Endl.) Krasser. This is the dominant tree of the summer green forests of south eastern Tierra del Fuego. The other trees also had a southern South American provenance. Of his 46 records of *N. pumilo*, 33 were from South Georgia (mainly from the southern shore line) and only two were from the South Shetland Islands which are just to the north west of the Antarctic Peninsula. None was recorded on the South Orkney Islands.

Thus there are suitable substrates in the form of wood, both as logs from South America and as pilings and jetties at Grytviken, which could be a source of inocula. While brown algae occur at Signy, they are more abundant at Grytviken, where the higher sea temperature is likely to be more conducive to the growth of both algae and fungi.

An examination of the stranded logs, and of the jetties is certainly

Figure 27.1 Sketch map to show the positions of the collecting sites relative to South America and the Antarctic Peninsula.

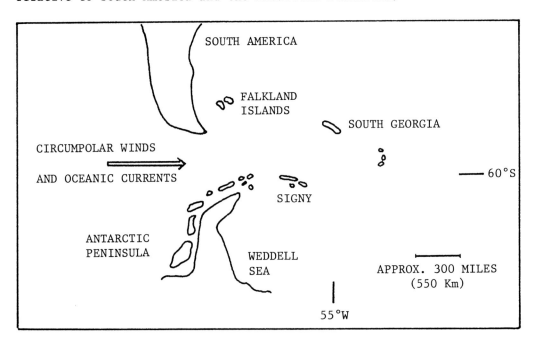

warranted to determine their fungal flora. At the same time the use of greenheart and southern beech wood blocks may be productive if the fungi are acclimatized to these timbers. Studies on rates of decomposition, particularly at the prevailing low temperatures are also needed. As some of the timbers may have been *in situ* for eighty years the decomposition rates must be slow, and are presumably limited by the low temperatures.

Table 27.3 lists the marine fungi known from Arctic and Antarctic zones. It highlights: 1) how few studies have been undertaken; 2) the very specialized nature of the collections to date. None of the fungi observed during this study has been reported by other workers from the Antarctic and Arctic geographical zones.

Of the species collected in this study *Ceriosporopsis circumvestita*, *C. tubulifera*, *Monodictys pelagica* and *Zalerion maritimum* can be regarded as cold water temperate species, while the remainder are cosmopolitan in their distribution. *Monodictys pelagica* has been reported as ubiquitous in cold coastal waters (Meyers & Reynolds 1960; Hughes & Chamut 1971). Few records exist of this species from tropical or sub-tropical zones (Kohlmeyer & Kohlmeyer 1979; Hyde unpublished). Similarly *C. circumvestita* (Koch 1982) and *Zalerion maritimum* (Zainal & Jones 1984, 1986) have been reported from warmer sub-tropical waters. Marine mycologists have discussed the geographical distribution of marine fungi but no clear picture has emerged. Further data on the effects of substrata, salinity, temperature requirements and intertidal/ subtidal immersion are required before our understanding of the biogeography of these fungi is complete.

ACKNOWLEDGEMENTS

One of us (G.J.F.P.) acknowledges with gratitude the hospitality of the British Antarctic Survey, Cambridge, and the financial help received from the Royal Society and the Transpolar Fund. The help of Dr. D. Allsopp and of Dr. J.L. Hurst in placing and recovering the wood blocks is gratefully acknowledged.

REFERENCES

Bahnweg, G. & Sparrow, F. (1972). *Aplanochytrium kerguelensis* gen. nov. spec. nov., a new phycomycete from subantarctic marine waters. Archiv. Mikrobiol., 81, 45-49.

Bahnweg, G. & Sparrow, F. (1974). Four new species of *Thraustochytrium* from Antarctic regions, with notes on the distribution of zoosporic fungi in the Antarctic marine ecosystem. Amer. J. Bot., 61, 754-766.

Bailey, A.D. & Wynn Williams, D.D. (1982). Soil microbiological studies at Signy Island, South Orkney Islands. Brit. Antarct. Surv. Bull., 51, 167-191.

Fell, J. (1976). Yeasts in oceanic regions. In Recent Advances in Aquatic Mycology, ed. E.B. Gareth Jones, pp. 93-113. London: Elek Science.

Headland, R.K. (1982). South Georgia: a Bibliography. Cambridge: British
        Antarctic Survey.
Heal, O.W., Bailey, A.D. & Latter, P. (1967). Bacteria, fungi and
        protozoa in Signy Island soils compared with those from a
        temperate moorland. Phil. Trans. Roy. Soc., Series B, 252,
        191-197.
Hughes, G.C. (1974). Geographical distribution of the higher marine
        fungi. Veröff. Inst. Meeresforsch. Bremerh., Suppl., 5,
        419-441.
Hughes, G.C. (1975). Studies of fungi in oceans and estuaries since
        1961. 1. Lignicolous, caulicolous and foliicolous species.
        Oceanogr. Mar. Biol. Ann. Rev., 13, 69-180.
Hughes, G.C. & Chamut, P.S. (1971). Lignicolous marine fungi from
        southern Chile, including a review of distributions in the
        southern hemisphere. Can. J. Bot., 49, 1-11.
Hurst, J.L., Pugh, G.J.F. & Walton, D.W.H. (1983). Fungal succession and
        substrate utilisation on the leaves of three South Georgia
        phanerogams. Br. Antarct. Surv. Bull., 58, 89-100.
Kerry, E. (1979). Microbiological studies of soils from Sabrina Island,
        Antarctic. In D. Lewis, Voyage to the ice. The Antarctic
        expedition of Solo, pp. 112-114. London: Wm. Collins.
Koch, J. (1982). Some lignicolous marine fungi from Sri Lanka. Nord. J.
        Bot., 2, 163-169.
Kohlmeyer, J. (1973). Spathulosporales, a new order and possible missing
        link between Laboulbeniales and Pyrenomycetes. Mycologia,
        65, 614-647.
Kohlmeyer, J. & Kohlmeyer, E. (1979). Marine Mycology: The Higher Fungi.
        New York: Academic Press.
Latter, P. & Heal, O.W. (1971). Psychrophiles in Antarctic soil. Soil
        Biol. Biochem., 3, 365-379.
Meyers, S.P. & Reynolds, E.S. (1960). Occurrence of lignicolous fungi in
        northern Atlantic and Pacific marine localities. Can. J.
        Bot., 38, 217-226.
Pugh, G.F.J. & Allsopp, D. (1982). Microfungi on Signy Island, South
        Orkney Islands. Brit. Antarct. Surv. Bull., 57, 55-67.
Smith, R.I.L. (1985). *Nothofagus* and other trees stranded on islands in
        the Atlantic sector of the Southern Ocean. Br. Antarct.
        Surv. Bull., 66, 47-55.
Tubaki, K. (1961). On some fungi isolated from the Antarctic materials.
        Publ. Seto Mar. Biol. Lab., 14, 3-9.
Tubaki, K. & Asano, I. (1965). Additional species of fungi isolated from
        the Antarctic materials. Jap. Sci. Reports, Biology, 27,
        1-12.
Zainal, A. & Jones, E.B.G. (1984). Observations on some lignicolous
        marine fungi from Kuwait. Nova Hedwigia, 39, 569-583.
Zainal, A. & Jones, E.B.G. (1986). Occurrence and distribution of
        lignicolous marine fungi in Kuwait coastal waters. Proc. 6th
        Inter. Biodeterioration Symp. Washington, in press.

K.D. Hyde

E.B.G. Jones

S.T. Moss

INTRODUCTION

The production of mucilaginous hyphal sheaths by fungi is well established (Szanislo *et al*. 1968; Davidson 1973; Palmer *et al*. 1983 a,b), however, the function of these sheaths is speculative and many roles have been suggested. An adhesive role for the sheath has been postulated by several workers (Akai *et al*. 1967; Kozar & Netolitsky 1978; Rees & Jones 1984; Hyde *et al*. 1986) although other functions have also been suggested (Davidson 1973; Palmer *et al*. 1983 a,b; Sutter *et al*. 1984). Hyphal sheaths are common in marine fungi and have been noted in *Leptosphaeria albopunctata* (Westend.) Sacc. (Szanislo *et al*. 1968), *Lulworthia medusa* (Ellis et Everh.) Cribb et Cribb (Davidson 1973), *Pleospora gaudefroyi* Patouillard and *Camarosporium roumeguerii* Sacc. (Crabtree & Gessner 1982), *Halosphaeria appendiculata* Linder and *Ceriosporopsis circumvestita* (Kohlm.) Kohlm. (Hyde *et al*. 1986). In this study 15 species of marine fungi germinating on wood veneers and/or polycarbonate membranes were observed to produce mucilaginous sheaths at the scanning electron microscope (SEM) level and micrographs to illustrate the hyphal sheaths of several of these species are presented. The major function of the sheath is thought to be one of attachment, but more than one role is not ruled out. These aspects are considered and quantitative evidence for mycelial adhesion (Hyde 1985) is discussed.

MATERIALS AND METHODS

Spore suspensions of test fungi were allowed to settle on wood veneers and/or polycarbonate membranes (placed on sea water corn meal agar) until germ tube initiation. The material was then prepared for observation in either a JEOL 35C or JEOL T20 scanning electron microscope by the method previously described by Jones *et al*. (1983).

RESULTS

Mucilage was produced by the germ tubes and young hyphae of 15 species of marine fungi growing on wood veneer and/or polycarbonate membrane surfaces (Table 28.1). The mucilaginous hyphal sheaths are illustrated in Figures 28.1-4. The texture, quantity and appearance of the sheaths differed between species. Little mucilage was produced by *Amylocarpus encephaloides* Currey (Figures 28.1a,b), whereas copious mucilage was produced by *Ceriosporopsis circumvestita* (Figures 28.1e,f), *Halosphaeria appendiculata* (Figures 28.4c,d) and *Leptosphaeria contecta*

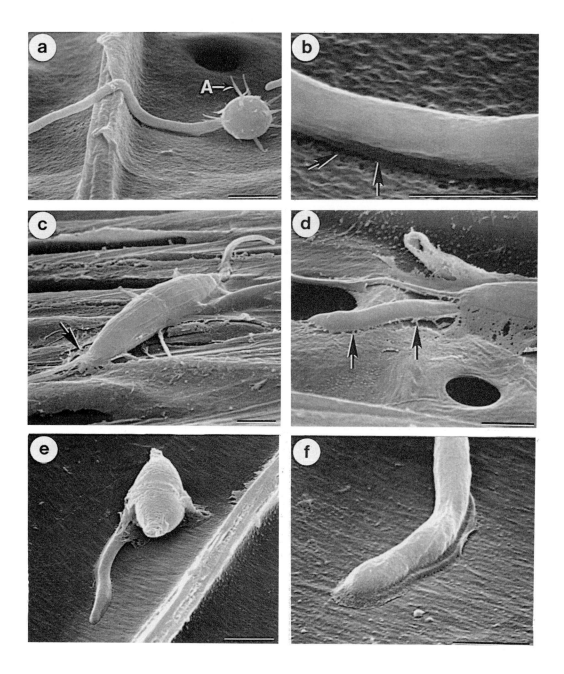

Figures 28.1a–f Scanning electron micrographs of ascospores, developing hyphae and hyphal tips of marine fungi on wood veneers. Bars = 5 μm. Figures 28.1a,b *Amylocarpus encephaloides*, germinated ascospore with spine-like appendages (A). Note the extracellular fibrous mucilage between the hypha and the substratum (arrowed). Figures 28.1c,d *Corollospora maritima*, both spores and germ tubes possess an associated mucilaginous sheath (arrowed).

Kohlm. (Figures 28.2c-d). In *C. circumvestita* (Figures 28.1e,f) and
Kohlmeyeriella tubulata Jones, R.G. Johnson et Moss (Figures 28.2a,b)
the mucilage was in the form of a pad, whereas in *H. appendiculata*
(Figures 28.4c,d), *L. contecta* (Figures 28.2c-e) and *Corollospora*
maritima Werdermann (Figures 28.1c,d) the hyphal sheath was thin and had
flattened over the substrate. The sheath texture was non-fibrous in
C. maritima (Figures 28.1c,d), *H. appendiculata* (Figures 28.4c,d) and
Halosarpheia viscosa (Schmidt) Shearer et Crane (Figure 28.4b), while in
C. circumvestita (Figures 28.1e,f), *K. tubulata* (Figures 28.2a,b) and
Lulworthia sp. (Figures 28.3a-c) the sheath was fibrous. In all the
species studied, however, there were discrete regions of attachment,
attachment points, between the mucilage sheath and the substrate
surface, which indicated an adhesive role.

In *C. maritima* the sheath was observed to enrobe the ascospore as well
as the germ tube and form attachment points with the substrata
(Figures 28.1c,d), whereas in other species mucilage did not appear to
be produced by the ascospore (e.g. *Lulworthia* sp., Figures 28.3a,c). In
K. tubulata the ascospore wall was smooth and devoid of mucilage with a
clear line of demarcation between the ascospore wall and the region of
mucilage production by the germ tube (Figures 28.2a,b). Again the
discrete attachment points between the mucilage sheath of *C. maritima*
and the substrate surface indicate an adhesive role.

DISCUSSION
The production of mucilaginous hyphal sheaths has been
demonstrated in many terrestrial fungi at both scanning (Palmer *et al.*
1983 a,b) and transmission (Wheeler & Gantz 1979; Hau & Rush 1982)
electron microscope levels and in marine fungi (Davidson 1973; Crabtree
& Gessner 1982; Rees & Jones 1984; Hyde *et al.* 1986). Fungi from filter
beds and other processes using microbial films are also known to produce
mucilaginous sheaths (Anderson & Blain 1980).

The functions of the sheaths of fungi are speculative. Akai *et al.*
(1967) and Hau & Rush (1982) concluded that sheaths play a role in the
attachment to substrata, thus aiding initial colonization while Clough
& Sutton (1978) showed that hyphae of the mycorrhizal fungus *Glomus*
adhere to sand grains and cause their aggregation. A role in attachment
is also apparent in the mycelial sheaths of fungi from fermenter
systems. However, attachment of fungi in fermenter systems is a source
of problems as mycelial growth on submerged surfaces may cause fouling
of the stirrer shaft, baffles and pH probes (Anderson & Blain 1980;
Solomons 1980).

A role other than attachment has been postulated by Palmer *et al.*

Figures 28.1e,f *Ceriosporopsis circumvestita*, apices of the germ tubes
are surrounded by extensive sheaths of mucilage which spread over the
substrata. The sheaths are fibrous and attach the germ tubes to the
substrata.

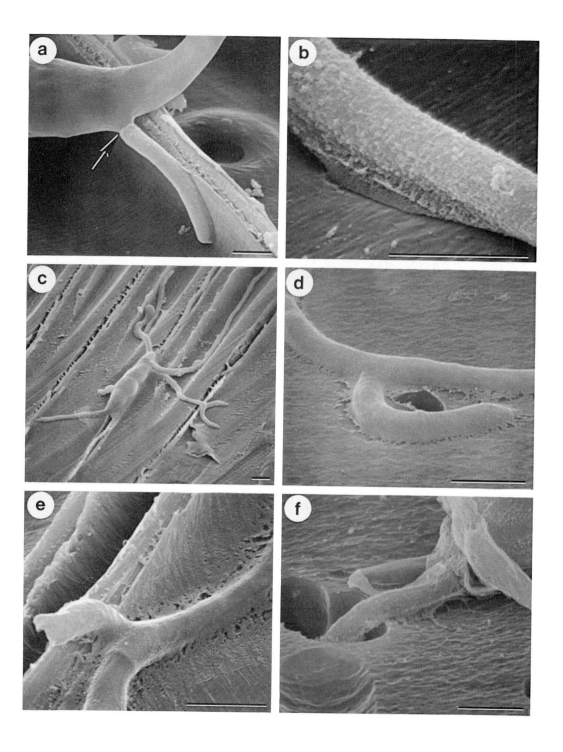

(1983 a,b), who suggested that the sheath may serve as a source of support and nutrition in addition to containing and transporting depolymerizing agents. Davidson (1973) indicated that the sheath of *Lulworthia medusa* modified the hyphal environment, caused adhesion of other organisms, retarded desiccation and provided a matrix to harbour and concentrate enzymes.

The possibility that mucilage production by germ tubes and hyphae of marine fungi was associated with attachment to surfaces was first postulated by Rees & Jones (1984) and later by Hyde *et al.* (1986). These workers observed mucilage production by the germ tubes of 8 species of marine fungi and mucilage mediated attachment was evident in 7 of these species. In the present study 15 species of fungi have shown mucilage mediated attachment, of which 4 species are confirmations of previous results and 11 species are demonstrated for the first time. Sheaths have also been shown on hyphae of the marine fungi *Leptosphaeria albopunctata* (Szaniszlo *et al.* 1968), *Lulworthia medusa* (Davidson 1973), *Pleospora gaudefroyi* and *Camarosporium roumeguerii* (Crabtree & Gessner 1982) and hyphopodia of *Buergenerula spartinae* Kohlm. et Gessner (Onyile *et al.* 1982), and are apparently common. The sheath is thought to adhere the mycelium during germination to aid initial colonization.

Quantitative evidence for mycelial attachment has been demonstrated by Hyde (1985). He settled ascospores of *Groenhiella bivestia* Koch, Jones et Moss on perspex discs attached to a Radial Flow Chamber and established the shear stress needed to remove the spores at time intervals up to 96 hours (Fowler & McKay 1980). His results showed that a greater shear stress was required to remove spores with increased time and that this was accompanied with germ-tube initiation. In species where germ-tube initiation did not occur there was no (*Ceriosporopsis circumvestita*) or little (*Amylocarpus encephaloides*) increase in strength of attachment. These results indicate that adhesion of *G. bivestia* was due to mucilage production by the germ tube (Figure 28.4a).

Figures 28.2a-f Scanning electron micrographs of ascospores, developing hyphae and hyphal tips of marine fungi on wood veneers (Figs 28.2a,c,e) or polycarbonate membranes (Figs 28.2d,f). Bars = 5 µm.
Figures 28.2a,b *Kohlmeyeriella tubulata*, germ tubes attached to the substrata by discrete pads of mucilage. Note the zone of demarcation between the mucilage produced by the germ tube and the ascospore wall (arrowed) which lacks mucilage.
Figures 28.2c-e *Leptosphaeria contecta*, a mucilaginous sheath surrounds the ascospore (Fig. 28.2c) and is continuous with the mucilaginous sheaths formed by the germ tubes. Germ tubes and germ tube tips are coated with fibrous mucilaginous material which forms attachment points with the surface.
Figure 28.2f *Halosphaeria salina*, germ tube coated with fibrous mucilaginous material which forms attachment points with the substratum.

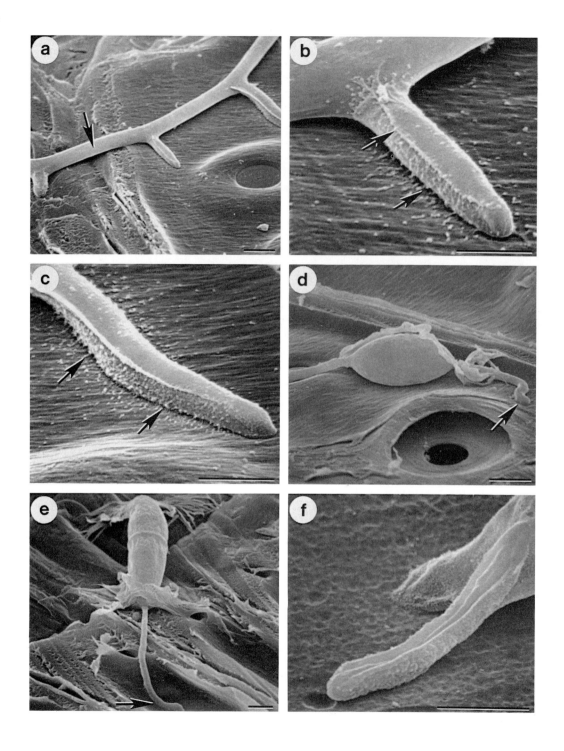

It is concluded that mucilaginous sheaths may have more than one
function and in marine fungi one of the functions is attachment.
Sheaths may also retard desiccation, supply nitrogen to the growing
hyphal tips, detoxify copper and other metallic ions, modify the hyphal
environment, cause adhesion of other organisms, provide a matrix for
harbouring, transporting and concentrating enzymes and may serve as a
source of nutrition and support (Davidson 1973; Palmer *et al.* 1983 a,b;
Rees & Jones 1984; Sutter *et al.* 1984).

REFERENCES
Akai, S., Fukutomi, M., Ishida, N. & Kunoh, H. (1967). An anatomical
         approach to the mechanism of fungal infections in plants. In
         The Dynamic Role of Molecular Constituents in Plant Parasite
         Interaction, eds. C.J. Mirocha & I. Uritani, pp. 1-18.
         St. Paul, Minnesota: American Phytopathological Society.
Anderson, J.G. & Blain, J.A. (1980). Novel developments in microbial
         film reactors. In Fungal Biotechnology, eds. J.E. Smith,
         D.R. Berry & B. Kristiansen, pp. 125-152. London: Academic
         Press.
Clough, K.S. & Sutton, J.C. (1978). Direct observation of fungal
         aggregates in sand dune soil. Can. J. Microbiol., 24,
         333-335.
Crabtree, S.L. & Gessner, R.V. (1982). Growth and nutrition of the
         salt-marsh fungi *Pleospora gaudefroyi* and *Camarosporium
         roumeguerii*. Mycologia, 74, 640-647.
Davidson, D.E. (1973). Mucoid sheath of *Lulworthia medusa*. Trans. Brit.
         mycol. Soc., 60, 577-601.
Fowler, H.W. & McKay, A.J. (1980). The measurement of microbial
         adhesion. In Microbial Adhesion to surfaces, eds.
         R.C.W. Berkeley, J.M. Lynch, J. Melling, P.R. Butler &

Figures 28.3a-f Scanning electron micrographs of ascospores,
developing hyphae and hyphal tips of marine fungi on wood veneers
(Figs 28.3a-e) or a polycarbonate membrane (Fig. 28.3f). Bars = 5 µm.
Figures 28.3a-c *Lulworthia* sp., the wall of the ascospore lacks
mucilage (Fig. 28.3a, arrowed), unlike the developing germ tube, which
is coated with fibrous material (Fig. 28.3b, arrowed). The mucilage is
restricted to the surface of the germ tube in contact with the veneer
and forms attachment points (Fig. 28.3c, arrowed).
Figure 28.3d *Haligena* sp. (?), mucilage is present as a discrete
pad-like layer between the germ tube tip and the veneer (arrowed).
Figure 28.3e *Remispora stellata*, during germination ascospore is attached
to the veneer by the polar appendages. Following germination mucilage
in the form of a pad-like layer is formed between the hypha and the
veneer (arrowed).
Figure 28.3f *Nimbospora bipolaris*, the surface of the germ tube in
contact with the membrane is coated with a layer of fibrous material,
whereas the upper surface lacks this layer. The polar appendage is
mucilaginous and attaches the ascospore to the polycarbonate membrane
during germination.

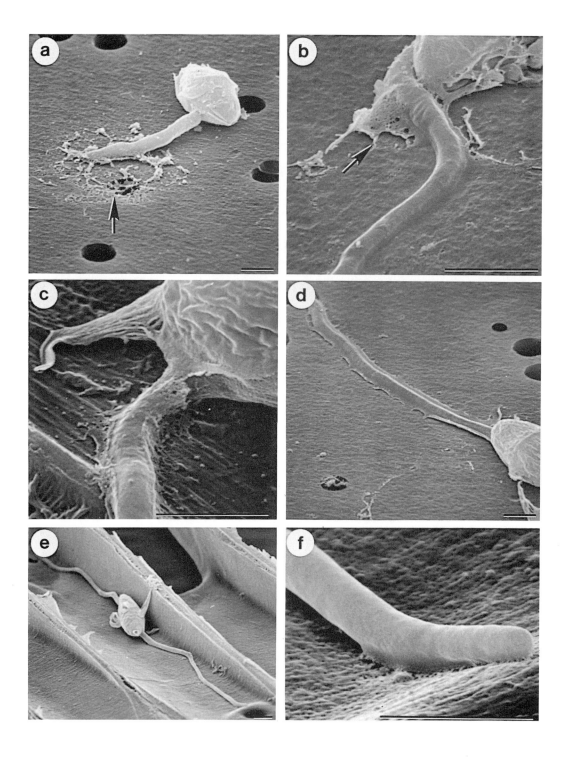

Table 28.1 Observations on mucilage production by germ
tubes and hyphae of marine Ascomycotina

Fungus	Quantity	Texture	Form
Amylocarpus encephaloides Currey	Small	Fibrous	Sheath?
Ceriosporopsis circumvestita (Kohlm.) Kohlm.	Copious	Fibrous	Pad
Corollospora maritima Werdermann	Copious	Non-fibrous	Sheath
Groenhiella bivestia Koch, Jones et Moss	Copious	Non-fibrous?	Sheath
Halosarpheia viscosa (Schmidt) Shearer et Crane	Copious	Non-fibrous	Sheath
Halosphaeria appendiculata Linder	Copious	Fibrous/non-fibrous	Sheath
Halosphaeria salina (Meyers) Kohlm.	Medium	Fibrous	Sheath
Halosphaeriopsis mediosetigera (Cribb et Cribb) T.W. Johnson	Medium	Non-fibrous	Pad
Kohlmeyeriella tubulata Jones, R.G. Johnson et Moss	Copious	Fibrous	Pad
Leptosphaeria contecta Kohlm.	Copious	Fibrous	Sheath
Lulworthia sp.	Copious?	Fibrous	Sheath
Nimbospora bipolaris Hyde et Jones	Medium	Fibrous	Pad?
Remispora stellata Kohlm.	Medium	Non-fibrous	Pad
Remispora sp.	Medium	Non-fibrous?	Pad
Torpedospora ambispinosa Kohlm.	Medium	Non-fibrous	Pad

Figures 28.4a-f Scanning electron micrographs of ascospores,
developing hyphae and hyphal tips of marine fungi on wood veneers
(Figs 28.4c,e,f) or polycarbonate membranes (Figs 28.4a,b,d).
Bars = 5 μm.
Figure 28.4a *Groenhiella bivestia*, germ tube is attached to the
polycarbonate membrane by mucilaginous material (arrowed).
Figure 28.4b *Halosarpheia viscosa*, a sheath of mucilaginous material is
visible on the polycarbonate membrane (arrowed) and is continuous with
the germ tube.
Figures 28.4c,d *Halosphaeria appendiculata*, germ tubes are attached to
the substratum by extensive mucilaginous sheaths.
Figures 28.4e,f *Halosphaeriopsis mediosetigera*, germinated ascospore
showing sub-polar origin of germ tubes (Fig. 28.4e). Note the tip of
the germ tube which is attached to the wood veneers by non-fibrous
mucilaginous material.

B. Vincent, pp. 143-161. Chichester: Ellis Horwood Ltd.

Hau, F.C. & Rush, M.C. (1982). Preinfectional interactions between *Helminthosporium oryzae* and resistant and susceptible rice plants. Phytopathology, 72, 285-292.

Hyde, K.D., Jones, E.B.G. & Moss, S.T. (1986). How do fungal spores attach to surfaces? In Proceedings 6th International Biodeterioration Symposium, in press.

Hyde, K.D. (1985). Spore settlement and attachment in marine fungi. Ph.D. thesis, C.N.A.A., Portsmouth Polytechnic, U.K.

Jones, E.B.G., Johnson, R.G. & Moss, S.T. (1983). Taxonomic studies of the Halosphaeriaceae: *Corollospora* Werdermann. J. Linn. Soc. (Bot.), 87, 193-212.

Kozar, F. & Netolitsky, H.J. (1978). Studies on hyphal development and appressorium formation of *Colletotrichum graminicola*. Can. J. Bot., 56, 2234-2242.

Onyle, A.B., Edwards, H.H. & Gessner, R.V. (1982). Adhesive material of the hyphopodia of *Buergenerula spartinae*. Mycologia, 74, 777-784.

Palmer, J.G., Murmanis, L. & Highley, T.L. (1983 a). Visualisation of hyphal sheaths in wood-decay Hymenomycetes. I. Brown rotters. Mycologia, 75, 995-1004.

Palmer, J.G., Murmanis, L. & Highley, T.L. (1983 b). Visualisation of hyphal sheaths in wood-decay Hymenomycetes. II. White rotters. Mycologia, 75, 1005-1010.

Rees, G. & Jones, E.B.G. (1984). Observations on the attachment of spores of marine fungi. Botanica mar., 27, 145-160.

Solomons, G.L. (1980). Fermenter design and fungal growth. In Fungal Biotechnology, eds. J.E. Smith, D.R. Berry & B. Kristiansen, pp. 55-80. London: Academic Press.

Sutter, P.H., Jones, E.B.G. & Wälchi, O. (1984). Occurrence of crystalline hyphal sheaths in *Poria placenta* (Fr.) Cke. J. Inst. Wood Science, 10, 19-23.

Szaniszlo, P.J., Wirsen, C. & Mitchell, R. (1968). Production of a capsular polysaccharide by a marine fungus. J. Bact., 96, 1474-1483.

Wheeler, H. & Gantz, D. (1979). Extracellular sheaths on hyphae of two species of *Helminthosporium*. Mycologia, 71, 1127-1135.

R. Mouzouras

INTRODUCTION

The degradative activities of marine fungi were reported by Barghoorn & Linder (1944) in their pioneer work on marine lignicolous fungi. They demonstrated typical soft rot bores in the middle layers of secondary walls of tracheids and vessels. Subsequently, several workers have studied the activities of marine fungi on cellulosic substrates (Kohlmeyer 1958 a,b; Meyers & Reynolds 1959 a,b, 1960; Meyers 1968). Soft rot activity and weight loss of wood samples caused by marine fungi under laboratory conditions were demonstrated by Kohlmeyer (1958 a), Jones (1971) and Eaton & Jones (1971). Henningsson (1976) reported fourteen fungi from the Baltic which were able to cause soft rot and a considerable loss in dry mass of hardwoods. Similarly, Curran (1979) reported on soft rot associated weight loss of wood blocks using two methods. Leightley (1980) showed that twenty five marine fungi caused soft rot of wood and one caused white rot. More recently, Hale (1983) found *Cirrenalia macrocephala* (Kohlm.) Meyers et Moore also to cause soft rot of wood in marine environments.

Jones (1976) listed 132 higher marine fungi on wood of which only three were species of the Basidiomycotina. Although Leightley & Eaton (1979) showed that the marine basidiomycete *Nia vibrissa* caused white rot of wood, the remaining two marine species, *Digitatispora marina* and *Halocyphina villosa*, have not previously been tested and this chapter concentrates on their wood decay abilities. Weight loss experiments were carried out with all three marine species of the Basidiomycotina and the deuteromycete *Monodictys pelagica* (Johnson) Jones. In addition numerous species of marine Ascomycotina were screened for their abilities to cause soft rot decay of wood.

MATERIALS AND METHODS

The ability of four marine fungal species to degrade wood was assessed by the use of weight loss tests. Three of these, *D. marina*, *N. vibrissa* and *M. pelagica*, were isolated from timbers of the Tudor ship Mary Rose. The fourth, *H. villosa*, was isolated from mangrove wood of the Seychelles. Species used for screening of soft rot cavity attack were isolated from wood exposed in various marine habitats (Table 29.1) and are part of the Portsmouth Polytechnic Culture Collection.

Table 29.1 Marine fungi tested for soft rot decay

Species	Cavity attack	Penetration attack	Source of Isolate
Aniptodera sp.	–	+	Seychelles
Ceriosporopsis circumvestita Kohlm.	+	+	Friday Harbor, USA
Groenhiella bivestia Koch, Jones et Moss	+	+	Denmark
Halosarpheia fibrosa Kohlm. et Kohlm.	–	+	Seychelles
H. retorquens Shearer et Crane	+	+	Friday Harbor, USA
H. spartinae (Jones) Shearer et Crane	+	+	Seychelles
Halosphaeria appendiculata Linder	+	+	Farlington, UK
Kohlmeyeriella tubulata (Kohlm.) Jones, Johnson et Moss	+	+	Denmark
Lanspora coronata Hyde et Jones	+	+	Seychelles
Leptosphaeria sp.	–	+	Seychelles
Lulworthia sp. 1	+	+	Seychelles
Lulworthia sp. 2	–	+	Thailand
Marinospora longissima (Kohlm.) Cavaliere	+	+	Denmark
Nautosphaeria cristaminuta Jones	–	+	Poole, UK
Nereiospora comata (Kohlm.) Jones, Johnson et Moss	+	+	Denmark
Nimbospora effusa Koch	+	+	Seychelles
Orcadia sp.	–	+	Seychelles
Pleospora sp. 1	–	+	Seychelles
Pleospora sp. 2	–	+	Seychelles
Remispora galerita Tubaki	+	+	Seychelles
R. maritima Linder	+	+	Friday Harbor, USA
R. pilleata Kohlm.	+	+	Denmark
R. stellata Kohlm.	+	+	Friday Harbor, USA
Savoryella lignicola Jones et Eaton	+	+	Seychelles

Weight loss tests were carried out using the medium described by Jones
(1971) except that seven 1 cm^3 blocks per wood species (*Ochroma
lagopus* Sw., *Pinus sylvestris* L. and *Fagus sylvatica* L.) and 100 cm^3 of
the medium were used. The flasks which contained the medium and wood
blocks were inoculated with the fungus mycelium and incubated on an
orbital shaker (96 rpm) at 10°C and 20°C, for 8, 16 and 24 weeks. Loss
in dry weight was obtained according to Henningsson (1977) except that
following incubation, wood blocks were leached with warm water (*ca* 60°C)
for 48 hours, to remove any salts absorbed from the culture medium. The
percentage loss in dry weight was estimated using five of the blocks
from each flask, the remaining two blocks were used for microscopic
examination.

Fungi screened for soft rot decay were cultivated using the balsa shake
broth method described by Hyde (1985) to promote sporulation by marine
fungi in axenic culture.

Specimens for scanning electron microscopy (SEM) were fixed with 4%
glutaraldehyde in 0.2M (pH 7.2) phosphate buffer for a minimum of
4 hours. Thereafter the procedure of Moss & Jones (1977) was followed
and specimens were examined in a JEOL T20 SEM at 20kV.

Cellulase and xylanase production by *D. marina* and *H. villosa* were
assayed using the method of Rautela & Cowling (1966), with the following
substrates added: (i) 0.125% Walseth cellulose prepared from Whatman
CF11 cellulose powder according to the procedure of Walseth (1952);
(ii) 0.125% Whatman CF1 cellulose powder; (iii) 0.125% larch xylan
(Sigma).

The ability of *D. marina* and *H. villosa* to degrade lignin was
investigated by the use of six phenolic compounds (Table 29.4)
following the method described by Käärik (1965). Tests were carried out
on cultures grown for 22 days at 22°C on both sea water cornmeal agar
and sea water 1.5% malt extract agar. Colonies of both fungi grew
submerged within the medium and to enhance diffusion of reagents, wells
were cut around the colony using a 6mm cork borer. White rot fungi were
distinguished from brown rot fungi using the method of Bavendamm (1928)
modified to be used with marine fungi by Leightley & Eaton (1979).

RESULTS AND DISCUSSION

Table 29.1 lists the fungi screened for soft rot decay.
Sixteen species were shown to cause soft rot cavity attack. The other
eight species showed penetration of cell walls but neither cavity nor
what Nilsson (1973) termed "erosion attack" was observed. Only four
of the species listed in Table 29.1 have been previously tested.
Leightley (1980) showed cavity attack in *Halosphaeria appendiculata*,
Nereiospora comata and *Savoryella lignicola* and this is confirmed in
the present study whereas Jones (1962, 1971) demonstrated cavity attack
for *Nautosphaeria cristaminuta* and this is not confirmed here.
Previously, 26 marine fungi have been shown to cause cavity attack
(Leightley 1980; Hale 1983) and this study increases the number to 42.

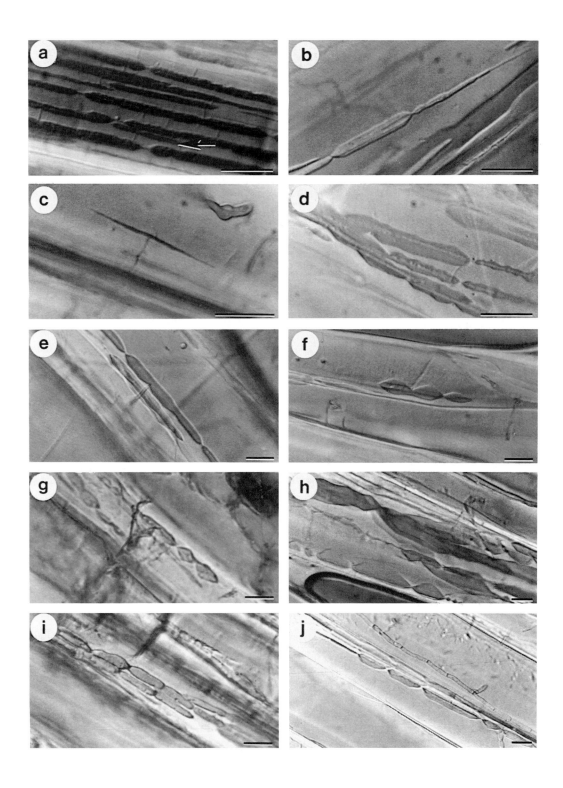

This indicates the importance of these fungi in degrading submerged
wood in the marine environment. Soft rot fungi may be better adapted
for this habitat, since they preferentially degrade cellulose through a
nonoxygen demanding process (Levi & Preston 1965).

During studies with soft rot fungi it was noted that the shape and size
of the cavity varied and this is illustrated in Figures 29.1a-j.
Figures 29.1a-d illustrate the cavities formed by *Monodictys pelagica*.
Typical cylindrical cavities with biconical ends (Figure 29.1a),
associated satellite cavities and cavities with proboscis hyphae
(Figure 29.1b) were common. Unusual cavities with a serrated appearance
were also observed (Figure 29.1d) associated with *M. pelagica* growing on
the same wood sample. A similar dimorphism of cavities formed by
Halosphaeria appendiculata is illustrated by Figures 29.1e,f in which
typical cylindrical and diamond shaped cavities are illustrated.
Diamond shaped cavities were also observed associated with hyphae of
Lanspora coronata (Figure 29.1g), irregular cavities with *Marinospora
longissima* (Figure 29.1h) and cylindrical cavities without biconical
ends associated with *Savoryella lignincola* (Figure 29.1i). The
successive cavities (Figure 29.1j) formed by *Groenhiella bivestia* were
also observed for all the species studied.

Observations from this study suggest that a particular fungus does not
necessarily produce a characteristic cavity shape. Such variation in
cavity shape has been documented in the past and the most probable
explanation is that shape is regulated by the cellulose structure
(Nilsson 1976). However, this does not explain the different cavity
shapes observed within close proximity of each other. Recently, Hale
(1983) showed similar observations to those made in this study and
discusses in detail the micromorphological mechanisms of soft rot
cavity formation.

The decay capability of *Monodictys pelagica*, *Nia vibrissa*, *Digitatispora
marina* and *Halocyphina villosa* was assessed using weight loss tests and
Table 29.2 presents the results obtained. Values for the four organisms
were adjusted to account for weight loss in the controls. It is
evident that *M. pelagica* is the most active degrader, since weight
losses of 41% were obtained after 24 weeks at 22° C. Activity is

Figures 29.1a-j Polarized light micrographs of soft rot cavities in
O. lagopus. Bars = 5 µm.
Figures 29.1a-d *Monodictys pelagica*; Fig. a, cylindrical cavities with
biconical ends and satellite cavities (arrow); Fig. b, proboscis hypha;
Fig. c, T-branch; Fig. d, serrated cavities.
Figures 29.1e,f *Halosphaeria appendiculata*; Fig. e, typical cylindrical
cavities; Fig. f, diamond shaped cavities.
Figure 29.1g *Lanspora coronata*; diamond shaped cavities.
Figure 29.1h *Marinospora longissima*; irregular cavities.
Figure 29.1i *Savoryella lignicola*; cylindrical cavities with rounded
ends.
Figure 29.1j *Groenhiella bivestia*; cavities in succession.

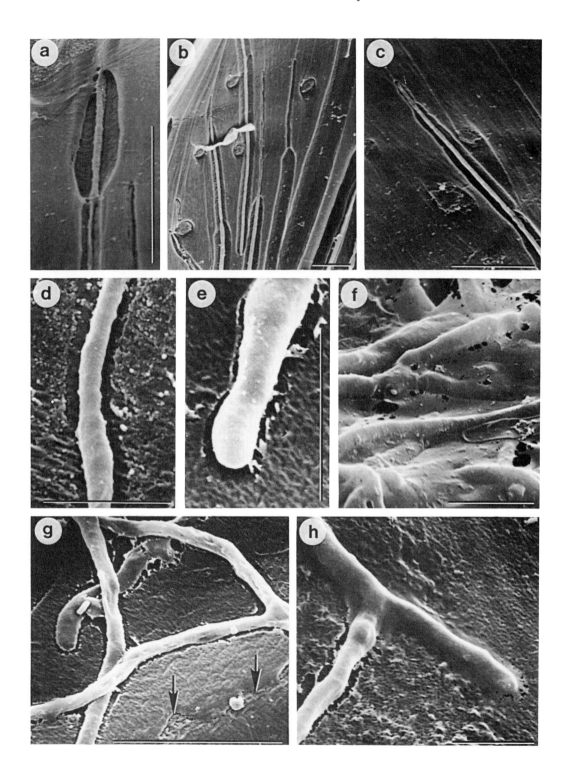

present in significant levels at 10° C, which is not surprising since
M. pelagica is widely distributed, from temperate regions (Jones 1968)
to the extremes of the Sub Antarctic (Pugh & Jones Chapter 27) and the
Seychelles (Hyde Chapter 26). Figures 29.2a-c illustrate the severity
of soft rot attack, by this species, after eight weeks incubation.

N. vibrissa showed considerable weight loss on blocks of *O. lagopus* at
all incubation periods but significant weight loss for *F. sylvatica* was
only recorded after 24 weeks. Leightley & Eaton (1979) failed to show
weight loss of *P. sylvestris* and *F. sylvatica* sapwood with this species

Table 29.2 Mean percentage loss in dry weight for *Ochroma lagopus*,
Pinus sylvestris and *Fagus sylvatica*

Fungus species	Timber species	% loss in dry weight*					
		8 weeks		16 weeks		24 weeks	
		10°C	22°C	10°C	22°C	10°C	22°C
CONTROLS	*O. lagopus*	2.28	5.33	5.05	4.92	3.62	4.83
	P. sylvestris	2.28	3.68	3.24	2.25	3.17	3.17
	F. sylvatica	2.04	2.60	2.81	2.70	0.58	2.04
Monodictys pelagica (Johnson) Jones	*O. lagopus*	11.30	12.74	18.32	24.24	28.30	40.87
	P. sylvestris	1.33	-0.52	-0.02	1.72	0.39	-0.68
	F. sylvatica	2.13	8.90	9.72	16.12	11.03	20.76
Nia vibrissa Moore et Meyers	*O. lagopus*	2.27	9.64	6.48	20.34	13.75	27.91
	P. sylvestris	0.53	-0.60	-0.40	1.65	-1.26	0.08
	F. sylvatica	0.13	-0.06	0.46	2.25	4.23	5.55
Digitatispora marina Doguet	*O. lagopus*	8.06	1.06	10.12	3.88	14.33	4.94
	P. sylvestris	0.60	-0.14	-0.09	0.36	-0.33	0.57
	F. sylvatica	0.07	-0.02	1.97	0.13	10.07	2.68
Halocyphina villosa Kohlm. et Kohlm.	*O. lagopus*	1.65	5.39	-0.84	17.58	-0.58	22.98
	P. sylvestris	0.26	-1.30	0.13	0.48	-0.59	-0.09
	F. sylvatica	-0.11	2.36	1.20	3.62	3.43	7.98

* adjusted to account for loss in dry weight of controls

Figures 29.2a-c Scanning electron micrographs of *Monodictys pelagica*
after eight weeks on *Ochroma lagopus*; Fig. a, cavity and hypha breaking
through the S3 layer; Figs b,c, proboscis hyphae. Bars = 5 μm.
Figures 29.2d-h Scanning electron micrographs of *Halocyphina villosa*
hyphae on *Ochroma lagopus* showing; Fig. d, erosion round hyphae;
Fig. e, erosion around hyphal tip; Fig. f, hyphae covered by mucilage;
Fig. g, hyphae with mucilage layer on the cell wall surface (arrows);
Fig. h, hyphae coated with mucilage. Bars = 5 μm.

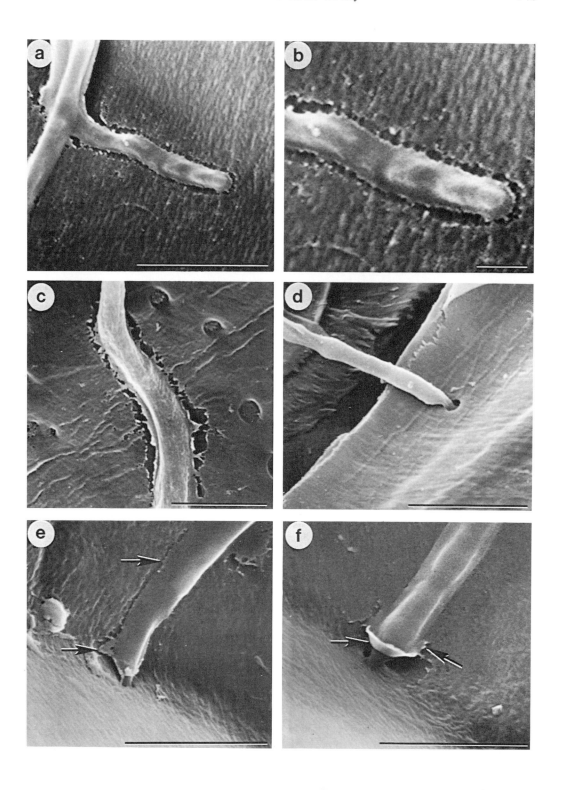

after 12 weeks and this is confirmed here, in that weight loss for
F. sylvatica was only recorded at 24 weeks, with no loss in
P. sylvestris. *D. marina* prefers low temperatures since significant
weight losses for *O. lagopus* were recorded at 10° C at all incubation
periods. In contrast to this, the tropical species *H. villosa* caused
the greatest weight loss at 22° C. The three species of Basidiomycotina
tested were only able to show significant weight losses on *F. sylvatica*
after 24 weeks, showing that *M. pelagica* is most active since it shows
weight loss on *F. sylvatica* throughout the test period. However, no
species showed significant weight loss for *P. sylvestris* which is in
agreement with previous reports (Eaton & Jones 1971; Nilsson 1973;
Henningsson 1976).

Examination of *H. villosa* and *D. marina* material with the SEM
revealed that, after 16 weeks incubation, degradation zones were found
to be present, forming erosion troughs around the hyphae of *H. villosa*
(Figures 29.2d,e) and *D. marina* (Figures 29.3a,b). Such discrete
erosion channels are considered to be diagnostic for white rot fungi
(Bravery 1971). Mucilagenous material was observed to be associated
with hyphae, especially those of *H. villosa* (Figures 29.2f-h). Large
amounts of slime were also recorded to be present during removal of
wood blocks from the culture flasks, after 16 and 24 weeks incubation.
Figure 29.2g illustrates the mucilage layer deposited over the wood cell
wall. A less conspicuous layer of mucilage was also found associated
with the hyphae of *D. marina* (Figure 29.3c). Similar observations were
made by Leightley & Eaton (1979) during their studies on *N. vibrissa*.
Bore holes were also observed for both species (Figures 29.3d-f) and the
involvement of mucilage was also noted for penetration hyphae
(Figures 29.3e,f). Crossley (1979) carried out ultrastructural studies
of mucilage in association with the hyphae of wood rotting fungi and
proposed mechanisms for cell wall lysis. A number of functions have
been attributed to slime sheaths (Palmer *et al.* 1983 a,b), e.g. to
prevent the hyphae and cell wall from evaporation and desiccation,
protect the fungus from chemical and enzymatic influences (Foisner
et al. 1984), for transportation of lytic enzymes (Green *et al.* 1980)
and the immobilization of heavy metals (Sutter *et al.* 1984).

Wood decay enzyme assays support the morphological observations
recorded for *D. marina* and *H. villosa*. Using the Rautela & Cowling
(1966) assay method production of cellulase and xylanase were detected
as shown in Table 29.3. Lignolytic enzymes were detected using the
method of Käärik (1965) and results for laccase and tyrosinase are
listed in Table 29.4. It is evident that laccase was present in both
species but tyrosinase was only detected in *D. marina*. Results obtained

Figures 29.3a-f Scanning electron micrographs of *Digitatispora marina*
on *Ochroma lagopus* showing; Figs a,b, hyphal branch causing erosion
across cellulose microfibrils; Fig. c, hypha and associated mucilage
layer, the discontinuity of mucilage around the hypha is a preparation
artefact; Fig. d, bore hole; Figs e,f, penetration hyphae with
mucilaginous sheaths (arrows). Figures 29.3a,c-f, bars = 5 μm;
Figure 29.3b, bar = 1 μm.

Table 29.3 Detection of cellulase and xylanase production by clearance
of agar columns by *D. marina* and *H. villosa*

Species	Temperature (° C)	Time (days)	Depth of clearance (mm)		
			Walseth	CFI*	Xylan
D. marina	10	14	0	3.5	0
		27	2	10	3
		40	4	28	6
	22	14	0	6	0
		27	3	19	4
		40	6	27	5
H. villosa	22	14	0	4	0
		27	1	18	3
		40	2	24	4

* Whatman cellulose powder (long fibres).

Table 29.4 Assay results for laccase and tyrosinase production by
D. marina and *H. villosa* after growth for 24 hours on sea water cornmeal
agar

Reagent	Colour change	Laccase reaction	Tyrosinase reaction	*D. marina*	*H. villosa*
1-napthol	violet to black	+	−	±	+
pyrocatechol	brown	+	+	+	+
p-cresol	orange red to brown	−	+	±	−
tannic acid	yellowish brown	+	+	+	+
gallic acid	yellowish brown	+	+	+	+
L-Tyrosine	brown to dark brown	−	+	−	−

+ = positive reaction.
− = negative reaction.
± = weakly positive reaction.

using sea water malt extract agar as the growth medium gave positive results for only three, i.e. 1-napthol, gallic acid, tannic acid, of the six phenolic compounds tested. However by the use of sea water cornmeal agar the results were more conclusive (Table 29.4). Confirmation of the presence of phenol oxidases using the method of Bavendamm (1928) was not successful, even though positive reactions were obtained for gallic and tannic acids when used according to Käärik (1965) (Table 29.4). However, the two fungi failed to grow on the Bavendamm medium.

The studies presented in this chapter show that a wide range of marine fungi can degrade wood via two types of decay processes, soft rot and white rot. Such organisms undoubtedly play an important role in the degradation of submerged wood and thus recycling of nutrients in the marine environment.

ACKNOWLEDGEMENTS
    I wish to thank the S.E.R.C. and the Mary Rose Trust for financial support and Prof. E.B.G. Jones for his invaluable advice and encouragement.

REFERENCES
Barghoorn, E.S. & Linder, D.H. (1944). Marine fungi: their taxonomy and
        biology. Farlowia, 1, 395–467.
Bavendamm, W. (1928). Uber das vorkommen und den Nachweis von oxydasen
        bei holzzerstörenden Pilzen. Z. Pflkrankh. PflPath.
        PflSchutz., 38, 257–276.
Bravery, A.F. (1971). The application of scanning electron microscopy
        in the study of timber decay. J. Inst. Wood Science, 5,
        13–19.
Curran, P.M.T. (1979). Degradation of wood by marine and non-marine
        fungi from Irish coastal waters. J. Inst. Wood Science, 8,
        114–120.
Crossley, A. (1979). The use of electron microscopy to compare wood
        decay mechanisms. Ph.D. thesis, Imperial College of Science
        and Technology, U.K.
Eaton, R.A. & Jones, E.B.G. (1971). The biodeterioration of timber in
        water cooling towers. I. Fungal ecology and the decay of
        wood at Connah's Quay and Ince. Mat. Org., 6, 51–80.
Foisner, R., Messner, K., Stachelberger, H. & Röhr, M. (1984). Electron
        microscopic detection and chemical analysis of three-
        lamellar structures in wood-destroying fungi. Int. Res.
        Group on Wood Pres., Doc. 1240.
Green, N.B., Dickinson, D.J. & Levy, J.F. (1980). A biochemical
        explanation for the observed patterns of fungal decay in
        timber. Int. Res. Group on Wood Pres., Doc. 1111.
Hale, M.D.C. (1983). The mechanisms of soft rot cavity formation in
        wood. Ph.D. thesis, Portsmouth Polytechnic, U.K.
Henningsson, B. (1977). Methods for determining fungal biodeterioration
        in wood and wood products. In Biodeterioration Investigation
        Techniques, ed. A.H. Walters, pp. 277–294. London: Applied
        Science.

Henningsson, M. (1976). Degradation of wood by some fungi from the
        Baltic and the west coast of Sweden. Mat. Org., 3, 509-519.

Hyde, K.D. (1985). Spore settlement and attachment in marine fungi.
        Ph.D. thesis, Portsmouth Polytechnic, U.K.

Jones, E.B.G. (1962). Marine fungi I. Trans. Br. mycol. Soc., 45,
        93-114.

Jones, E.B.G. (1968). The distribution of marine fungi on wood submerged
        in the sea. In Biodeterioration of Materials: Proc. 1st Int.
        Biodet. Symp., Southampton, eds. A.H. Walters &
        J.S. Elphick, pp. 460-485. London: Elsevier.

Jones, E.B.G. (1971). The ecology and rotting ability of marine fungi.
        In Marine Borers, Fungi and Fouling Organisms of wood, eds.
        E.B.G. Jones & S.K. Eltringham, pp. 237-258. Paris: OECD.

Jones, E.B.G. (1976). Lignicolous and algicolous fungi. In Recent
        Advances in Aquatic Mycology, ed. E.B. Gareth Jones,
        pp. 1-49. London: Elek Science.

Kaärik, A. (1965). The identification of the mycelia of wood decay fungi
        by their oxidation reactions with phenolic compounds. Studia
        Forestalia Suecica, No. 31.

Kohlmeyer, J. (1958 a). Holzzerstörende Pilze im meerwasser. Holz Roh-u.
        Werkstoff, 16, 215-220.

Kohlmeyer, J. (1958 b). Beobachtungen über mediterrane meerespilze
        sowie das Vorkommen von marinen moderfaule-Errengen in
        Aquariumszuchten holzzerstörender Meerestiere. Ber. Dt. bot.
        Ges., 71, 98-116.

Leightley, L.E. (1980). Wood decay activities of marine fungi. Botanica
        mar., 23, 387-395.

Leightley, L.E. & Eaton, R.A. (1979). Nia vibrissa - a marine white rot
        fungus. Trans. Br. mycol. Soc., 73, 35-40.

Levi, M.P. & Preston, R.D. (1965). A chemical and microscopic
        examination of the action of the soft rot fungus Chaetomium
        globosum on beech wood. Holzforschung, 19, 183-190.

Meyers, S.P. (1968). Degradative activities of filamentous marine fungi.
        In Biodeterioration of Materials: Proc. 1st Int. Biodet.
        Symp., Southampton, eds. A.H. Walters & J.S. Elphick,
        pp. 594-609. London: Elsevier.

Meyers, S.P. & Reynolds, E.S. (1959 a). Growth and cellulolytic activity
        of lignicolous deuteromycetes from marine localities. Can.
        J. Microbiol., 5, 493-503.

Meyers, S.P. & Reynolds, E.S. (1959 b). Cellulolytic activity in
        lignicolous marine ascomycetes. Bull. mar. Sci. Gulf and
        Caribbean, 9, 441-455.

Meyers, S.P. & Reynolds, E.S. (1960). Occurrence of lignicolous fungi
        in Northern Atlantic and Pacific marine localities. Can. J.
        Bot., 38, 217-226.

Moss, S.T. & Jones, E.B.G. (1977). Ascospore appendages of marine
        Ascomycetes: Halosphaeria mediosetigera. Trans. Br. mycol.
        Soc., 69, 313-315.

Nilsson, T. (1973). Studies on wood degradation and cellulolytic
        activity of microfungi. Studia Forestalia Suecica, No. 104.

Nilsson, T. (1976). Soft rot fungi - decay patterns and enzyme
        production. Mat. Org., 3, 103-112.

Palmer, J.G., Murmanis, L.L. & Highley, T.L. (1983 a). Visualisation of hyphal sheath in wood decay Hymenomycetes. I. Brown rotters. Mycologia, 75, 995-1004.

Palmer, J.G., Murmanis, L.L. & Highley, T.L. (1983 b). Visualisation of hyphal sheath in wood decay Hymenomycetes. II. White rotters. Mycologia, 75, 1005-1010.

Rautela, G.S. & Cowling, E.B. (1966). Simple cultural test for relative cellulolytic activity of fungi. Appl. Microbiol., 14, 892-898.

Sutter, H.P., Jones, E.B.G. & Wälchli, O. (1984). Occurrence of crystalline sheaths in Poria placenta (Fr.) Cke. J. Inst. Wood Science, 10, 19-23.

Walseth, C.S. (1952). Occurrence of cellulases in enzyme preparations from microorganisms. Tappi, 35, 228-233.

R.A. Eaton

INTRODUCTION
The protection of wood in the sea is primarily dependent on impregnation with toxic wood preservatives. The fixation and permanence of preservatives are essential prerequisites for the effective performance of chemical treatment in submerged marine wooden structures. Preservative treatment is used to protect wooden boats, exposed piling, planking, decking, braces and fenders and has also been applied to log rafts in transit to sawmills. The natural durability of certain timber species enhances resistance to degradation, and the heartwood of many tropical species is particularly resistant to attack. High natural durability is sometimes associated with high silica content in certain species but extractives, e.g. alkaloids, which are present in some species also impart greater resistance. However, the heartwood of less durable timbers can be difficult to treat with preservatives and satisfactory retention levels may be difficult to achieve.

The most important agents of biodeterioration of wood in the sea are the marine wood-boring animals. The speed and severity of attack by these organisms varies with wood species, water temperature and local conditions, the distribution of wood borer species and preservative treatment of the wood. Additionally, some wood-boring animals are tolerant to preservatives used in marine timbers. Microbial degradation of timbers in the sea is superficial and the major agents of decay are the soft rot fungi of the Ascomycotina and Deuteromycotina. Three lignolytic marine Basidiomycotina are also recognized whilst terrestrial species cause decay of marine piling above the waterline and decay of wood in small boats. Lignolytic bacteria have been found to attack the surface layers of wood exposed in the sea and form part of the microbial flora on decayed wood.

The superficial infestation of timbers in sea water by microorganisms is considered to be of minor importance, but surface softening of wood through microbial decay is believed to enhance settlement of marine borer larvae leading to subsequent attack at deeper levels in the wood.

THE BIOLOGY OF MARINE WOOD BORING ANIMALS
Molluscs and crustaceans are the major agents of timber degradation in the sea. Attack of submerged timber by molluscs can result in extensive tunnelling of both the outer and inner regions of

the wood.

The molluscs are represented by the Teredinidae and Pholadidae. The teredinids or shipworms comprise many genera and species including *Teredo*, *Lyrodus* and *Bankia* spp. and their taxonomy has been extensively studied by Turner (1966, 1971). The shipworms excavate tunnels in wood through the rasping action of two anterior valves, and the wood which is ingested provides food for the worm. A calcareous deposit lines the surface of the tunnel which allows the shipworms within wood to be observed using x-radiography. The posterior end of the animal has two siphons (excurrent and incurrent) which normally extend from a small hole, up to 2 mm diam, in the wood surface. When wood is not submerged, the siphons are retracted and the aperture is plugged with the pallets. The form of the pallets is a useful feature in the identification of shipworms.

The pholads or piddocks do not have the same diversity in genera and species as the teredinids and perhaps the most important genus is *Martesia*. They are bivalve organisms and species can be distinguished on the basis of the shape of the calcareous valves, the accessory plates between the valves and the siphons. Unlike the teredinids, the pholads do not feed on the wood which they excavate in creating their burrows, but they feed on plankton in the surrounding water. Nevertheless they cause considerable breakdown of timbers particularly in sub-tropical and tropical waters and survive in a wide range of salinity, water pH and temperature extremes. *Martesia* spp. are believed to be most active in timber piling at the low to mid-tide position (Figure 30.1a). The size of tunnels produced by wood-boring molluscs varies a great deal depending on species, age, wood species and local conditions. Teredinids have been reported to produce tunnels up to 2 metres long and 15 mm in diameter. The pholads do not deposit a calcareous lining on the surface of the tunnel. The diameter of tunnel apertures can be 5 mm but pholads are able to excavate tunnels up to 8 times their own length. Infection of new timber surfaces is achieved by larval settlement; adult wood-boring molluscs remain in the same piece of wood throughout their lives.

The wood-boring crustaceans are members of the Isopoda and Amphipoda. The genera *Limnoria* (gribble) and *Sphaeroma* (pill-bug) are the most important isopods but they exhibit different modes of attack. *Limnoria* spp. degrade the surface regions of timber by excavating a network of fine galleries (1-3 mm diam). Three species cause the greatest damage - *L. lignorum* Rathke (cold temperate), *L. quadripunctata* Holthuis (temperate) and *L. tripunctata* Menzies (temperate/tropical). Elevated sea water temperatures near to industrial warm water outfalls can provide conditions for greater species diversity in temperate regions. The animals are small, generally no more than 2.5 mm long, and move about freely within the excavated galleries. Adult animals may also move to fresh timber surfaces and when established, excavate galleries along the grain of the wood.

The pill-bugs or *Sphaeroma* spp. usually attack wood by burrowing across the grain producing a "honeycomb" of tunnels which are roughly 10-15 mm long and 5-10 mm diam. Species of *Sphaeroma* are a major problem in tropical and warm temperate waters and degrade wood in the upper tidal regions (Figure 30.1a). Animals can also survive extremes of salinity and temperature.

The Amphipod *Chelura* is widespread and generally associated with *Limnoria* infestations of wood. Species of *Chelura* excavate and enlarge the galleries made by the limnorids and also feed on their faecal pellets.

NUTRITION IN MARINE BORERS
In general, the feeding habits of marine borers have great bearing on the use of appropriate toxicants to protect timber against attack. Considerable research has been directed towards understanding the interrelationships between marine borers, their substrate and associated microorganisms in order to define specific targets which may be disturbed and thus provide avenues for formulating new control measures. However, the digestive processes of those animals which use wood as a nutritional source are not fully resolved and work continues in this area (Turner 1984). The presence of cellulolytic microbial symbionts in the gut of teredinids has been thought to be significant in the digestion of wood particles, but cellulolytic marine fungi are believed to be of little importance in nutrition (Morton 1978; Mann 1984). Bacterial cellulolysis is more widely accepted and bacteria isolated from the gland of Deshayes in six teredinid species have nitrogen-fixing and cellulolytic abilities (Waterbury *et al.* 1983). These authors noted the potential biotechnological significance of these two characteristics in the same bacterial cell.

The demand for nitrogen by wood-boring animals may not be met solely by the limiting levels of nitrogen in wood. Nitrogen supplements may be provided to teredinids in the form of marine phytoplankton, dissolved organic matter and through the activities of nitrogen-fixing symbiotic bacteria. The role of microorganisms in the nutrition of limnorids has been studied extensively. Ray & Stuntz (1959) believed that fungi were unimportant in the feeding activities of limnorids but the observations of other workers indicate that microorganisms are a necessary component of their diet. Because microorganisms are absent from the gut surface of limnorids (Boyle & Mitchell 1978, 1980) it has been suggested that these arthropods utilize wood already infected with bacteria and fungi. This provides a diet of partially digested wood containing cellulase enzymes and microbial nitrogen as well as vitamins and organic micronutrients derived from the external wood-inhabiting microflora. The growth of marine soft rot fungi and lignolytic bacteria in superficial wood elements supports the grazing limnorid flora.

Selective feeding experiments on *Limnoria* were carried out using cultures of marine soft rot fungi (Geyer 1980). Although the animals were found to show some selective feeding preference for particular fungi, the overall conclusion drawn from this and previous work

(Kohlmeyer *et al.* 1959) was that *Limnoria* species are dependent on a diet of wood which is infected with marine microorganisms and that marine fungi contribute significantly to this total microbial biomass.

In contrast to the surface feeding activities of *Limnoria* spp. and *Chelura* spp., the crustacean borers – *Sphaeroma* spp. excavate galleries deep into timber. There is some debate about the nutritional behaviour of *Sphaeroma*. It has been suggested that *Sphaeroma* spp. require a supplemented diet and feed on plankton and microorganisms which inhabit the surfaces of tunnels produced by the animals.

MICROBIAL DECAY OF WOOD

Bacteria and fungi have only a small direct part to play in the destruction of submerged wood in the sea. Decay is superficial and is attributed to the cellulolytic ability of microorganisms which attack the wood cell walls in cells adjacent to the surface of the timber. Bacteria are observed in the cell lumina and are able to breakdown cell walls by releasing lignolytic enzymes causing erosion of the cell wall around individual bacteria. Bacteria have also been shown to penetrate the inner S_3 layer of the cell wall and excavate fine tunnels within the secondary cell wall (Nilsson & Daniel 1983). A radiating network of tunnels characterizes the early stages of decay leading to ultimate collapse of the wood cell wall producing a granular appearance under the light microscope (Figures 30.1c,d).

Soft rot decay of wood by marine Ascomycotina and Deuteromycotina has been known since the work of Barghoorn & Linder (1944). Soft rot decay results in cellulolysis without lignin breakdown, creating a dark, spongey texture to the wood surface. On drying, severely decayed wood shows characteristic cross-cracking. Hyphal penetration of the cell wall from the wood cell lumen is the first stage of the decay process. Hyphae branch in the S_2 region and follow the helical orientation of the cellulose microfibrils. Hyphal growth within the wood cell wall is not continuous as observed in the cell lumen, but is oscillatory (Hale & Eaton 1985 a). This start-stop mode of growth was first recognized in *Halosphaeriopsis mediosetigera* (Cribb et Cribb) T.W. Johnson (Leightley & Eaton 1977) but has since been observed in other soft rot fungi. The cessation of hyphal tip growth is followed by enzyme release over the hyphal surface and the formation of cylindrical or diamond-shaped cavities within the S_2 cell wall region. During cavity formation a fine

Figure 30.1a Marine piling (*Syncarpia* sp. - turpentine) at Cairns, Australia showing attack by *Martesia*, mid to low tide position, and *Sphaeroma*, upper tide position (photograph courtesy of J. Barnacle, CSIRO).
Figure 30.1b Chains of fungal soft rot cavities in tracheid wall of *Pinus sylvestris* L. Bar = 10 μm.
Figures 30.1c,d Light micrograph (Fig. c) and scanning electron micrograph (Fig. d) of bacterial decay of tracheid walls of *Pinus sylvestris*. Bars = 20 μm.

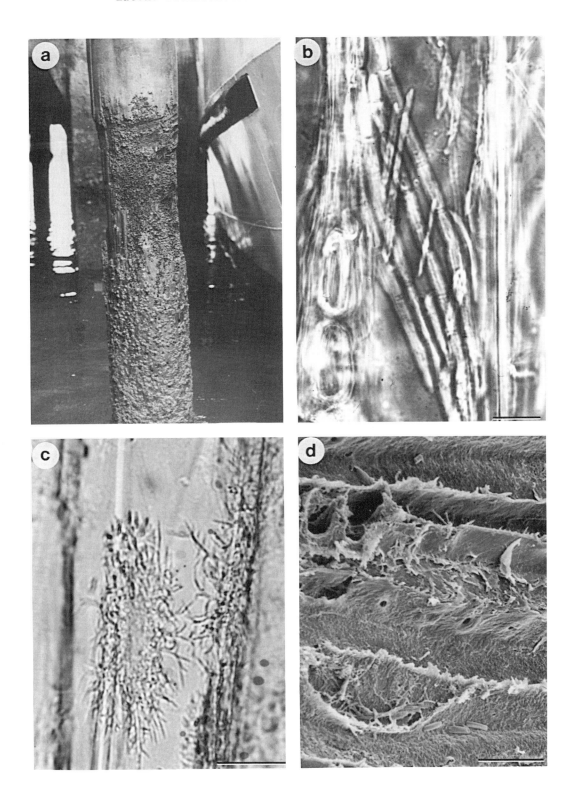

hypha forms from the hyphal tip region within the cavity and this grows
in a helical orientation within the wood cell wall. A further cessation
of apical growth is followed by cavity elaboration and hyphal widening.
This pattern of growth is repeated to form a helical chain of cavities
(Figure 30.1b). Studies on the fine structure of hyphae during growth
and cavity formation included investigations with the marine isolate
Humicola alopallonella Meyers et Moore (Hale & Eaton 1985 b,c).

Basidiomycete decay of wooden structures for marine use is a particular
problem in the cut surfaces of pile tops where the instances of decay
are concentrated in the heartwood. This form of internal decay is
observed in piling timbers treated with preservative where untreated
heartwood is exposed to infection when the ends of piling timbers are
sawn off. Remedial fumigation treatments using vapam and chloropicrin
have given encouraging results in controlling both decay and marine
borer attack (Helsing 1979). Decay of timber by terrestrial
Basidiomycotina was also recognized inside wooden fishing boats (Savory
& Eaves 1965). The cause of decay was attributed to poor ventilation
within the fish-hold.

TRADITIONAL WOOD PRESERVATION
Timber species with high natural resistance to
biodeterioration in the sea have been used for centuries, but in those
species with a lower durability the earliest forms of protection
afforded to wood exposed in sea water were surface barriers and coatings
(Barnacle 1976). Such methods are still in use and include the coating
of fishing vessels with animal oils and fats and plant oils and resins
sometimes mixed with lime. Regular application of these coatings is
necessary for reasonable protection but frequent beaching of small boats
to allow drying out is a simple and effective way of limiting the
establishment of marine boring animals.

Creosote
The introduction of wood-tar creosote for the preservation
of timbers in naval ships began during the eighteenth century. However,
surface application of creosote did not offer long-term protection and
the pressure treatment of timbers, established in the nineteenth
century, was necessary to achieve preservative retentions suitable to
prevent attack. Full-cell pressure treatment with creosote should
provide retention levels ranging from 160–400 kg m^{-3} depending on the
timber species. This is the recommendation of the British Wood
Preservers' Association and is appropriate for exposure in temperate
waters. In such circumstances adequate protection of softwoods is
generally achieved although instances of failure in Douglas fir have
been reported (Bramhall 1968).

Pressure treatment of hardwoods to yield satisfactory retention levels
and good distribution of creosote is variable and dependent on species
density and anatomy. Problems of treated hardwood failure are more
frequently recorded in tropical waters although instances of attack of

creosoted softwoods are not uncommon. McQuire (1971) suggested that
poor penetration, loss and degrade during service, and the variable
nature of creosote could account for premature failure of creosoted
marine timbers. McQuire pointed out also that the type of marine borer
present would have an important effect. In 1951 Menzies had identified
Limnoria tripunctata as an organism which was able to attack creosoted
piling. The tolerance to creosote by *L. tripunctata* was aggrevated in
timbers exposed in warmer waters. The fact that bacteria degrade
creosote in treated wood indicated that an association might exist
between creosote-tolerant *L. punctata* and its own gut microflora.
However, examination of the gut surfaces of *Limnoria* spp. has shown them
to be clear of microorganisms (Boyle & Mitchell 1984) although
microorganisms can be found in the gut contents and faeces. Since
modification and breakdown of creosote by microorganisms is known to
occur prior to attack of treated wood by *L. tripunctata* it seems likely
that synergism between the ingested microflora and the limnorids is a
key factor in creosote tolerance.

Water-borne preservatives
The development of water-borne preservatives during the
first half of this century led to their use in wooden marine
installations. Copper-chrome-arsenic (CCA) and ammoniacal copper
arsenate (ACA) are widely used preservatives in the marine environment
and recommended pressure treatment aims for total sapwood penetration to
give dry salt retention values ranging from 32-48 kg m $^{-3}$. The
retention values are greater in warm waters but brittleness in timber
treated with salt preservatives has given concern where strength
characteristics are important, e.g. harbour structures including fenders
and pilings.

Copper-chrome-boron (CCB) preservatives have also been tested for marine
use and in an international trial conducted over several years CCB
treatment was found to perform at least as effectively as CCA. This
trial included a range of treated wood species and it emerged from this
and other work that although satisfactory performance can be expected
from softwoods treated to high loadings, protection of hardwoods is
less consistent (Eaton 1985).

Copper-based preservatives are generally effective against the limnorids
but Barnacle *et al.* (1983) reported attack of CCA-treated piling by
L. tripunctata and *L. quadripunctata* in Australia. Other marine borers
which exhibit tolerance to copper-containing preservative treatments
are *Martesia striata* L. and species of *Sphaeroma*. In the case of
Martesia, the toxic effect of preservative impregnation may be
considerably nullified because wood particles are not ingested during
tunnel excavation. The question of wood particle ingestion by boring
Sphaeroma spp. is also not fully resolved and may have a bearing on
their ability to attack CCA-treated wood. In addition, copper
accumulation in the caecae of *Sphaeroma* specimens attacking CCA-treated
piles may in part explain the tolerance to CCA treatments, but no
chromium or arsenic accumulations were reported (Cragg & Icely 1982).

Dual treatments
        The susceptibility of creosote-treated timbers to attack by
limnorids led to the development of pressure treatment with copper-based
preservative followed by a drying stage and further pressure treatment
with creosote. Such dual treatment of softwoods in the United States
has been shown to be as effective as high preservative salt treatment
alone and is recommended for high hazard sites. Nevertheless, *Limnoria*
and pholad (*Martesia*) attack of timber installed in long-term trials
(Richards 1977) has been reported but so far trials using this method of
treatment are limited, outside the United States.

NOVEL PRESERVATIVE CHEMICALS
        In recent years the potential use of chemicals with specific
toxic effects against marine borers has been investigated.

Synthetic pyrethroid insecticides have low water solubility and a high
level of permanence in wood. These properties plus the accumulated
toxicity data against *Limnoria* spp. (Carter 1984) suggest that these
chemicals may have value as components of new marine preservative
systems. Bultman & Parrish (1983) investigated the molluscicide
N-tritylmorpholine. In short-term trials at a high hazard marine site
in Central America, no attack by pholads or teredinids was recorded but
damage by limnorids was severe. The development of new preservatives
composed of chemicals which are environmentally acceptable and which
have selective toxic action against marine borers are currently being
sought. However, the permanence and stability of these compounds in
sea water are necessary pre-requisites for their long-term commercial
use.

It has been known for some time that certain timber species are
particularly resistant to damage by marine borers. Extractives removed
from *Dalbergia retusa* Hemsl. and *Tabebuia guayacan* (Seem.) Hemsl.,
species which exhibited these characteristics, were impregnated into
pine (Bultman 1976). Following exposure, inhibition against certain
teredinids (Turner 1976) and marine fungi (Furtado & Jones 1976) was
observed and the compound obtusaquinone afforded reasonable protection
in the field. Laboratory experiments showed that obtusaquinone appeared
to inhibit the enzyme phenoloxidase which is involved in the initial
stages of shell formation in settled larvae. The larvae were unable to
penetrate the wood and did not survive. Further work with tannins has
shown that they have similar inhibitory effects to the quinones and
patents have been registered on these treatments (Turner 1984).

PHYSICAL PROTECTION
        Physical barriers applied to the surface of marine timbers
have been used for many years. These barriers include metals, concrete
and plastic materials. Malleable metals such as lead, zinc and copper
have been applied in sheets nailed to timbers. Concrete casings are
also effective except where cracking takes place. More recently plastic
sheathing (polyvinyl chloride and polyethylene) has been employed by

in situ wrapping of piles using divers or zippering plastic tubing. Preventative methods include heat-shrunk polyethylene film around piling timber before driving (Steiger & Horeczko 1981) and currently, investigations are underway into polyurethane coating systems.

The sea is a particularly hazardous environment for the successful protection of wood. The wide variety of organisms which are able to infest timbers used in the sea, their incidence and the environmental factors which determine their virulence throughout the world affect the long term performance of preserved wood. Traditional wood preservatives have proved satisfactory in many sites but the greater hazard in the tropics requires further research into new chemicals and the use of physical barriers (Eaton 1985).

REFERENCES

Barghoorn, E.S. & Linder, D.H. (1944). Marine fungi: their taxonomy and biology. Farlowia, 1, 395-467.

Barnacle, J. (1976). Wood and its preservation in the sea - a resumé. Proc. 4th Int. Congress Marine Corrosion and Fouling, 57-66.

Barnacle, J., Cookson, L.J. & McEvoy, C.N. (1983). *Limnoria quadripunctata* Holthuis - a threat to copper-treated wood. Int. Res. Group on Wood Pres., Doc. 4100.

Boyle, P.J. & Mitchell, R. (1978). Absence of microorganisms in crustacean digestive tracts. Science, 200, 1157-1159.

Boyle, P.J. & Mitchell, R. (1980). Interactions between microorganisms and wood-boring crustaceans. In Biodeterioration: Proc. 4th Int. Symp. Berlin, eds. T.A. Oxley, D. Allsopp & G. Becker, pp. 179-186. London: Pitman.

Boyle, P.J. & Mitchell, R. (1984). The Microbial Ecology of Crustacean Wood Borers. In Marine Biodeterioration: an interdisciplinary study, eds. J.D. Costlow & R.C. Tipper, pp. 17-23. Washington, D.C.: U.S. Naval Institute.

Bramhall, G. (1968). Comparison of boring patterns in the inspection of Douglas fir marine piling. Proc. Am. Wood Pres. Assoc. Conv., 64, 45-49.

Bultman, J.D. (1976). Research at the Naval Research Laboratory on bioresistant tropical woods: an overview. In Biodeterioration of Tropical Woods, ed. J.D. Bultman, pp. 1-6. Washington, D.C.: Naval Research Laboratory, Department of the Navy.

Bultman, J.D. & Parrish, K.K. (1983). N-tritylmorpholine as a potential marine wood protectant against teredinids and pholads - a preliminary evaluation. Int. Res. Group on Wood Pres., Doc. 497.

Carter, S.W. (1984). The use of synthetic pyrethroids as wood preservatives. Proc. Brit. Wood Pres. Assoc. Conv.: 32-41.

Cragg, S.M. & Icely, J.D. (1982). An interim report on studies of the tolerance of *Sphaeroma* (Crustacea: Isopoda) of CCA-treated timber. Int. Res. Group on Wood Pres., Doc. 491.

Eaton, R.A. (1985). Preservation of marine timbers. In Preservation of Timber in the Tropics, ed. W.P.K. Findlay, pp. 157-191. Dordrecht, Netherlands: Martinus Nijhoff/Dr. W. Junk.

Furtado, S.E.J. & Jones, E.B.G. (1976). The performance of *Dalbergia* wood and *Dalbergia* extractives impregnated into pine and exposed in a water cooling tower. In Biodeterioration of Tropical Woods, ed. J.D. Bultman, pp. 41-56. Washington, D.C.: Naval Research Laboratory, Department of the Navy.

Geyer, H. (1980). Influence of marine fungi present on wood on the feeding activity and propagation of *Limnoria tripunctata* (Crustacea: Isopoda). Z. Angew. Zool., 67, 79-100.

Hale, M.D. & Eaton, R.A. (1985 a). Oscillatory growth of fungal hyphae in wood cell walls. Trans. Br. mycol. Soc., 84, 277-288.

Hale, M.D. & Eaton, R.A. (1985 b). The ultrastructure of soft rot fungi. I. Fine hyphae in wood cell walls. Mycologia, 77, 447-463.

Hale, M.D. & Eaton, R.A. (1985 c). The ultrastructure of soft rot fungi. II. Cavity forming hyphae in wood cell walls. Mycologia, 77, 592-605.

Helsing, G.G. (1979). Controlling wood deterioration in waterfront structures. Sea Technology, June, 20-21.

Kohlmeyer, J., Becker, G. & Kampf, W.D. (1959). Versuche zur Kenntnis der Ernahrung der Holzbohrassel *Limnoria tripunctata* und ihre Beziechung zu holzzerstorenden Pilzen. Z. Angew. Zool., 46, 457-489.

Leightley, L.E. & Eaton, R.A. (1977). Mechanisms of decay of timber by aquatic microorganisms. Proc. Brit. Wood Pres. Ass. Conv., 221-250.

McQuire, A.J. (1971). Preservation of timber in the sea. In Marine borers, fungi and fouling organisms of wood, eds. E.B.G. Jones & S.K. Eltringham, pp. 339-346. Paris: OECD.

Mann, R. (1984). Nutrition in the Teredinidae. In Marine Biodeterioration: an interdisciplinary study, eds. J.D. Costlow & R.C. Tipper, pp. 24-29. Washington, D.C.: U.S. Naval Institute.

Menzies, R.J. (1951). A new species of *Limnoria* (Crustacea: Isopoda) from Southern California. Bull. South Calif. Acad. Sci., 50, 86-88.

Morton, B. (1978). Feeding and digestion in shipworms. Oceanogr. Mar. Biol. Ann. Rev., 16, 107-144.

Nilsson, T. & Daniel, G. (1983). Tunnelling bacteria. Int. Res. Group on Wood Pres. Doc. 1186.

Ray, D.L. & Stuntz, D.E. (1959). Possible relation between marine fungi and *Limnoria* on submerged wood. Science, 129, 93-94.

Richards, B.R. (1977). Comparative values of dual-treatment and water-borne preservatives for long-range protection of wooden structures from marine borers. Proc. Am. Wood Pres. Assoc., 73, 128-131.

Savory, J.G. & Eaves, A. (1965). Decay in Scottish fishing boats. Princes Risborough, U.K.: Forest Products Research Laboratory.

Steiger, F. & Horeczko, G. (1981). Controlling marine borer attack of timber piles with plastic wraps. Int. Res. Group Wood Pres., Doc. 479.

Turner, R.D. (1966). A survey and illustrated catalogue of the Teredinidae (Mollusca: Bivalvia). Cambridge, Mass.: Museum of Comparative Zoology, Harvard University.

Turner, R.D. (1971). Identification of marine wood boring molluscs. In Marine borers, fungi and fouling organisms of wood, eds. E.B.G. Jones & S.K. Eltringham, pp. 17-64. Paris: OECD.

Turner, R.D. (1976). Search for a weak link. In Biodeterioration of Tropical Woods, ed. J.D. Bultman, pp. 31-40. Washington, D.C.: Naval Research Laboratory, Department of the Navy.

Turner, R.D. (1984). An overview of research on marine borers: past progress and future direction. In Marine Biodeterioration: an interdisciplinary study, eds. J.D. Costlow & R.C. Tipper, pp. 3-16. Washington, D.C.: U.S. Naval Institute.

Waterbury, J.B., Calloway, C.B. & Turner, R.D. (1983). A cellulolytic nitrogen-fixing bacterium cultured from the gland of Deshayes in shipworms (Bivalvia: Teredinidae). Science, 221, 1401-1403.

INDEX